Ecological and Environmental
Physiology of **Birds**

Ecological and Environmental Physiology of **Birds**

J. Eduardo P. W. Bicudo
*Departamento de Fisiologia, Instituto de Biociências,
Universidade de São Paulo*

William A. Buttemer
*School of Biological Sciences,
University of Wollongong*

Mark A. Chappell
University of California, Riverside

James T. Pearson
Department of Physiology, Monash University

Claus Bech
Department of Biology, NTNU, Norway

OXFORD
UNIVERSITY PRESS

*This book has been printed digitally and produced in a standard specification
in order to ensure its continuing availability*

OXFORD
UNIVERSITY PRESS

Great Clarendon Street, Oxford OX2 6DP
United Kingdom

Oxford University Press is a department of the University of Oxford.
It furthers the University's objective of excellence in research, scholarship,
and education by publishing worldwide.

Oxford is a registered trade mark of Oxford University Press in the UK
and in certain other countries

British Library Cataloguing in Publication Data
Data available

Library of Congress Cataloging in Publication Data
Data available

ISBN 978-0-19-922844-7

Ecological and Environmental Physiology Series (EEPS)
Series Editor: Warren Burggren, University of North Texas

This authoritative series of concise, affordable volumes provides an integrated overview of the ecological and environmental physiology of key taxa including birds, mammals, reptiles, amphibians, insects, crustaceans, mollusks, and fish. Each volume provides a state-of-the-art review and synthesis of topics that are relevant to how that specific group of organisms have evolved and coped with the environmental characteristics of their habitats. The series is intended for students, researchers, consultants, and other professionals in the fields of physiology, physiological ecology, ecology, and evolutionary biology.

A Series Advisory Board assists in the commissioning of titles and authors, development of volumes, and promotion of the published works. This Board comprises more than 50 internationally recognized experts in ecological and environmental physiology, providing a combination of both depth and breadth to proposal evaluation and series oversight.

The reader is encouraged to visit the EEPS website for additional information and the latest volumes (http://www.eeps-oxford.com/). If you have ideas for new titles in this series or just wish to comment on EEPS, please do not hesitate to contact the Series Editor, Warren Burggren (University of North Texas; burggren@unt.edu).

Volume 1: Ecological and Environmental Physiology of Amphibians
Stanley S. Hillman, Philip C. Withers, Robert C. Drewes, Stanley D. Hillyard

Volume 2: Ecological and Environmental Physiology of Birds
J. Eduardo P.W. Bicudo, William A. Buttemer, Mark A. Chappell, James T. Pearson, Claus Bech

Contents

Acknowledgements

We would like to thank all those who, through their work on birds, have been important sources of inspiration to us for undertaking the task of writing a book on Ecological and Environmental Physiology of Birds. They are too numerous to list comprehensively, but our mentors, George A. Bartholomew, William R. Dawson, Dick Grau, Craig Heller, Kejll Johansen, Knut Schmidt-Nielsen, and Roger Seymour have been particularly inspiring, not only to us but to the whole community of avian physiologists (indeed, of animal physiologists in general). Their guidance in our early careers, and their deep enthusiasm and encouragement for our different pursuits are much appreciated.

Writing a book like this is a very lengthy endeavor, and many people have contributed directly or indirectly towards its completion. Certainly our families, colleagues, and students tolerated the sometimes intense focus on the work with encouragement and good humor. One of us (Eduardo Bicudo) benefited from the hospitality of the staff of the Animal Demography Unit (former Avian Demography Unit; ADU) at the University of Cape Town (UCT), South Africa, during the early stages of writing. Special thanks are due to Les Underhill (ADU chairman), John Cooper, Chris Lotz, and Sue Kuyper, who were just superb hosts, and also to Margaret Koopman, librarian at the Niven Library, Percy FitzPatrick Institute of African Ornithology at UCT. Eduardo, who also spent some time at the University of Wollongong, Australia, during the early stages of writing, had the privilege to engage in stimulating discussions with Lee Astheimer, Bill Buttemer, and Terry Dawson. Lee contributed substantially to early discussions that shaped the book contents, but other obligations prevented her further participation. During this period, Eduardo was supported by the University of Wollongong as a Visiting Professorial Fellow. Jim Jones, at the University of California, Davis, also hosted Eduardo, and kindly made available to him all the facilities at UCDavis so he could continue to write the book while on his sabbatical leave.

During much of his work on this volume, Mark Chappell was hosted by Roberto Nespolo at the Universidad Austral de Chile in Valdivia, and by the Polish Academy of Sciences Mammal Research Institute in Bialowieza, under a Marie Curie Fellowship.

Other people have also helped us, and José Guilherme Chauí-Berlinck, Bill Karasov, Joe Williams, and Gabrielle Nevitt kindly offered their invaluable advice and assistance at different stages.

Foreword

Birds are a truly special group of vertebrates in many ways. They fly, mostly, and while we easily accept this because of its familiarity, it is a very special lifestyle and we are still uncertain as to how it evolved. Mammals are often considered to be the vertebrates that dominate the continents but birds occur over more of the Earth's surface and play important roles in ecosystems on a global scale. They are excellent bio-indicators in our changing world and improving our knowledge of them is crucial to our understanding of these global environmental developments. The importance of this new volume on the *Ecological and Environmental Physiology of Birds* sits squarely within this greater context.

From a personal perspective, a casual foray into bird ecophysiology markedly enhanced my insights into the adaptive responses of birds in a variety of challenging environments. Birds are generally one of our first interfaces with wild animals, be they city pigeons, garden sparrows or the gulls at the beach. These encounters can lead to a continuing fascination with birds, perhaps as amateur bird watchers or even as professional biologists, such as the contributors to this volume. My career in ecological physiology grew from such early experiences. A collection of bird eggs obtained during my pre-teen wanderings in the dry bush near my home in the Australian rangelands is my first remembered step along this path. A curiosity about the odd Australian mammals (such as kangaroos) that were also encountered in this environment has dominated much of my research over the years but my interest in birds continued and increasingly I realised that birds were functioning in extreme environments in ways that far exceeded their physiological abilities, as then understood. It was the Emu (*Dromaius noveahollandia*) a supposedly primitive, giant bird that really awoke me to the very different physiologies of birds and mammals and the adaptive potential of birds. While the kangaroos at our field site hid in the shade to avoid the summer sun the Emus spent the day foraging in the open desert, yet their reported urine concentrating abilities were far lower than those of the kangaroos.

These observations resulted in a major excursion into avian ecological physiology, with a necessary steep learning curve for my students and me to gain insights into the unexpected complexity of the descendants of the dinosaurs. Luckily, we had support from colleagues, several of whom are contributors to this volume. William Calder III, a friend from my postdoctoral days in Knut Schmidt-Nielsen's lab at Duke University, visited and gave me my first lessons on avian metabolism. William Dawson from the University of Michigan also visited and he recommended the sturdy Bill Buttemer as someone with knowledge of bird thermal biology who might also handle large Emus. We were doubly rewarded when Bill also brought Lee Astheimer to the University of New South Wales. Erik Skadhauge, a friend from my Duke University days came on sabbatical from Denmark and provided much on-site training in bird excretory physiology to my students, Shane Maloney and Robert Herd and me. Mark Chappell came to look at Australian birds and gave much help, as did Eduardo Bicudo when he was consulted on bird energetics. Eventually we came to understand the fascinating features of the physiological ecology of the Emus, which also provided a refreshing insight into the biology of kangaroos.

When Eduardo Bicudo was planning this book he asked if I would like to contribute. I declined, however, since my knowledge is rather narrow and bird comparative physiology has progressed significantly in the last few years. Fortunately, I was able to make some appropriate suggestions leading to some of those now contributing to this volume. Eduardo has kindly let me read chapters as they have progressed and my education has continued apace.

Much knowledge across the spectrum of bird physiology has been gained in the years since the Emu studies, but the authors of this volume have avoided an exhaustive "laundry list" approach. Instead, they focus on our current understanding of a set of topics in ecological and environmental physiology that are of particular interest to ornithologists, but also have broad biological relevance. The introductory chapter covers the basic body plan of birds and their still-enigmatic evolutionary history. The focus then shifts to a consideration of the essential components of that most fundamental of avian attributes: the ability to fly. The emphasis here is on feather evolution and development, flight energetics and aerodynamics, migration, and as a counterpoint, the curious secondary evolution of flightlessness that has occurred in several lineages. This sets the stage for subsequent chapters, which present specific physiological topics within a strongly ecological and environmental framework. Chapter 2 covers gas exchange and thermal and osmotic balance, together with the central role of body size. Chapter 3 addresses "classical" life history parameters – male and female reproductive costs, parental care and investment in offspring, and fecundity *versus* longevity trade-offs – from an ecophysiological perspective. Chapter 4 – "*Adaptations and specializations*" – offers a comprehensive analysis of feeding and digestive physiology, adaptations to challenging environments (high altitude, deserts, marine habitats, cold), and neural specializations (notably those important in foraging, long-distance navigation, and song production). Throughout the book, classical studies are integrated with the latest research findings.

Of course, numerous important and intriguing questions await further work, and the book concludes with a discussion of research methods and approaches – emphasizing cutting-edge technology – and a final chapter on future directions that should help point the way forward for both young and senior scientists. This volume is broad and incisive; a similar one would have been invaluable when my group started researching Emus in the 1980s. An informed colleague was not always on hand.

Finally, returning to a more personal and subjective level, I believe it is important for even the most dedicated and dispassionate of scientists to acknowledge and, indeed, embrace the special emotional appeal of birds. The contributors to this book are a diverse group of professional biologists from across the globe, each with their own particular specialized training and expertise. Each is thoroughly versed in the rigorous, disciplined, and objective methods of science and has used them to build a career, in whole or in part, probing physiological questions and adding to our understanding of how avian biology "works" at a mechanistic level. But from a more emotive and aesthetic perspective, all have a boundless enthusiasm for, and appreciation of birds simply as the beautiful, elegant, and utterly fascinating creatures that they are. I hope the contents of this volume will add to that special sense of wonder engendered by the sight of a bird in the sky.

Terence J. Dawson
Emeritus Professor of Zoology, University of New South Wales
Sydney, Australia.

1

Introduction – Blueprint of a Bird (*Bauplan*/Body plan)

1.1 What is a Bird?

Birds are mostly recognized by their ability to fly, even though not all avian species are capable of flight. However, the body plan of birds is essentially the same irrespective of their mode of locomotion. Generally speaking, a typical bird possesses a streamlined and compact body covered with feathers, a pair of wings (forelimbs), a pair of hind limbs and a pointed beak.

A closer look reveals that birds have a series of specific adaptations that distinguish them from other vertebrates. Many elements of the skeleton have been modified; the major skull bones are fused, the wing and leg elements are reduced in number, many bones are hollow and air-filled, the skull shows a unique form of cranial kinesis (movement of all or part of the upper jaw relative to the braincase), the sternum possesses a carina (an expanded area for the attachment of the two muscles associated with flight—the pectoral and the supracoracoid muscles), and the clavicles have been fused into a furcula. Other changes include an exclusively oviparous (egg laying) mode of reproduction, loss of teeth, development of a muscular gizzard, air sacs (allowing a unidirectional flow of air in the parabronchi), loss of one ovary in most species and a great reduction in the mass of the skin (which is paper-thin in most species of flying birds). This body plan may seem to have imposed a number of constraints on birds, but on the other hand it has allowed them to become one of the most successful of animal taxa (see e.g., Chatterjee 1997; Feduccia 1996; Padian 1986).

Birds occur over more of the Earth's surface than any other vertebrate group and play an important role in ecosystems on a global scale. Because of this, we must understand how birds have evolved, and we also must know what environmental and ecological circumstances existed at the time of their early adaptive radiation.

With approximately 10,000 species catalogued to date, extant birds constitute a highly successful vertebrate class. For example, major lineages of the approximately 5739 species of passerine birds (perching birds) have diversified in all continents and now occupy nearly all terrestrial ecosystems. The songbirds (oscines, suborder Passeri) alone comprise nearly half of all extant birds, encompassing a staggering ecological and behavioral diversity (Barker *et al.* 2004).

The most remarkable trait found in the majority of existing birds is, of course, their ability to fly. Although we find species which can fly, or at least glide, within other vertebrate taxa (fish, amphibians, reptiles, and mammals), for more than 90% of the existing species of birds, flight is their major and most important mode of locomotion. Therefore, there seems to be a strong and continual selection pressure for the maintenance of the flight apparatus in birds. Flightlessness, on the other hand, is a pervasive phenomenon. If there is not a strong and continual selection for the maintenance of the flight apparatus, it tends to disappear (Feduccia 1996).

1.2 Evolution of Birds

How birds evolved is still a matter of great controversy. It is not the purpose of this volume to enter into the debate that pervades this important and intriguing topic. However, it is relevant to summarize it here, especially because new fossil records have been discovered recently in China (Liaoning, north-east China).

Archaeopteryx, from the Jurassic, is still considered to be the oldest and most primitive bird. However, well preserved and abundant new fossils of birds and dinosaurs (Fig. 1.1) have provided unprecedented new evidence on the dinosaurian origin of birds, the arboreal origin of avian flight (not widely accepted), and the origin of feathers prior to flapping flight (more commonly accepted) (Clarke *et al.* 2006). The Mesozoic avian assemblage mainly comprises two major lineages: the prevalent extinct group Enanthiornithes, and the Ornithurae, which gave rise to all modern birds, as well as several more basal taxa. Cretaceous birds radiated into various paleoecological niches, particularly fish- and seed-eating. Significant changes in size, morphology, ecology and diet, and variation in flight capabilities ranging from gliding to strong muscle-powered flight, highlight the diversification of the avian lineage in the early Cretaceous (145.5 million years ago). There is little evidence, however, to support a Mesozoic origin of modern avian groups (see e.g., Zhou 2004; Zhou and Zhang 2005).

The most recent discoveries in China (Yon *et al.* 2006) reveal that extant birds (Neornithes) may have originated from a goose-like ancestor, *Gansus yumenensis*, an amphibious, flight-capable bird from the Early Cretaceous.

Paleoecological evidence from the early Cretaceous points to the existence of a warmer climate than at present and probably warmer on a worldwide basis than at any other time during the Phanerozoic. The temperature difference from the poles to the equator was about one-half the present gradient. Floral evidence suggests that a tropical to subtropical climate prevailed over much of the planet (Heimhofer *et al.* 2005). It is also likely that birds evolved in an O_2 rich atmosphere correlated with a time of increasing air density, probably in the late Jurassic.

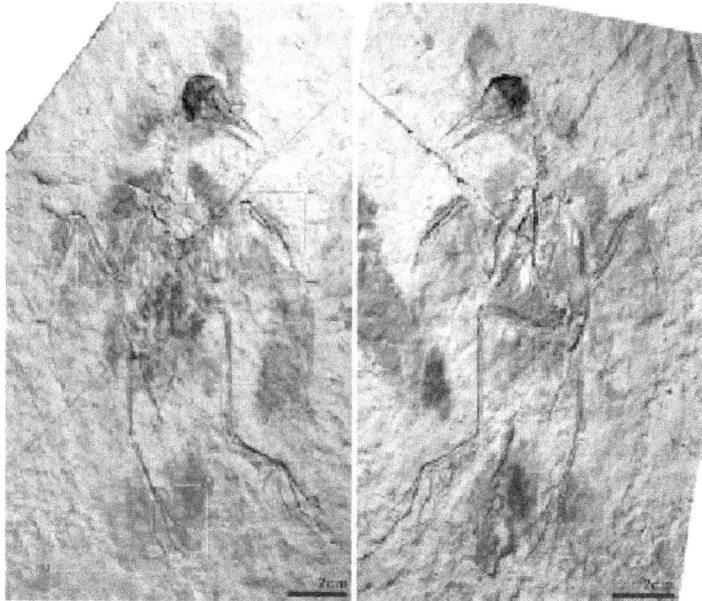

Fig. 1.1 Holotype of ornithurine bird *Hongshanornis longicresta* gen. et sp. nov. from the Lower Cretaceous of Inner Mongolia, China (IVPP V14533). (*Left*) Part. (*Right*) Counterpart. (Zhou and Zhang, 2005; with permission from the National Academy of Sciences).

Recent studies have shown that during the late Jurassic oxygen concentrations rose to current levels, or slightly higher (Falkowski *et al.*, 2005). The three independent origins of vertebrate flapping flight, in pterosaurs, bats, and birds, may have also been facilitated by a hyperdense (denser than early Jurassic levels, mainly due to the higher molecular weight of oxygen versus nitrogen) and a hyperoxic atmosphere (reviewed by Dudley, 1998). This combination would provide increased lift per unit of wing area, and a greater ability to power the oxidative metabolism that provides the large flux of ATP necessary for flapping flight. In addition to the potential role of hyperoxia in the evolution of animal flight, enhanced oxidative metabolism contributes to augmentation of heat production, allowing maintenance of the high levels of performance seen during powered flight. To picture the precise environmental conditions which may have favored the evolution of avian flight is not easy because most likely a mosaic of interplaying factors were involved. At present, it is safe to say that the whole scenario seems to have been very favorable in terms of energy and nutrient supplies, atmospheric physical conditions, and the spreading of a new ecological niche to be exploited, i.e., the concomitant expansion of the angiosperms (Heimhofer *et al.* 2005).

Paleontological studies show that modern avian groups probably first appeared in the Paleocene or Eocene and experienced an explosive radiation in the early

Cenozoic. Molecular studies have, however, favored a much earlier origin of modern avian groups, nearly 50 million years earlier than inferred from the fossil record (reviewed by Zhou 2004).

Because of a putative explosive radiation, deep avian evolutionary relationships have been difficult to resolve. In a recent phylogenomic study of birds, in which ~32 kilobases of aligned nuclear DNA sequences from 19 independent loci for 169 species, representing all major extant groups were examined, Hackett *et al.* (2008) were able to recover a robust avian phylogeny from a genome-wide signal, which included multiple loci with diverse rates of evolution, in particular the rapid evolving introns.

Their major findings can be summarized as follows: The largest clade in Neoaves was a well-supported land bird clade (node F, Fig. 1.2) that contained the Passeriformes (perching birds, representing more than half of all avian species), which is allied with several morphologically diverse orders. These included Piciformes (woodpeckers and allies), Falconiformes (hawks and falcons), Strigiforms (owls), Coraciiformes (kingfishers, hornbills, rollers and allies), Psittaciformes (parrots), Coliiformes (mousebirds), and Trogoniformes (trogons). One of the most unexpected findings was the sister relationship between Passeriformes and Psittaciformes (node A, Fig. 1.2), with Falconidae (falcons) sister to this clade. Sister to the land birds is the Charadriiformes (shorebirds, gulls and alcids; yellow, node G, Fig. 1.2). According to Hackett and colleagues, regardless of the exact placement of Charadriiformes in their analysis, they consistently support that this order is not basal within Neoaves (Paton *et al.* 2002), and refute the hypothesis that transitional shorebirds gave rise to all modern birds (Feduccia 1995). Their phylogeny also revealed a highly supported water bird clade (node H, Fig. 1.2), including members of the Pelecaniformes (totipalmate birds), Ciconiiformes (storks and allies), Procellariiformes (tube-nosed birds), Sphenisciformes (penguins), and Graviiformes (loons). Basal to the water birds were two clades of terrestrial and arboreal taxa (node J, Fig. 1.2): Musophagiformes (turacos) and a clade (node I, Fig. 1.2) including core Gruiformes (rails, cranes and allies), Cuculiformes (cuckoos), and Otididae (bustards, typically considered as belonging to Gruiformes). One of the most important findings reported by Hackett and colleagues was that several well-accepted orders were not monophyletic. Their analysis provided strong support that Tinamiformes (tinamous) are found within Struthioniformes (ostriches and allies; node Q, Fig. 1.2), Apodiformes (hummingbirds and swifts) are found within Caprimulgiformes (nightjars and their allies; node L, Fig. 1.2), and Piciformes are found within Coraciiformes (node C, Fig. 1.2). Typical Pelecaniformes and Ciconiiformes were intermixed in a clade that excludes one traditional pelecaniform family: the Phaethontidae (tropicbirds). The Gruiformes represented at least four distinct clades in their tree. Falconidae and Accipitridae (falcons and hawks) formed distinct clades, rather than a monophyletic Falconiformes (Fig. 1.2).

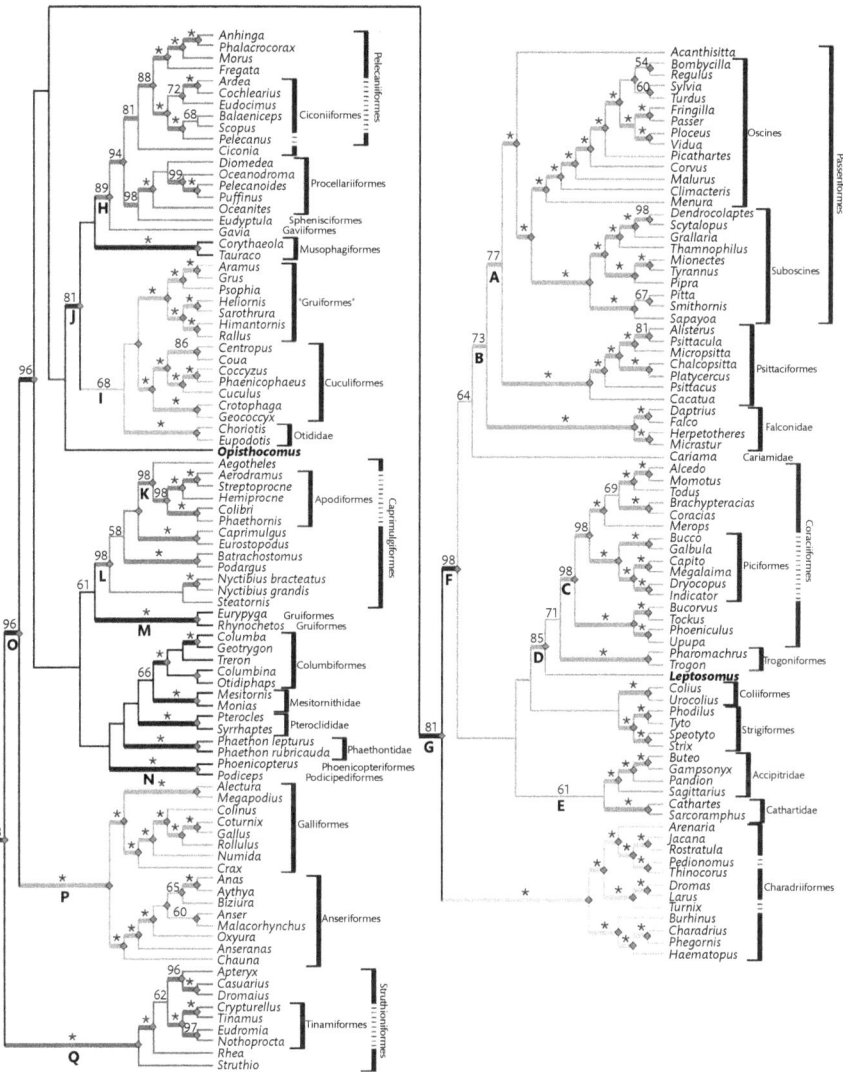

Fig. 1.2 Phylogenetic tree with major groups of birds (Hackett *et al.*, 2008; with permission from AAAS). See text for details.

The broad structure of Hackett and colleagues' phylogeny suggests diversification along general ecological divisions, such as water birds, shorebirds, and land birds. Adaptations to these environments arose multiple times (Fain and Houde 2004), because many aquatic birds were not part of the water bird clade (e.g., tropicbirds, flamingos, and grebes) and terrestrial birds were found outside of the land bird clade (e.g., turacos, doves, sandgrouse, and cuckoos). Their phylogeny also indicated that several distinctive lifestyle niches, such as nocturnal (owls, nightjars, and allies), raptorial (falcons, hawks, eagles, New World vultures, seriema, and owls), or pelagic (tubenosed birds, frigatebirds, and tropicbirds), have evolved multiple times. Furthermore, their results reinterpret the evolution of various adaptations (e.g., the diurnal Apodiformes evolved from nocturnal/crepuscular Caprimulgiformes, and the flighted Tinamiformes arose within the flightless Struthioniformes) and biogeographic patterns (e.g., the New Caledonian Kagu and Subtropical Subittern are sister taxa).

1.3 Summary of Passerine Bird Phylogeny

The diversity of extant species of the order Passeriformes ("perching birds") is comparable to that of all living mammals combined. A strongly supported phylogenetic tree based on single-copy nuclear gene sequences (RAG-1 and RAG-2) for the most complete sampling of passerine families to date has been proposed recently (Barker et al., 2004). It suggests multiple waves of passerine dispersal from Australasia into Eurasia, Africa and the New World, beginning as early as the Eocene period. In essence, it reverses the classical scenario for oscine biogeography. A reassessment of comparative analyses of passerine diversification and adaptation is therefore required as the new data gathered by Barker and colleagues (2004) implies a revised history of the group (Fig. 1.3).

In summary, at the base of the revised phylogenetic tree, proposed by Barker and colleagues (2004), the New Zealand endemic family Acanthisittidae (New Zealand wrens) is the sister group of all other passerines, and the later clade can be divided into suboscines and the oscine songbirds. Within the suboscines, there is additional support for a division into a New World and Old World assemblages. The phylogenetic structure among basal oscine lineages is according to Barker and colleagues more complex than previously realized.

Quantitative biogeographic methods to infer dispersal patterns, combined with molecular clock methods (Reisz and Müller, 2004) to infer the absolute timing of the radiation of passerine lineages across the globe, have provided a set of spatial and temporal constraints on the rates and patterns of passerine diversification.

Basal divergences among *Acanthistitta* and the suboscine passerines are trans-Antarctic, with profoundly diverging lineages distributed in South America, New Zealand, and Australasia, suggesting a Gondwanan influence on the distribution of these taxa (Cracraft, 2001). The earliest fossil passerine is from the early Eocene of Australia (Boles, 1995), and the next known fossil is from late Oligocene of Europe (Mourer-Chauviré, 1989). In contrast, molecular results obtained by Barker and colleagues indicate that passerines began their diversification in the

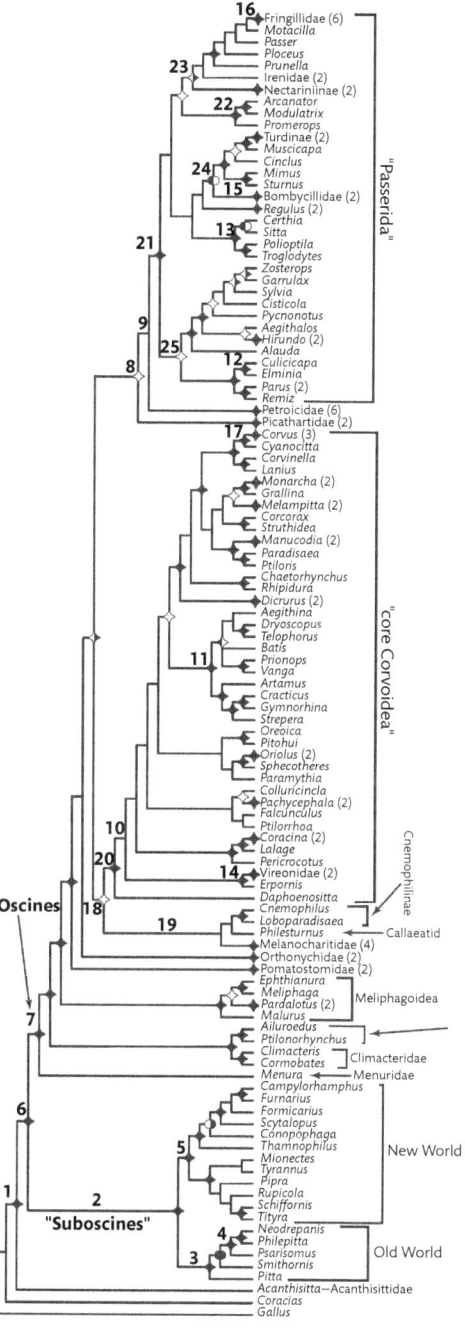

Fig. 1.3 Relationships among passerine birds based on analysis of combined RAG-1 and -2 sequences (146 taxa, 4,126 aligned nucleotide positions). The topology presented is the best ML (maximum likelihood) estimate, rooted using *Gallus* and *Coracias*. Multiple exemplars of genera and certain higher taxa have been collapsed for clarity (number of species indicated in parentheses after genus or group). Nodal support is indicated by symbols and symbolfills. The quotes around the taxon "Passerida" indicate that Sibley and Ahlquist's (1990) definition of the group excludes the underlined taxa within this clade and includes the underlined taxa shown here as falling outside the clade. Their Corvida includes all oscines (node 7) not in the Passerida. Likewise, the quotes around the taxon "Suboscines" indicate that this group traditionally contains the Acanthisittidae, here shown as basal within passerines. (From Barker *et al.*, 2004; with permission from the National Academy of Sciences).

Late Cretaceous, approximately 30 million years before their first appearance in the Southern Hemisphere fossil record, and 60 million years before their first appearance in the north. Barker and colleagues (2004) proposed that the large gap in the northern record is probably a function of the southern origins of passerines and the temporal history of their radiation out of Gondwana.

In their study, Barker et al. (2004) suggest that the highly speciose New World suboscines began their diversification in South America near the Cretaceous-Tertiary boundary, indicating provincialism and vicariance before continental breakup of South America, Antarctica, and Australia. Cracraft (2001) proposes that the biogeographic pattern of the remaining suboscines, the relatively species-poor broadbills and pittas, currently found in Madagascar, continental Africa, Asia, Australia and New Guinea, seems to implicate tectonic vicariance involving India, Madagascar, and Antarctica followed by diversification in Asia.

Diversification of oscines thus appears to have involved dispersal of multiple lineages from Australasia, a single one of which, the Passerida, represents the massive diversification of passerines (Barker et al., 2004). The establishment of divergent oscine lineages in Asia and Africa was probably mediated by the presence of mesic forests throughout northern Africa and Eurasia before Miocene desertification (Axelrod and Raven, 1978), which in turn generated divergent patterns of distribution in many bird groups and also in other vertebrate lineages (Cracraft, 1973).

Due to the appearance of oceanic barriers from continental drift, in combination with climatic worsening conditions, it is unlikely that dispersal of most oscines into the New World occurred via the North Atlantic (McKenna, 1975) or Antarctica (Askin and Spicer, 1995). This argument is in agreement with broad-scale spatial and temporal analyses of the extant Holarctic fauna (Sanmartín et al., 2001) and with the timing and pattern of mammalian dispersal indicated by the fossil record (Webb and Opdyke, 1995): both suggest that dispersal into the New World in the late Oligocene/early Miocene periods was predominantly through Beringia (Barker et al., 2004). This, in turn, is consistent with the hypothesis of relatively recent emberezine dispersal and diversification in South America, together with the strongest push of Andean uplift (Burns, 1997).

1.4 Evolution of Feathers

The recent discovery of feathered dinosaurs from Liaoning Province in China (Xu *et al.* 2001) seriously challenges the notion that feathers are unique to birds. Feathers are characteristic of extant birds, but they might have originated earlier than the group descended from the most recent common ancestor of *Archaeopteryx* and extant birds.

It is important to bear in mind that the origin of groups, the origin of a structure and the origin of a behavior or function are fundamentally different questions.

Also, because the fossil record still has many gaps, it is not possible to assert whether the origin of feathers was concomitant to the origin of flight. Nevertheless, over the past decade, every new discovery made in Mesozoic strata around the world bolsters the dinosaurian nature of birds and the non-volant origin of feathers.

More recently, a selective regime favoring a streamlining of body contours and surfaces has been proposed by Homberger and de Silva (2000) as being instrumental in driving the morphological and functional transformation of an unfeathered reptilian integument into a feather-bearing avian one. This hypothesis is consistent with a new structural and functional analysis of the microanatomy of the avian feather-bearing integument. This considers the skin and feathers as a complex inte-

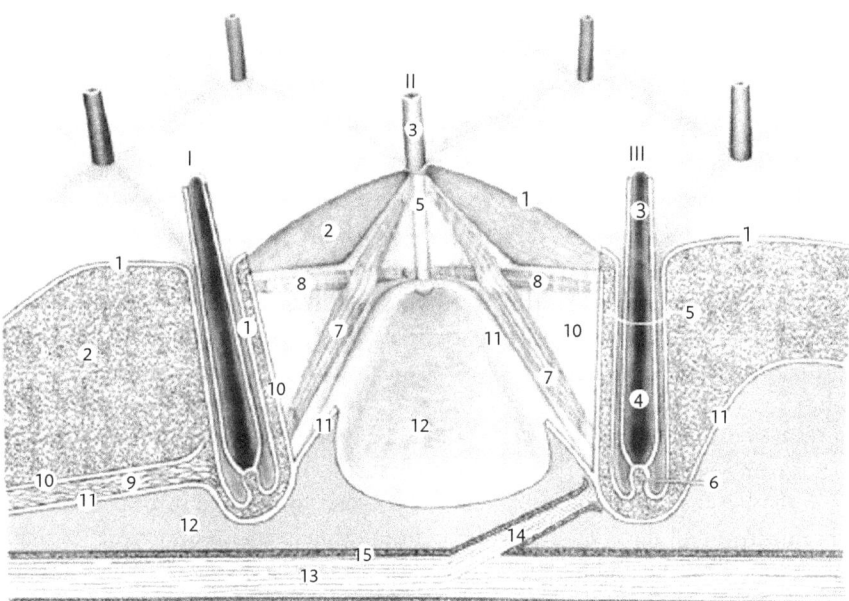

Fig. 1.4 Diagrammatic transverse section through the feather-bearing integument of the domestic turkey, *Meleagris gallopavo*, at about mid-length of the neck. The view is towards caudal of the bird, the left side of the figure is towards dorsal of the bird, and the right side of the figure is towards ventral of the bird. A wedge of tissue, comprising the epidermis, dermis, elastic membrane, and a superficial portion of the *Fascia superficialis*, was removed from the center of the section to reveal the obliquely arranged feather muscles. 1 epidermis, 2 dermis; 3 rachis of a feather (cut), 4 calamus of a feather, 5 feather follicle, 6 feather papilla, 7 smooth erector feather muscle, 8 smooth depressor feather muscle, 9 smooth apterial muscle, 10 elastic epimysium of the apterial and feather muscles, 11 *Lamina elastica*, 12 *Fascia superficialis*, 13 striated subcutaneous muscle (*M. constrictor colli*), 14 *Pars pennae* of the striated subcutaneous muscle, 15 collagenous epimysium of the striated subcutaneous muscles, I–III numbered feathers (Homberger and de Silva, 2000; with permission).

grated organ system that includes an intricate hydraulic skeleto-muscular apparatus of the feathers, a dermo-subcutaneous muscle system of the integument, and a subcutaneous hydraulic skeletal system formed by fat bodies (Fig. 1.4). Key elements of the evidence supporting this view are: i) the presence of depressor feather muscles that are not needed as antagonists for the erector feather muscles but can counteract external forces such as air currents; ii) the fact that the highly intricate feather-bearing integument constitutes a machinery to move feathers or to stabilize them against external forces; iii) the role of feathers in streamlining the body contours and surfaces of birds; iv) the role of feathers as controllable temporary turbulators, i.e., creating turbulence in the boundary layer and preventing the detachment of the air flow from the body surface under certain circumstances, such as during landing; and v) the critical role that a streamlined body plays in avian flight. The last factor in particular is likely to have guided evolutionary transformation from ecologically versatile quadrupedal reptiles with diverse locomotor patterns to volant bipedal birds without passing through parachuting or gliding stages. These transformations are likely to have occurred more than once.

The evolution of a coat of feathers allowed fusiform body contours without adding considerable weight in the form of fat, which is a serious handicap to airborne organisms. Even though fat tissue is an integral part of the hydraulic skeleto-muscular apparatus of the feathers and of the subcutaneous hydraulic skeletal system, the amount needed for this purpose is a small fraction of the amount of subcutaneous fat that would be needed to streamline an unfeathered body. Streamlining with fat would also reduce the mobility and flexibility of the neck. A fusiform body shape is especially significant for drag reduction in small birds because of their relatively larger surface area. Thus the selective pressure towards a fusiform body shape is bound to be greater for small than larger species (small birds are generally more completely fusiform than larger birds), so that it is likely that the characteristic fusiform body shape originated in a small ancestor (Homberger and de Silva, 2000). In addition, smaller size is also more forgiving of an imperfect flight apparatus.

In modern birds, feathers serve a vast number of functions. Primary and secondary wing feathers create the flight surface, and contour feathers are essential for streamlining. Down is one of the best insulation substances found within Animalia, and extant birds use feathers, along with their own body heat, to incubate eggs. In both terrestrial and aquatic birds, feathers form a waterproof (or water-resistant) coat. Social signaling is very important to birds, and a variety of specialized plumages of diverse form, color, and pattern are central to this.

1.5 Developmental Biology of Feathers

The origin and evolution of feathers have been of great interest (Prum 1999; Prum and Brush 2002; Sumida and Brochu 2000), particularly with the recent discoveries of feathered dinosaurs in Northern China (Zhou *et al.*, 2003; Xu *et al.*, 2004). From the many intermediate forms, we have learned that feathers evolved through stepwise evolutionary novelties to produce diverse morphology with new functions. Today's birds have evolved anatomically-specific feather forms, ranging from radially symmetric downy feathers to bilaterally symmetric flight feathers to radically modified feathers used as display ornaments.

The existence of primitive feathers, or protofeathers, in theropod dinosaurs such as *Sinosauropteryx* has been challenged. The integumental structures proposed as protofeathers are instead, according to recent findings, the remains of structural fibers that provide toughness and bear striking similarity to the structure and levels of organization of dermal collagen (Feduccia *et al.* 2005; Lingham-Soliar *et al.* 2007).

The feather has a folicullar structure with the dermal papilla at its base. The feather filament is a cylindrical structure with mesenchymal pulp inside. Moving up the developing feather shaft from the proximal to the distal end, there are the collar, where cells actively proliferate, the ramogenic zone, where barb ridges start to form, and maturing feather branches, where barb ridges form the ramus and barbule plates that keratinize to become barbs. In downy feathers, all barb ridges are parallel to the long follicular axis. In flight feathers (remiges), bilaterally symmetric barb ridges converge obliquely toward the anterior follicle, leading to fusion and the creation of the shaft (Fig. 1.5).

The evolution of bilaterally symmetric feathers is a fundamental process leading toward flight. One major mystery is how the genetics and developmental processes can form both radially symmetric downy feathers and bilaterally symmetric flight feathers.

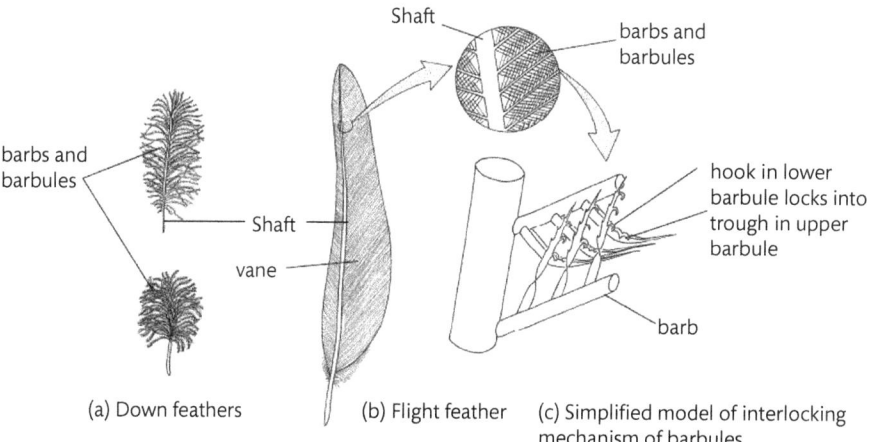

Fig. 1.5 Feather structure (© Copyright D. G. Mackean; with permission).

In developing downy feather follicles, barb ridges are organized parallel to the long axis of the feather follicle. In developing flight-feather follicles, the barb ridges are organized in a helix toward the anterior region, leading to the fusion and creation of a rachis. An anterior-posterior molecular gradient of the gene *wingless int* (Wnt3a) in flight feathers but not in downy feathers has recently been discovered (Yue *et al.* 2006). Global inhibition of the Wnt gradient transforms bilaterally symmetric feathers into radially symmetric feathers. Production of an ectopic local Wnt3a gradient reoriented barb ridges toward the source and created an ectopic rachis. The orientation of the Wnt3a gradient is dictated by the dermal papilla. Swapping dermal papillae between wing covert and breast downy feathers demonstrates that both feather symmetry and molecular gradients are in accord with the origin of the dermal papilla.

Feather epidermal stem cells are in a ring-configured niche, horizontally placed in the radial symmetric feathers and tilted anteriorly-posteriorly in bilateral symmetric feathers. These epidermal stem cells are true stem cells that can be modulated into distinct symmetric forms by the different microenvironments created by the dermal papilla. Furthermore, an anterior-posterior Wnt gradient is involved in the property of this micro-environmental niche. Therefore, the fates of feather epidermal cells are not predetermined through some molecular codes but can be modulated (Yu *et al.* 2002). Data recently obtained suggest that feathers are shaped according to the following sequence of events:

> Dermal papilla → Wnt gradient → helical barb ridge organization → creation of rachis → bilateral symmetry

Diverse feather forms can be achieved by adjusting the orientation and slope of the molecular gradients, which then shape the topological arrangements of the feather epithelia, thus linking molecular activities to organ forms and novel function.

1.6 Origin and Evolution of Flight

The new arguments for the cursorial origin of flight of birds are almost all based on the assumption that bird-related Theropods were cursorial animals, and on the recent discoveries of feathered dinosaurs from Early Cretaceous of China (125 million years ago). The cursorial theory assumes that ancestral birds developed powered flight from the ground up without a transition stage involving an intermediate arboreal or gliding phase (Padian and Dial, 2005). Most of the fossil evidence, however, points to an arboreal scenario for the origin of avian flight (e.g. Feduccia, 1996). The so-called arboreal theory proposes that flight evolved from tree-dwelling ancestors, and predicts an intermediate gliding phase (Long et al., 2003).

The evolution of powered flight in birds is recognized as the key adaptive breakthrough that contributed to the biological success of the group. We are now starting to understand how the transformation of wing design from non avian dinosaurs to early birds may have occurred, mainly based on a wealth of newly-discovered fossil from China (see Section 1.2). Among the most recent findings, *Microraptor gui* offers perhaps the best evidence that arboreal dromaeosaurs might have acquired powered flight through a gliding stage where both forelimbs and hindlimbs were involved (Chatterjee and Templin, 2007).

According to Chatterjee and Templin (2007), the most unusual feature in *Microraptor* is the presence of long, asymmetric flight feathers on the entire length of the metatarsus, which are not present in *Archaeopteryx*, feathered dromaeosaurs, and modern birds. Also, according to these authors, because *Microraptor* could not extend its hindwings directly behind the forewings in the same plane, it probably held its feet lower than its arms, a configuration that would most likely have offered them a more stable anatomical and aerodynamical configuration. Based on anatomical and aerodynamic modeling, and once the parasagittal posture and feather orientation of the hindlimb are understood, the wings of *Microraptor* resemble those of a staggered biplane when viewed from the side, where the forewing forms the dorsal wing and the metatarsal wing forms the ventral one (Chatterjee and Templin, 2007).

Xu and colleagues (2003) presented anatomical evidence indicating that *Microraptor* was not capable of direct lift-off from the ground or from a running start, because it lacked the supracoracoideus pulley to elevate its wings. Furthermore, a running takeoff would damage the ventral metatarsal wings. Instead, by using *phugoid* gliding (when birds take off from a perch, they do not seem to use excess power; they loose height at first and then swoop up with a large-amplitude undulation to swing between two perches), *Microraptor* could potentially have traveled form one tree to another tree by undulating flight covering a horizontal distance of more than 40 meters. With the presence of tibial feathers, *Microraptor* could eliminate 40% of drag when the legs were held in a z-fashion (Chaterjee and Templin, 2007).

Support from current fossil and phylogenetic evidence indicates a gradual shift from the hindlimb to forelimb as the main locomotor structure during the evolution of avian flight. According to Chaterjee and Templin (2007), extant raptors (hawks and falcons), differently from other birds, maintain their hindlegs in a z-configuration during preparation for aerial attack and carrying prey, hanging loosely their tibiae in a vertical plane (Fig. 1.6). The so-called feathered "trousers" are a quite noticeable feature found in predatory birds, keeping their prey-catching legs streamlined during aerial attack. *Microraptor*, apparently, constitutes one of the most important pieces of the puzzle to find the answer to the role of leg feathers in the flight of extant raptors (Chaterjee and Templin, 2007).

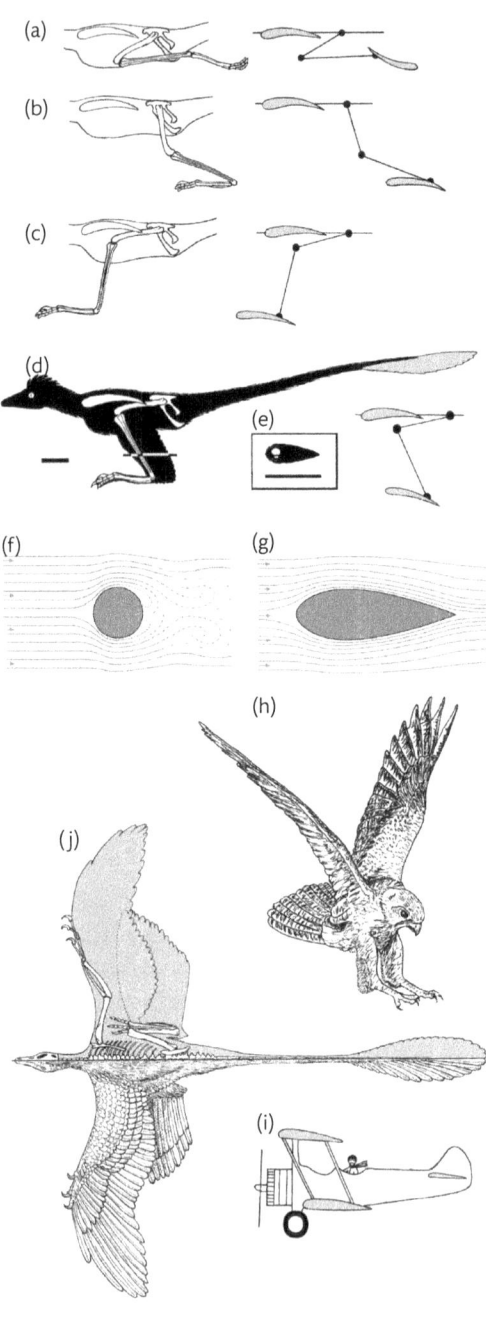

Fig. 1.6 Wing planform of *Microraptor*. (*A–D*) Different possible hindlimb postures during flight. (*A*) Hindlimb backwardly directed as in modern birds. (*B–D*) Biplane configuration. (*B*) Hindlimb backwardly sloping position. (*C*) Hindlimb forwardly sloping in predatory strike position. (*D*) Hindlimb in z-fashion with a body silhouette showing the animal in lateral view with an upwardly tilted tail for pitch control. (*E*) Cross-section of the tibia –fibula showing a streamlining and stretching effect of the cylindrical tibia by adding feathers caudally. (*F*) Cylindrical structure offers maximum resistance to the airstream as the airflow behind it becomes broken up into eddies, creating turbulence. (*G*) Filling the spaces in a cylindrical structure in front and behind improves streamlining, as in the case of the feathered tibia of raptors. (*H*) Pouncing posture of a raptor, *Falco*. (*I*) A typical staggered biplane for comparison with *Microraptor*; in biplane aircraft of the 1920s, there was a large additional drag of wires, struts, etc. between the two wings, which eventually made the biplane obsolete except for a niche application; such drag-induced structures were absent in *Microraptor*. (*J*) Life reconstruction of *M. gui* (IVPP V13352) in dorsal view showing the morphology and distribution of hindlimb feathers (*Left*) and orientation of the hindlimb (*Right*) during gliding; proximal feathers on the humerus and femur are inferred (data are from ref. 12). (Scale bar, 5 cm.) (Chaterjee and Templin, 2007; with permission from the National Academy of Sciences).

Finally, a new concept, 'wing-assisted incline running' (WAIR; Dial, 2003), has provided an interesting explanation for the origin and evolution of avian flight. It argues that the incipient wings of feathered Theropod dinosaurs probably played a role similar to that of modern birds which employ wing-assisted incline running, rather than flying, to reach elevated refuges such as cliffs, trees and boulders. This hypothesis appears to combine aspects of the arboreal and cursorial scenarios.

The WAIR hypothesis has interesting implications when we consider the ontogeny of flight in extant birds. There are large differences between precocial and altricial birds with regards to their locomotor behavior and performance (which will be explored further in this volume), in particular those aspects which are related to their ecology (Dial, 2003; Dial et al., 2006).

Precocial and altricial birds Recent birds have modes of development, which vary from highly altricial to highly precocial modes. Most tree-nesting and cavity-nesting birds have altricial chicks, which require long periods of feeding before they are strong enough to leave the nest on their own. In contrast, most shorebirds, game birds, and ducks have precocial chicks, which can walk away from the ground nest within hours of hatching.

Although it is not the purpose of this volume to discuss in depth the origin of flight, it is important to emphasize that this matter is far from being completely settled. New discoveries will most likely be made in the near future, especially from the fossil record, which in turn may or may not challenge our current views of the facts.

1.7 Energy Requirements for Flight

As we described earlier, birds evolved an array of special morphological characteristics which allowed them to exploit the aerial stratum with enormous success. When we think of bird flight, powered flapping flight is probably what first comes to mind. As flapping flight is the most predominant type of flight in extant birds, we will first discuss the energy requirements involved in this type of locomotion. Requirements for the other types of flight (gliding, soaring, and hovering) will be discussed later on together with their ecological implications.

Since rapid wing movement is intrinsic to flapping flight, a flying bird resembles a helicopter more than it does an airplane. The wings push an air stream slightly downward at the fastest forward speed. Except during brief periods of upward or downward movement or acceleration, the rate of creation of downward momentum must equal body weight (see e.g., Alexander 2004; Pennycuick 1975 and 2008; Vogel 2003; Videler 2006).

In airplanes, functions such as staying aloft and progressing forward with decent speed and economy, are divided, with fixed wings responsible for the first and a propeller or jet thrust responsible for the second and only indirectly for the first. Nature's fliers, however, are able to effectively combine both in a single – albeit complex – structure. Scaling provides the key. Wing loading is the ratio of weight (mass X gravitational force, or m·g) to wing surface area (S):

$$\textbf{Wing loading} = \textbf{m·g/S} = \rho\textbf{·V·g/S} \tag{1}$$

If we ignore the density of the bird (ρ) and gravity (g), which are both size-independent, we obtain a residual dimension of length from equation 1. Therefore, wing loading will be proportional to length or the cube root of body mass. So a smaller flier (assuming isometry) has a lower wing loading, quite aside from aero-dynamic variables such as Reynolds number (the ratio between inertia and the viscosity of air). In other words, the smaller flier is lighter relative to its wing area, and thus it needs less lift relative to that area. Lift depends on speed, in particular on the square of speed (Lift = $1/2\ C_1 \cdot \rho \cdot S \cdot v^2$; where C_1 is coefficient of lift, ρ is density, S is surface area, and v is speed). Accordingly, the smaller flier does not have to fly as fast to achieve sufficient lift to sustain flight. The larger birds hedge a bit, bending the wings in midspan as they beat them, so the inner part of each wing works mainly as a fixed wing and the outer part as a propeller. And the smaller the flying bird, the better it gets at zero-speed flight – important for take-off and for hovering. Big birds typically need a running takeoff and cannot hover; smaller ones can hover very briefly; very small birds can perform sustained hovering.

Active flight has a high power requirement, but efficiency is speed-dependent and hence a flying animal has the option of choosing the most economical speed. What matters for economy in flapping, forward flight is drag. Drag increases with the square of speed, and power is drag times speed, hence power require-ments increase with the cube of speed. So, for the best economy per unit distance it is better to fly slowly. However, things are more complicated due to the aero-dynamics of flight. Not only does the body of the bird suffer drag, but so do the wings, as a price for making lift. One component, called "profile drag", behaves in such a way that wings cannot avoid surface friction. Profile drag increases with speed and is relatively low at low Reynolds numbers. Wings, additionally, suffer "induced drag" because they are not infinitely long, and they thus have to give passing air some finite deflection. Induced drag increases as flying speed drops. At the same time, higher speed does increase the drag on the bird's body. What is worse, in practice, is that it gradually reduces the animal's advance ratio (how far the animal moves forward relative to how large is the amplitude of the wingbeat; the latter also called stroke angle) and limits the power available from beating wings (see Vogel 2003). Because induced drag increases as flying speed drops, the power requirement for forward flight increases at slower speeds than the most economical speed, and therefore it is not always better to fly at slow speeds.

The muscular power available may be some fairly fixed multiple of basal metabolic rate (BMR: the rate of metabolism in the zone of thermoneutrality (see Chapter 2) when the animal is at normal body temperature, is post-absorptive, and is quiescent; e.g., Aschoff and Pohl 1970b). However, the aerodynamic power available is not. Aerodynamic power is maximal at some forward velocity intermediate between zero and top speed. Multiplying drag by forward velocity (not the speed of local flows over the wings) gives the necessary power input. The flight muscles must generate that power and transmit it outward across the hinge points of the wings. Birds can sustain flight wherever power available equals or exceeds power required. Permissible speeds are not necessarily the most economical speeds, in the context of energy cost or achievable flight range on a given amount of metabolic fuel. For example, for a starling weighing 70 g the minimum power speed (i.e., where energy cost is lowest) is 7.8 m sec^{-1} (about 13 km hr^{-1}) while the speed providing maximum flight range is 13.3 m sec^{-1} (about 48 km hr^{-1}).

The relationship between mechanical power output (rate at which mechanical work is done) and forward velocity in bird flight is still controversial (Ellington 1990; Rayner 1999). For flying birds, aerodynamic theory predicts that mechanical power should vary as a function of forward velocity in a U-shaped curve. Empirical tests of this theory, using the Black-billed Magpie (*Pica pica*), suggest that the mechanical power curve is relative flat over intermediate velocities (Dial *et al.* 1997). More recently, Tobalske *et al.* (2003) obtained mechanical power curves for Cockatiels (*Nymphicus hollandicus*) and Ringed Turtle-doves (*Streptopelia risoria*). In contrast to the curve reported for magpies, the power curve for Cockatiels is acutely concave as predicted by theory, whereas that of doves is intermediate in shape and shows higher mass-specific power output at most speeds. Tobalske and colleagues also found that wing-beat frequency and mechanical power output do not necessarily share minima in flying birds. Thus, aspects of morphology, wing kinematics, and overall style of flight can greatly affect the magnitude and shape of species' power curves (Hedrick *et al.* 2002; Hedrick *et al.* 2004). The mechanical power output required for flight is only a fraction of the total metabolic energy needed by birds to fly, and the ratio of mechanical over metabolic power is a measure of efficiency. For flapping flight efficiency increases from a few percent in small birds to about 30% in large birds.

The minimum metabolic power required for flight has been calculated for a size range from 2–3 g hummingbirds to 19 kg bustards (Fam. Otididae). When the basal metabolic rate (BMR) has been added to the minimum metabolic power, we have an estimate for the total power required to fly. A hummingbird with a body mass of 4 g must increase its metabolic rate above the resting level by only a factor of about 3.3 to fly, while a 7.3 kg griffon vulture has to raise its metabolic rate to 20 times the resting level to stay airborne. Obviously the largest birds are expending energy at rates close to the upper limits of their abilities while sustaining level flight in still air. Therefore it is not surprising that most large flying birds spend more time in low-cost gliding flight than in expensive flapping flight (McMahon and Bonner 1983). Smaller birds, by comparison, have a surplus of power which can be used to improve acceleration, load-carrying ability, maneuverability, or other aspects of flight performance.

Many small birds, including Skylarks (*Alauda arvensis*), swallows and martins, use an intermittent mode of flight in which they alternately flap and then fold their wings. Whenever the drag due to the air friction on the outstretched wings exceeds the drag due to lift, there can be an energy saving for covering a given distance when the wings are folded a part of the time. This can be true only for birds that fly at speeds well in excess of their minimum-drag speed (a speed somewhat greater than the minimum-power speed), and therefore it can be true only for small birds with plenty of power to spare (see e.g., Videler, 2006). As an example, consider a small bird traveling at an average airspeed 50% greater than its minimum-drag speed. If it flapped its wings half the time and kept them folded the rest of the time, the average drag (and, therefore, the energy consumption for traveling a given distance) could be reduced by 17%.

Waterfowl such as swans and geese are excellent fliers and can cruise at relatively high speeds, but they also have high take-off speeds because of their high wing loadings (body mass/wing area). They can reach their take-off speeds by running along the water, using their big feet as paddles. The large birds of prey – hawks and eagles – often use another technique to launch themselves: they jump off tall trees to get up to flying speed. However, large eagles are quite capable of taking off from the ground without a running start, using their powerful legs to jump into the air.

The fact that the largest birds are at the ragged edge of not having enough power for sustained level flapping flight leads them to search for free rides in ascending currents of air. Ascending air can be found at low altitudes on the windward sides of cliffs and mountains. It can also be found at intermediate altitudes in the standing waves of rising and falling air downwind of uneven topography. Many soaring birds take advantage of "thermals": columns of warm air heated by hot ground and rising through cooler air. This phenomenon can occur over perfectly flat terrain and the use of it is commonly referred to as thermal soaring. Dynamic soaring, on the other hand, is practiced by many marine birds flying over ocean waves (Pennycuick 2002). They take advantage of the fact that winds experience drag at the water surface. Horizontal wind speeds are therefore lower near the surface and increase with altitude. Birds alternately ascend and descend, extracting energy from the velocity gradient. Albatrosses and other large seabirds are believed to make use of speed reductions in the wind gradient (see e.g., Vogel 2003; Videler 2006).

In aircraft, a long, narrow wing has a lower induced drag than a short, stubby one, even though both may have the same lift. This is because a long wing can spread the downwash over a large area behind the aircraft, with the result that the downwash velocity is reduced substantially by comparison with a short wing. Many soaring birds also obey this rule. For example, albatrosses have extremely long, narrow wings and resemble long-winged sailplanes. From perches on high cliffs, many seabirds merely jump out into updrafts when they want to soar. The updraft is always there, provided a sea breeze is blowing, and the bird rarely has to flap its wings.

As the sun warms the ground in the morning, the intensity of the vertical updrafts (thermals) slowly increases, reaching a peak in the mid-afternoon. The speed of the smallest vertical updraft necessary for soaring is a fixed fraction of the minimum-power airspeed for the bird in question and therefore is proportional to the square root of the bird's wing loading (body mass/wing area). Because wing loading increases with the size of the bird, the minimum conditions for thermal soaring are reached earlier in the morning for the smaller birds. Kites soar earlier in the day than vultures. This aerodynamic fact gives the smallest soaring birds the first crack at whatever food opportunities have accumulated during the preceding night.

The largest birds capable of hovering at a spot for appreciable periods are hummingbirds weighing about 20 g. They do this standing on their tails in the air and moving their wings back and forth. By creating a downward jet of the air, a hovering animal gives itself an upward thrust. To sustain hovering, the rate at which the downward airflow carries away momentum must be equal to the animal's weight (see e.g., Weis-Fogh 1972; Vogel 2003; Altshuler et al. 2004a; Videler 2006). The power per unit body mass required to generate lift is directly proportional to the square root of the mass and inversely proportional to the wingspan. The mass-specific power required for hovering increases by almost a factor of 6 in going from tiny cabbage butterflies to hummingbirds. A 20 g hummingbird with a wingspan of 30 cm requires about 130 watts kg^{-1} to hover, a muscular power close to the maximum sustainable power for an animal of this size (by comparison, the maximum power output of a human distance runner is about 15 watts kg^{-1}).

Variation in wing disc loading (WDL), which is based on the area "swept" by the beating wing (i.e., wing length and the subtended angle of the complete wing stroke, which must be 180° or less), has been used to elucidate differences either among hummingbird species in nectar-foraging strategies (e.g., territoriality, traplining) and dominance relations or among gender-age categories within species. Altshuler et al. (2004b), however, consider that variation in hummingbird flight and behavior, including hovering, cannot be easily classified using WDL, and instead is correlated with a diversity of morphological and physiological traits (Feinsinger and Colwell 1978; Stiles 1995). Hummingbird hovering flight approaches in many aspects that of insects, yet remains distinct because of effects resulting from an inherently dissimilar – avian – body plan (Weis-Fogh 1972; Warrick et al. 2005).

1.8 Migration

Migration is perhaps the most impressive aspect of the aerial life style of many birds. Birds are preadapted to migrate through their capacity for active flight as well as their species richness coupled with great ecological differentiation.

Birds have conquered virtually all terrestrial and oceanic biomes, and their migration routes encompass most of the earth's surface like a mesh. In extreme cases birds cover distances equivalent to the circumference of the planet; they cross oceans, deserts, mountains, and ice-fields, and there is no time of year in which one could not find birds migrating somewhere over the globe's surface. To make this system of migration possible the development of many sophisticated flight and fuel-management adaptations, sensory, and control mechanisms was necessary – a development that is still in process today (see e.g., Berthold 1993; Dingle 1996). Swift and continual changes in the environment, now mainly due to human influence (Drent *et al.* 2006), require ever-changing capabilities in contemporary migratory species (Berthold *et al.* 1992; Pulido and Berthold 2004).

1.8.1 Modes of Locomotion

Although flight is the main locomotory activity used by birds during migration, all other kinds of locomotion of which they are capable are also sometimes used to cover migratory distances. The flightless *Hesperornis* of the Cretaceous period had to migrate by swimming, and many recent, mainly marine, species also swim or dive for at least part of the journey. This applies, for example, to the flightless penguins as well as to species from Central European latitudes such as gannets, ducks, geese, grebes, loons, and alcids which may cover over 1,000 km during this process (Berthold 1993). Another locomotor activity related to swimming is also employed by some penguins. By "rowing" actions of their fin-like wings they "snow-paddle" on their bellies, occasionally switching to running and hopping, across snow and ice-fields to and from their breeding places, some thus covering several hundred kilometers. The high cost of pedestrian locomotion in emperor penguins, for example, may have important ecological implications in terms of their energy budget (Halsey *et al.* 2007). Apart from this, migration on foot is rare, and only occasionally occurs on a large scale.

1.8.2 Molting

Many bird populations that undergo a complete flight-feather molt over a short period of time, often leading to a temporary loss of flying or swimming ability, migrate to special molting areas that provide favorable protection and/or food supplies. Molt-related migrations differ considerably with regard to species, populations, and age- and sex-groups; ideal areas are nonetheless traditionally visited regularly (Jehl 1990).

Occasionally, fat deposits are accumulated for the molting period, which often coincides with complete flightlessness. However, the food resources are often so rich in some of the molting sites that body mass can be held or even increased despite restricted mobility. At times, birds may loose a considerable amount of body mass. During the period of flightlessness temporary atrophy of flight muscles is common, between 15 and 20% of a bird's body mass. Before departure, however, the reduced volume flight muscles are strengthened, a process frequently accompanied with exercising wing-beats. Reductions of kidney and pancreas mass during the transition from molt to migration that occur in Garden Warblers (*Sylvia borin*) are considered to be related to the demands of molt, while increased flight muscle maybe due to molt, migration or both (Bauchinger and Biebach 2006). As well as storing considerable fat, penguins seem to undergo a muscle hypertrophy prior to molt, apparently to provide the necessary protein for feather synthesis (they typically molt all feathers at once and cannot enter the water and feed during that period).

1.8.3 Migration Flocks

Flight formations (migration flocks) are common during migration, and at times are even made up of individuals of different species. Migratory birds can often be observed in flocks of varying density (starlings, finches) or in long broad rows (lapwings). Bow-shaped, V-shaped, double or multiple V-shaped formations or even simple oblique lines can also be seen, in which individuals fly to one side of, but closely behind, each other. In all these formations individuals frequently shift positions, and the leading birds periodically swing out from their position and rejoin the line further back. This indicates that leading birds face higher energy demands or that following birds conserve flight energy, presumably by taking advantage of the aerodynamic "wake" of other individuals. Total energy gain and its distribution among flock members vary according to the size and shape of the formation, lateral distance, and other factors. Theoretically, energy saving could amount to 50% of solitary flight costs, though this only applies to ideal formations, which are unlikely to occur frequently (e.g., Alerstam 1990).

1.8.4 Nocturnal Migration

It is not surprising that mainly nocturnal species also migrate at night. However, the majority of day-active migrants, especially small species, also migrate at night. If nocturnal migration is such a widespread trait, especially for small long-distance migrants of many different taxonomic groups, it must have clear advantages over diurnal migration (Berthold 1993). Such advantages could be: i) reduced pressure from predators; ii) gaining time: the use of night for migration provides more freedom to forage during the day; iii) saving energy: flying during the night is less energy-consuming because of the cooler and denser air, wind speeds are lower at night, the range of variable wind directions is lower at night, and vertical turbulences are less common at night; iv) reduced physiological stress: dangers of hyperthermia and dehydration are reduced in the lower ambient temperatures encountered during nocturnal migrations; and v) the use of a star-compass for orientation. Recent data point towards gaining time and saving energy as having the greatest significance among these factors.

In a world-wide context, bird migration takes place for all kinds of reasons. It is astonishing how much of the year individual species and populations spend on the move.

1.8.5 Evolutionary Aspects

The innumerable cases of breeding, resting, and stop-over site fidelity, clearly demonstrates that migrants can navigate precisely between specific places, even over distances of thousands of kilometers. Philopatry and fidelity to breeding areas, as well as birth-site fidelity, have been established as normal behavior for many species, and can been seen as prerequisites for the widespread formation of different races that is often apparent among migrants (Berthold and Querner 1981).

Although still a matter of debate, the origin of avian migration identifies the tropics (Rappole and Jones 2002) as the original "home" of the majority of long-distance migrants utilizing temperate breeding seasons. If this view has wide acceptance, then "traveling to breed" assumes a major role in building a scenario as to how the migratory strategy arose, based on a universal template (Piersma *et al*. 2005). Arctic-breeding shorebirds offer an interesting case for such speculations (Drent *et al*. 2006). The reconstruction of Red Knots' (*Calidris canutus*) "palaeo-flyways", by combining the dating of population divergence based on molecular technology with what is known about the distribution of suitable habitat at the time, reverses common knowledge on how this system spanning the polar region today actually arose (Buehler *et al*. 2006). Divergence dates are recent (20,000 years ago and less) and the evidence points to the migration linking breeding stations of *Calidris canutus islandica* in the high arctic of Canada to maritime Western European wintering areas as a recent "engineering feat" originating from an eastward expansion from the New World. This subspecies (the only representative electing to winter in temperate instead of tropical regions) likely represents a recent innovation, and the molecular data rule out a close relationship to the *Calidris canutus canutus* population with which it shares use of the Wadden Sea nowadays (*canutus* traveling on to winter on tropical African shores).

Temperate-tropical migrants are integral parts of tropical communities (e.g., Keast and Morton 1980) and their ecological attributes in the tropics may have influenced the evolution of their migratory behavior (Rappole 1995). Levey and Stiles (1992) proposed that a continuum of seasonal movements exists among Neotropical birds in response to resource fluctuations and that the evolutionary endpoint of such continua is Neotropical-Nearctic migration. Most Neotropical resident species that engage in small-scale movements are frugivores or nectarivores of edge, canopy, or open habitats – traits that entail reliance on variable resources (e.g., Croat 1975; Levey 1988; Blake and Loiselle 1991). Avian taxa which undertake seasonal migrations out of the Neotropics were hypothesized to have most likely evolved from lineages with these dietary and habitat characteristics for tracking highly variable resources across space and time (Levey and Stiles 1992). Based on these arguments, Levey and Stiles (1992) hypothesized that a nonrandom group of Neotropical lineages – those with at least partially frugivorous or nectarivorous species of edge, canopy, and open habitat – would be the most likely candidates to contain Neotropical-Nearctic migrants (species that breed north of the Tropic of Cancer and spend the non-breeding season to its south).

To evaluate this hypothesis, Levey and Stiles (1992) categorized passerine families of presumed Neotropical origin according to their habitat, diet, and migration. Their hypothesis was generally supported, i.e., Neotropical-Nearctic migration tended to develop in families with the hypothesized evolutionary precursors to migration. Chesser and Levey (1998) further exploited the relationship with diet, habitat, and migration by examining a separate migratory system – austral migration in South America – and by incorporating the avian phylogeny of Sibley and Ahlquist (1990) into their analysis. Austral migrants breed in southern South America during the austral summer and migrate north, toward or into the Neotropics, for the austral winter. They are drawn from the same taxonomic pool as Neotropical-Nearctic migrants but otherwise represent a largely independent migratory system. So, according to Chesser and Levey (1998), the two groups of migrants have presumably faced similar selection pressures and evolutionary constraints, and, therefore, any theory on the evolution of migration out of the Neotropics should hold for both the austral and Nearctic systems. The results obtained by Chesser and Levey showed that habitat is a likely contributing factor in the evolution of migration out of the Neotropics, but that diet appears to be only weakly related to migration. Their results are consistent at the family/subfamily level in both migration systems.

1.8.6 Adaptations

Migrating birds fly over a huge range of altitudes, from somewhat below sea level (passerines migrating on the lee-side of dykes or across natural depressions) up to

9,000 m (in species crossing the Himalayas). Most overland migration usually proceeds at intermediate altitudes.

Many requirements for successful migration are met by means of temporary migratory adjustment of a multitude of physiological and behavioral traits. A number of adaptations are, however, built into the morphology of migrants. By contrast with related bird groups, migrants are generally equipped with longer and more pointed wings, which for a variety of aerodynamic reasons, cause less air resistance than rounded wings. Depending on their way of life, migrants' wings may have one or the other form, but usually the former.

Further important adaptations include the construction of the "flight motor", the large pectoralis muscles. These can constitute up to 35% of total body mass in migrants. Migration is obviously an activity requiring great endurance capacity, and hence it relies on oxidative metabolism for energy. Long-distance migrants invariably possess rapidly oxidative glycolytic muscle fibers, whereas short-distance migrants possess fibers with lower oxidative capacity in addition to their fast-acting glycolytic fibers. In addition, long-distance migrants possess the fibers with the smallest diameters and the greatest capillary density, and therefore relatively shorter O_2 diffusion times (Altshuler and Dudley 2006). These features may be enhanced during migration by muscular hypertrophy and modifications of the muscles. Anaerobic metabolism apparently only plays a major role during take-off, landing, or fast directional changes: situations where brief but very intense bursts of power are needed.

An enhanced substrate delivery to working aerobic skeletal muscle is also important, particularly during heavy and strenuous exercise. Maillet and Weber (2007) tested the oxidative capacity (Krebs cycle and β-oxidation) of the flight muscles of Semipalmated Sandpipers (*Calidris pusilla*) during their fall migration from the Arctic to South America. Oxidative capacity increases rapidly during refueling in the Bay of Fundy (east coast of Canada), before the birds fly non-stop for approximately 4,500 km across the ocean. This is achieved by feeding on *Corophium volutator*, an amphipod containing high amounts of n-3 polyunsaturated fatty acids (n-3 PUFA), particularly eicosapentaenoic (20:5) and docosahexaenoic acid (20:6). In mammals, high dietary intake of n-3 PUFA is known to increase the capacity for oxidative metabolism (Jump 2002; Jump and Clarke 1999; Lapillone et al. 2004). In the sandpiper, according to Maillet and Weber (2007), flight muscle oxidative capacity and β-oxidation are stimulated through increases of citrate synthase and 3-hydroxiacyl dehydrogenase (HOAD) activities, respectively, and their activities are correlated with dietary n-3 PUFA content in phospholipids (22:6 for CS; 20:5 for HOAD. Dietary n-3 PUFA are thus used as molecular signals to prime flight muscles for extreme exercise during long distance flights, which in many cases may also occur at high altitudes.

1.9 Flightlessness

As mentioned before, flightlessness is an uncommon but pervasive avian phenomenon. If there is no strong and continual selection for the maintenance of the flight apparatus, it tends to disappear. In the context of this book, it is important to explore this topic. Recent studies carried out with the group Rallidae (rails) illustrates the problems involved in the evolution of flightlessness in extant birds and its ecological implications (Trewick, 1997).

The Rallidae have a broad geographic distribution across continental and oceanic islands (Olson, 1973a), and more than one quarter of extant or recent species – particularly island forms – are flightless (Olson, 1973a; Diamond, 1981). Many of them have recently been extirpated by humans and various introduced predators which have colonized islands on which they once inhabited (Steadman and Olson, 1985; Steadman, 1995). For example, the Pacific region was once very rich in flightless rails, as evidenced by extensive subfossil bones found in the Hawaiian archipelago, (Olson and James, 1991), New Zealand (Steadman, 1995), and also in other islands. It is likely that additional recently extinct species have yet to be accounted for (Pimm et al., 1994). The majority of the world's extant flightless rails persist in the New Zealand region (Trewick, 1997).

According to Trewick (1997), the devastating impact of introduced predators (including humans) on secondarily flightless avifauna is a strong indication that the earlier predator-free status of oceanic islands was of particular significance in the evolution of these species. Given this scenario, the frequency with which flightlessness evolved among rails is likely a function of both the behavioral flightlessness exhibited by many volant species (Ripley, 1977), and the apparent paradoxical frequency with which these species reach distant predator-free islands (Olson, 1973a; Ripley, 1977; Olson and James, 1991). The current view is that flightless rails are derived from self-colonizing volant ancestors. However, a panbiogeographic analysis suggests that the flightless weka, for example, (*Gallirallus australis*) and its ilk may have evolved from a common flightless ancestor (Beauchamp, 1989).

Trewick (1997) remarks that given the fact that gross skeletal features such as relative dimensions of leg and wing bones tend to be convergent amongst flightless species, analysis reveals significant differences between birds that have at one time been classified as the same species. Trewick (1996) showed that differences in the relative dimensions of bones from two morphotypes of the tahake (*Porphirio mantelli*) are consistent with the view that these two forms were independently derived from separate but similar volant ancestors, rather than having diverged after loosing the ability to fly.

The implications of the independent origin hypothesis are, according to Trewick (1997), considerable. They suggest an unexpected level of evolutionary repeatability and thus contradict the assumption that morphologically similar, and geographically proximal "forms," are necessarily each other their closest relatives. More importantly, the hypothesis emphasizes that as a factor in speciation, the loss of flight is associated with a process of morphological parallelism rather than divergence (Trewick, 1997). Flightless endemic species are probably recently derived, and monophyletic with their volant counterparts, despite the existence of sometimes profound differences in gross morphology (Olson, 1973b).

Trewick (1997) found little evidence to support a vicariant origin for any of the flightless rails he investigated. All of them have volant relatives which tend to be behaviorally flightless, and this is a possible explanation for the apparent frequency with which flightless rails may have evolved. Similarity of form is the expected result of convergence and parallelism. Allopatric taxa that are similar to one another in gross morphology are apparently more closely related to their respective volant ancestors than to one another, and may have originated over a broad timescale (Trewick, 1997). For example, even the two forms of tahake studied by Trewick, which appear quite similar, and which at one time existed in such close proximity to one another, are apparently not directly related. According to Trewick (1997), such coincidence of form within this taxon is further supported by subfossil evidence of other tahake-like birds around the world. According to Livezey (1993), this form of parallelism has also been proposed as one plausible explanation for the origin, on separate islands, of two forms of the most famous of all flightless birds, the dodo (*Raphus cucullatus*).

Evolution of flightlessness on islands can be viewed as an extreme form of allopatric speciation because flightlessness fixes the species in space and prevents gene flow (it prevents emigration, but not necessarily immigration) (Trewick, 1997). The occurrence of dispersal and colonization may be frequent, but not so frequent as to maintain gene flow preventing speciation, and the almost universal occupation of habitable islands by land birds is a very good indication that the frequency of founding dispersal is high (Trewick, 1997). Evidences thus indicate the occurrence of repeated dispersal events to some islands, leading to the conclusion that speciation through loss of flight is very rapid (Olson, 1973a; Feduccia, 1980). The process by which this happens, however, has not thus far been entirely elucidated (Trewick, 1997).

The study conducted by Trewick (1997) shows that in the right circumstances, evolution of flightlessness is both rapid and frequent. "It is a process which is repeated as and when suitable colonizers reach islands." The observed changes in morphology associated with the evolution of flightlessness constitute an adaptive response to selection pressures. They also have the effect of rapidly isolating island endemic species from genetic introgressions with closely related volant species, despite the possibility of the occurrence of ecological overlap, as it has been clearly evidenced from the incidence of repeated colonization episodes (Trewick, 1997).

1.10 Overview

In this introductory chapter, the main general design plan ("blue print") of birds, which makes them in many ways so distinct from the other animal taxa, has been depicted. Although there are still several gaps with regards to the underlying mechanisms involved in bird evolution and the evolution of flight in birds, the most recent findings have been presented, not only to bring updated information but also to emphasize that further research is much needed to fill those gaps towards a better understanding of the avian group as a whole.

The importance of feathers has been emphasized since birds are the only extant group within Animalia which possess them. The importance of body size with respect to energy expenditure during flight has been discussed with quantitative examples, and also how migratory birds can make use of energy saving strategies during long duration flights.

Finally, the evolution of migration, and its importance for the widespread distribution of birds around the earth have been presented, and also the importance of flight as a strong selective pressure without which flightlessness would instead prevail in the group.

The subsequent chapters will deal with specific topics of bird physiology within an ecological and environmental context—how specific adaptations have evolved, and the implications of such adaptations to the life style of extant birds.

2

General Physiological Principles

Birds show a number of structural and functional adaptations that are characteristic of the group as a whole and independent of the ability to fly. In this chapter, we will explore some basic physical and physiological principles and structural variables that will help us understand how birds "work". Of course, many of these physiological principles are fundamental to all vertebrates (and in some cases to all metazoans) but some are unique to the avian group.

We will focus our discussion on gas and heat exchanges, energy flow, and water and ion fluxes. Budget modeling and analysis of the effects of body size on function are indispensable tools to understand some basic principles. When pertinent, and whenever we can establish relationships with environmental adaptations, we will include typical bird examples to illustrate some of the topics treated in this chapter.

2.1 Gas Diffusion

For many practical problems, it is easier to calculate transport when diffusion is examined from a macroscopic point of view. In the respiratory organs of animals (e.g., gill and lungs), gases move between the environment and the organism entirely by diffusion so that oxygen (O_2) enters and carbon dioxide (CO_2) leaves the animal. It is therefore of interest to understand what factors affect how fast these gases diffuse, i.e. their rate of diffusion.

One factor that affects the rate of diffusion of a gas is its molecular weight: diffusion rates are inversely proportional to the square root of molecular weight. For example, CO_2 is heavier than O_2 and therefore diffuses more slowly. Other factors being equal, the rate of diffusion of CO_2 is 0.86 that of O_2. The higher solubility of CO_2 in water, with respect to O_2, makes it appear to diffuse faster than O_2, but this is not the case.

Another factor is the medium through which diffusion is occurring: diffusion through a gas (such as air) is enormously faster than diffusion through liquids, other factors being equal. Typically (depending on temperature), diffusion rates are six to ten thousand times faster in air than in water.

If we assume that A is the area of a plane through which particles pass, we can express this process as a *flux density*, J_x, i.e., the net number of particles crossing the plane per time per area. The product of the area and the distance traveled by a particle from one side of the plane to the other side is a volume. Therefore, the number of particles per volume at a given position is a measure of concentration C. The change in concentration per change in distance, say along an x axis, is the concentration gradient.

The flux of particles in the x direction is proportional to the gradient in concentration in the x direction and to the diffusion coefficient.

$$J_x = -D \cdot [\partial C / \partial x] \tag{1}$$

The negative sign tells us that transport proceeds from a position of higher concentration to one of lower concentration. D is the diffusion coefficient (e.g., of O_2 or CO_2, in air or water; unit: $cm^2 \cdot s^{-1}$). This differential equation is better known as *Fick's first equation of diffusion*, where D can be expressed as:

$$D = [J_x / \Delta C] \cdot [\Delta x / A]$$

We can then simplify Fick's equation as:

$$J_x = [A / \Delta x] \cdot \Delta C \cdot D$$

This allows stronger emphasis on the critical importance of the inverse square (area A) relationship with diffusion distance (Δx), implying that diffusion can only take place effectively over very short (i.e., cellular) distances.

2.2 Gas Exchange

Given the inverse-square relationship between diffusion rate and distance, the effective diffusion of respiratory gases is restricted to short traveling distances. Many small organisms obtain O_2 by diffusion through their body surfaces, without having any special respiratory organs and without circulating blood. Larger and more complex aerobic organisms like birds must also exchange gases with the environment by diffusion, but they have specialized surfaces for gas exchange and also a blood system to transport respiratory gases by convection, which is far more rapid than diffusion over any distance more than about a millimeter.

Among the vertebrates, birds and mammals have the highest rates of aerobic metabolism, and convective respiratory structures have evolved to rapidly move the air outside to gas exchange organs (lungs) located deep in their bodies. In both birds and mammals, the microstructure of the lung presents a huge surface area of thin, highly vascularized membrane for rapid diffusional movement of O_2 and CO_2 between air and blood. However, the structure and function of bird and mammal lungs differ in several important ways.

2.2.1 The Bird Lung

Birds possess highly complex respiratory or "plumbing" compared to mammals. The major differences between the two systems are that birds have large air sacs which connect with the bronchi and the compact lung by a series of passageways. The voluminous, thin-walled air sacs and air spaces are extensively ramified within the body cavity and in many species even extend into the bones of the extremities and the skull.

The volumes of the respiratory system of birds and mammals have conspicuous differences. The lung volume of a typical bird is only a little more than half that of a mammal of the same body size. In contrast, the tracheal volume of a bird is much larger than that of a mammal. The air sacs of a bird are several times as large as the lung. The total volume of the respiratory system of a bird is some three times that of a mammal.

In birds, gas exchange occurs in the finest branches of the bronchial system in the lung, known as parabronchi. In contrast with the tidal flow encountered in the lung of mammals, air in the avian lung flows continuously past the parabronchial gas exchange surface in one direction. This is the most striking structural and "mechanical" difference between the respiratory systems of birds and mammals, and it has profound physiological implications. At the least, this unidirectional parabronchial ventilation avoids the mixing of "old" and "new" air within the lung that reduces lung O_2 concentrations in tidal ventilatory systems, but it has other advantages as well.

To understand how the avian respiratory system works we need to know a few additional facts about its complex anatomy. The air sacs, anatomically and functionally, form two major groups, i.e., a posterior or caudal group that includes the large abdominal sacs, and an anterior or cranial group that consists of several somewhat similar sacs.

The trachea divides into two bronchi, with each bronchus running to and through one of the lungs, and terminating in the abdominal air sac. The cranial sacs connect to this main bronchus in the anterior part of the lung, and furthermore, some of the air sacs directly connect to the lung tissue. Air flows through this complex anatomy in an equally complex manner.

One of the most interesting achievements with regards to the avian respiratory system was the discovery of how it works. Investigators from four different laboratories using different techniques, and roughly at the same time, reached the same results (Bretz and Schmidt-Nielsen 1971; Bouverot and Dejours 1971; Brackenbury 1971; Scheid and Piiper 1971). According to the model conceived by these investigators, air in the avian respiratory system flows as follows: (i) during inhalation most of the air flows directly to the caudal air sacs; (ii) the cranial air sacs also expand on inhalation, however they do not receive any of the directly inhaled outside air; instead, they receive air from the lung; (iii) on exhalation, air from the caudal sacs flows into the lung; and iv) on the following inhalation, air from the lung flows to the cranial air sacs, and on the second exhalation, air from the cranial air sacs flows directly to the outside. Therefore, two full ventilation cycles are required to move a single volume of gas through the avian respiratory system. The two cycles are identical, each volume of gas being followed by another similar volume on the next cycle. In this pattern, air flows through the lung *in the same direction, posterior to anterior*, during both inhalation and exhalation. Aerodynamic valves (Jones *et al.* 1981; Banzett *et al.* 1987, 1991; Wang *et al.* 1988; 1992) are probably responsible for controlling flow direction.

Unidirectional flow through the parabronchi has important consequences for gas exchange between air and blood. Ventilation systems can be characterized by the relative orientation of the flow of the external medium (air or water) and of the internal perfusing fluid (blood). If the two flows are parallel to each other the flow is known as "cocurrent" and the respiratory gas partial pressure (O_2 or CO_2, for example) of the blood leaving the gas exchanger cannot exceed that of the external medium leaving the exchanger; at best the two fluids will equilibrate, with 50% exchange occurring. If, on the other hand, the two flows are arranged as a "countercurrent", the internal perfusing fluid may reach almost the same level of respiratory gases partial pressures as is present in the incurrent bathing medium. In this case, the effluent perfusate is in contact with the incoming and, let us say, maximally oxygenated external medium, so that even if the perfusate has already picked up a good loading of O_2 within the exchanger the gradient of this gas is still such that it will continue to load up until an almost complete exchange between the two is reached. Countercurrent exchange also reduces the "work" of breathing by reducing the volume of external fluid that must be moved through the gas exchange organ to provide a given amount of O_2 (or eliminate a given amount of CO_2).

According to the currently accepted model, the respiratory system of birds effectively functions as a crosscurrent gas exchanger (Scheid and Piiper 1972). A "crosscurrent" exchanger has properties that lie between cocurrent and countercurrent exchangers, and the effluent blood will load to a partial pressure of O_2 intermediate between those of the incoming and outgoing external media. The blood in the avian lung does not flow in parallel to air movement, but rather in an irregular complex capillary network. Blood leaving the lung is a mixture of blood from different parts of the lung and having different degrees of oxygenation.

The effectiveness of both unidirectional air flow and crosscurrent flow in the avian lung are of particular importance at high altitude. In experiments in which mice and sparrows were exposed to an atmospheric pressure of 350 mm Hg, corresponding to 6,100 m altitude, the mice were lying on their bellies and barely able to crawl, while the sparrows were still able to fly (Tucker, 1968a). Mice and sparrows have the same body mass, their blood has the same affinity for O_2, and their metabolic rates are similar, thus the difference cannot be explained in terms of their metabolic rates or their blood chemistry. Therefore, the air flow pattern in the avian lung as well as the crosscurrent type of gas exchange between blood and respiratory gases are the most plausible explanations for these results. They are consistent with the fact that birds in nature have been seen in the high Himalayas, flying easily at altitudes where human mountain climbers can barely walk without breathing supplementary O_2.

2.2.2 The Bird Egg

A bird's egg contains all the nutrients that the embryo needs to grow to hatching, except O_2. In general, the egg's temperature is maintained within narrow limits, about 34 to 38°C, by the incubating parents. Its only exchanges with the outside during the period of embryonic development are the influx of O_2 for its metabolism and effluxes of CO_2, produced by aerobic respiration, and water. Exchange occurs though tiny pores that penetrate the eggshell and bring the surrounding atmosphere into contact with the respiratory membranes of the embryo. Gases pass through these pores by diffusion. The egg must gain O_2 from the environment and eliminate CO_2. Loss of water by the same route is usually unavoidable because water vapor moves through the same pores as O_2 and CO_2, and the outside air has lower vapor pressure than the moist egg contents. Thus the egg is faced with the challenge of accomplishing the necessary gas exchange while limiting water loss to a tolerable level. Gas exchange is a flux determined by surface area, concentration gradient and conductance (the ease with which energy or material crosses the surface barrier). Wangensteen *et al.* (1971; 1974) discovered that the movement of gas through the pores of the eggshell is governed by the simple laws of physical diffusion, and most of the resistance to diffusion resides in the hard shell of the egg (Lomholt 1976).

As we saw earlier, Fick's law concerns the diffusion of a gas across a constant concentration gradient. Accordingly, the rate of diffusion (V) of gas across an eggshell, measured in cubic centimeters per second ($cm^3 \cdot s^{-1}$) is the product of pore area, A (cm^2), the diffusion coefficient of the gas, D ($cm^3 \cdot cm^{-1} \cdot s^{-1}$), the inverse of the length L of the pore (cm), i.e., shell thickness, and the difference in the concentration of gas (C) between the egg and the surrounding air ($cm^3 \cdot cm^{-3}$).

Because gases occur in an aqueous phase inside the egg, the difference in vapor pressure (P) between the inside and outside of the egg can be substituted for the concentration difference (C) with an appropriate conversion factor. Therefore, gas exchange may be described by the equation $V = (A/L)DP$ (Fick's equation). Experimentally determined values of D ($cm^3 \cdot cm^{-1} \cdot s^{-1}$) in air at 38°C (e.g., the incubation temperature of the chicken egg) are 0.27 for water, 0.23 for O_2 and 0.18 for CO_2.

The shell of a typical chicken egg is perforated by about 10,000 pores. Because the surface area of the egg is about $70\,cm^2$, there is an average of 1.5 pores per square millimeter of shell. The average pore diameter is about 0.017 mm. The idealized chicken egg has a pore area of $2.3\,mm^2$ and a shell thickness of 0.3 mm. The vapor pressure of water at 38°C is 50 torr. That is, the tendency of water to evaporate from a liquid surface at 38°C is exactly balanced by the tendency of water to move from the gas to the liquid phase when the pressure of water vapor at the surface is 50 torr, or about 7% of the total pressure of the atmosphere at sea level (760 torr).

Rahn *et al.* (1977a) used a simple device, called an "egg hygrometer", to determine the vapor pressure in the immediate vicinity of the egg under the incubating parent (for a complete description of the method, see Rahn *et al.* (1997a)). By placing calibrated egg hygrometers in the nests of many species of birds, Rahn and colleagues determined that the vapor pressure of water in the nests of most species is maintained between 18 and 26 torr regardless of the temperature and water content of the surrounding air. This regulation is achieved by the parent's adjusting the water permeability of the nest and the parent's temporal pattern of incubation.

Because the pressure gradients of O_2 and CO_2 are primarily controlled by the metabolic activity of the embryo, they change during the course of incubation. The partial pressure of O_2 in sea level atmosphere is approximately 150 torr, and that of CO_2 is close to zero. Through the incubation period (about 21 days in a chicken), O_2 consumption and CO_2 production increase as the embryo grows. As the embryo approaches hatching, the partial pressure of gases within the egg decreases to 104 torr for O_2 and increases to 37 torr for CO_2 (Rahn *et al.* 1974). Thus, as the requirement for O_2 increases, the partial pressure gradient across the shell also increases, enabling oxygen to enter the egg more rapidly. Similarly, as the embryo produces CO_2 more rapidly, its concentration within the egg increases, with a corresponding rise in CO_2 flux from the egg. Gas conductances place an upper limit on the embryo's rate of metabolism, and above this limit, higher concentrations of CO_2 would become toxic for the embryo. Increasing the gas conductance of the shell to alleviate this problem would result in excessive water loss.

Different environments place different demands on egg gas exchange. At high altitude, atmospheric pressure is reduced. Wangensteen *et al.* (1974) found that chickens kept at 3,800 m elevation (O_2 partial pressure = 90 torr) adapt to low O_2 partial pressure by reducing embryonic metabolism and prolonging the incubation period slightly. Because air pressure (hence density) at 3,800 m is 39% lower than at sea level, gas conductances are correspondingly higher. In order to prevent excessive water loss at high altitude, bird eggs there have reduced pore area (Rahn *et al.* 1977b; Carey 1980). At very high altitude, however, there must be a compromise between reducing water loss on the one hand and provision of the required O_2 supply to the embryo on the other (Monge *et al.* 1988).

Seabirds like the Wedge-tailed Shearwater (*Puffinus pacificus*) that breed on remote oceanic islands have eggs with reduced pore area to prevent excessive water loss from their prolonged incubation periods (Ricklefs 1984). The egg of the Wedge-tailed Shearwater is about the same size of a chicken egg. However, its incubation period lasts approximately 52 days. Thus, if the shearwater egg lost water at the same rate as the chicken egg, the embryo would die of dehydration before the end of the incubation period. Because water vapor pressure in the nests of shearwater and chickens are basically the same, the only solution to overcome this problem is a reduction in pore area, and hence gas conductance of the egg-shell. This solution, however, also reduces the embryo's ability to acquire O_2. Ackerman *et al.* (1980) observed that the shearwater embryo overcome this other problem by piping (i.e., it breaks a tiny hole through the eggshell) several days before it hatches and before its requirements for O_2 reach a maximum value. By doing this, the embryo is able to breathe air directly, avoiding the diffusion limitations of the eggshell.

Another example that illustrates the responses of organisms to their physical environment is given by Brush Turkeys (*Alectura lathami*). These birds belong to a chickenlike group of Australian birds known as the megapodes. Their eggs are not incubated directly by the parents, but are laid within mounds of vegetation built by the male. The eggs are warmed by the heat produced as the vegetation decomposes. The result of this unusual style of incubation is O_2 scarcity and high levels of CO_2 and water vapor surrounding the developing eggs. Seymour and Rahn (1978) measured partial pressures of 48 torr for water vapor, 100 torr for O_2 and 62 torr for CO_2 in Brush Turkey nests. The most critical problem is elimination of CO_2. To increase the flux of CO_2 without increasing its concentration within the egg, Brush Turkeys have greatly increased their eggshell gas conductance, which is 2.6 times higher than that of a typical egg of similar size. They can do this without risking egg dehydration because of the very humid nest environment within the incubation mounds, and the high eggshell conductance also facilitates O_2 uptake.

2.3 Temperature and Thermal Exchanges

To maintain a constant body temperature, a bird must satisfy the steady state condition in which the rate of heat gain equals the rate of heat loss. Typically the external environment is colder than body temperature, which means that birds constantly loose heat from their bodies. A major source of the heat needed to maintain this gradient is from metabolic heat production.

Metabolic heat production can easily increase more than 10-fold above resting levels with activity, and unless heat transfer to the outside is increased in the same proportion, body temperature will rise rapidly. Furthermore, the conditions for metabolic heat transfer vary tremendously with body size, as well as external factors such as ambient air temperature and wind conditions. Considerable study has been devoted to understanding the thermal relationships between birds and their environments.

Temperature exchange between birds and their environment is governed by basic thermodynamic principles. Whenever physical materials are at different temperatures, heat flows from a region of higher temperature to one of lower temperature. This transfer of heat takes place by *radiation* and by *conduction*. A body cannot transfer heat unless its surrounding environment, or some part of it, is at a different temperature than the surface of the body. There is, however, another way for terrestrial organisms to transfer heat: the *evaporation* of water (or in a few cases, the *condensation* of water vapor into liquid water).

2.3.1 Radiation

Heat transfer by radiation takes place in the absence of direct contact between objects. Energy is exchanged by means of photons that travel through space with the speed of light. All bodies at temperatures greater than 0 K radiate energy. As the temperature of a body increases, the total amount of energy radiated increases. Electromagnetic radiation passes freely through a vacuum, and for practical purposes atmospheric air can be regarded as fully transparent to radiation. The intensity of radiation from an object is given by the Stefan-Boltzman law, written as

$$Q = \sigma \cdot A \cdot T^4 \qquad (3)$$

where Q is the amount of energy radiated in joules per hour, σ (Stefan-Boltzmann constant) equals 2.04×10^{-8} J/(cm²·h·K⁴), A is surface area of the body in square centimeters (cm²), and T is its surface temperature in Kelvin. A body conforming to this equation is considered a *black body*. Many real bodies emit less energy than is described by this equation. A dimensionless quantity, called *emissivity*, is used to describe the degree to which a particular body conforms to black-body radiation. Emissivity varies from 0.0 to 1.0 and

$$Q = \varepsilon \cdot \sigma \cdot A \cdot T^4 \qquad (4)$$

If a body had an emissivity of 0.0, it would radiate no energy, and if it had an emissivity of 1.0, it would radiate the maximum amount of energy dictated by temperature (i.e., it would be a black body). Objects that are close to physiological temperature emit most of the radiation in the middle infrared ($T \approx 300$ K; peak at about 10,000 nm). The surfaces of most organisms have emissivities close to 1.0 for these wavelengths; i.e., they are functionally blackbody radiators.

If two surfaces are in radiation exchange, each emits radiation according to Stefan-Boltzmann law, and the net radiation transfer is given by

$$Q = \sigma \cdot \varepsilon_1 \cdot \varepsilon_2 \cdot (T_2^4 - T_1^4) \cdot A \tag{5}$$

As long as a temperature difference exists between two bodies, or between a body and its environment, a net radiative flux of energy will proceed from the warmer to the cooler body.

An interesting observation demonstrating the selective use of radiation was made by Calder (1973, 1974) in a study of the positioning of hummingbirds nests. Incubating females at high elevations in summer may be exposed to radiant night sky temperatures as low as −20°C. Yet, the total estimated rate of energy expenditure for thermoregulation was within basal expenditures (BMR) because the heavily insulated nest obscured three-fourths of an incubating female's surface, and because the nest was placed under branches, thus blocking radiation from the bird's exposed back to the open sky. The predawn radiant temperatures of the branches varied from 3.5 to 6.5°C, i.e. some 25°C above those of the open sky.

2.3.2 Conduction

Heat transfer by conduction takes place between substances that are in contact with each other, whether they are solids, liquids or gases. Heat is transferred from one molecule to another without material exchange. It is an important means of heat transfer within bodies, depending on their thermal conductivity, and between bodies, depending on their area of contact.

For most terrestrial tetrapods, including birds, conduction is the least important means of thermal exchange with the environment, mainly because conduction to air is low and the principal areas of contact with the solid or liquid parts of the environment are the small foot areas upon which tetrapods stand (McNab 2002). At times of rest or during egg incubation in the case of birds, the area of contact and therefore conductive exchange may be significantly greater. Conductive heat transfer also occurs between a body and the boundary layer of the atmosphere at the solid-fluid interface, but heat transfer in this case is normally determined by the bulk movement of molecules in the fluid beyond the boundary layer.

Thermal conduction varies with the surface area (A) of contact between bodies, the temperature gradient between the two bodies, and a constant (thermal conductivity, k) that measures the ease of transfer of heat:

$$Q = -k \cdot A \cdot (T_2 - T_1) \tag{6}$$

where $(T_2 - T_1)$ is the thermal gradient. Thermal conductivity (units: J cm/cm²·h·°C) is an expression for how easily heat flows in a given material, and it reflects its molecular makeup. For example, thermal conductivity through still air (i.e., air trapped within feather layers) is very low, while conductivity through water is very high.

2.3.3 Convection

The transfer of heat in fluids (gases and liquids), especially at an interface with a solid, is almost invariably accelerated by the process of convection, i.e., fluid movement. This form of heat transfer is a function of several parameters that include surface area and the temperature difference between the fluid and the surface of the solid:

$$Q = h_c \cdot A \cdot (T_s - T_a) \tag{7}$$

where h_c is the convective coefficient (units: $J/cm^2 \cdot h \cdot °C$). The convective coefficient is a complex variable that is strongly dependent on the velocity of fluid movement in relation to the exchange surface.

Convection in a fluid may be caused by temperature differences or by external mechanical forces. Cooling or heating a fluid usually changes its density, and this causes mass flow. If a warm solid surface is in contact with a cold fluid (liquid or gas), the heated fluid expands and rises, being replaced by cool fluid. In this case, the bulk flow, or convection, is caused by the temperature difference and is called free or natural convection. Fluid movement can also be caused by external forces, such as wind, water currents, animal movement, or an electric fan. This type of convective movement is referred to as forced convection.

Because convection depends on mass transfer in fluids, the process is governed by complex laws of fluid dynamics, and includes variables such as fluid viscosity and density, in addition to its thermal conductivity. Convective heat transfer does not depend only on the area of the exposed surface. Variables like the curvature and the orientation of the surface give rise to complex mathematical expressions, which can cause great difficulties in the analysis of heat transfer from an animal.

An example of the use of these principles was shown by Chappell *et al.* (1989) in a study of the differential impact of convective cooling in Adélie Penguins (*Pygoscelis adeliae*) and Blue-eyed Shags (*Phalacrocorax atriceps*), in Antarctica. At 20°C and low wind speed (2.6 m·s^{-1}), the smaller (2.6 kg) shags had a total rate of metabolism 14% greater than that of the larger (4.0 kg) Adélies. When the air temperature dropped to −20°C at the same wind speed, the rate of metabolism in the Adélies increased, and was 13% greater than that measured in the shags. If in that air temperature, wind speed increased to 5.7 m·s^{-1}, the penguins rate of metabolism was 31% greater than that found in the shags, even though the shags' rate increased by 12%. The sensitivity of the Adélies to convective heat transfer apparently results from their poorly insulated flippers, whose structure facilitates propulsion in water, not heat conservation in air. When Adélies face an ambient air temperature below 0°C, heat transfer through the flippers may be high, particularly when wind speeds are very high. This is because penguins cannot hide their flippers under body feathers, and skin temperature in the exposed flippers must be kept above freezing. Freezing is not an issue when swimming, because water temperature is never cold enough to pose a significant frostbite risk.

2.3.4 Evaporation

The evaporation of water requires a significant amount of heat transfer. To evaporate 1 g of liquid water at room temperature (~20 °C) to water vapor at the same temperature requires 2443 J. The rate at which heat is transferred by evaporation equals LE, where L is the latent heat of vaporization ($\approx$2400 J/g H_2O at 40°C) and E (g/h) is the rate at which water is vaporized. The amount of water that can be evaporated from a surface, and hence the amount of heat transfer, is dependent on air movement. The rate of evaporative water loss can be described by Fick's law, and it depends on surface area, temperature of the surface (and boundary layer), and the difference in water vapor density between the boundary layer and the external air mass:

$$E = h_d \cdot A \cdot (\rho_o - \rho_a), \tag{8}$$

where h_d is a mass transfer coefficient; A, effective surface area; ρ_o, saturated water vapor density (g.cm^{-3}) at surface temperature; and ρ_a, water vapor density of the air mass (g·cm^{-3}). The latter factor is strongly affected by air movement, with wind generally reducing the water vapor density in the air surrounding the animal, thereby increasing evaporation (convection is a faster process than evaporative diffusion). The saturated water vapor density increases with temperature (T, in Kelvin) according to the relation

$$\rho_o = (9.16 \times 10^8) \cdot e^{-(5218/T)} \tag{9}$$

where e is the base of the natural logarithms, the constant 5218 is approximately H_{evap}/R, H_{evap} is the molar heat of vaporization, and R is the universal gas constant (Atkins 1992).

For respiratory water loss, Hutchinson (1955), King and Farner (1964), and Salt (1964) described the coefficient in the above equation as a function of respiratory volume. This converts the equation for passive evaporation into one for forced convection. Therefore, the equation becomes

$$E = k \cdot V \cdot A \cdot (\rho_o - \rho_a), \tag{10}$$

where V is ventilation rate and k is the slope of the water loss-ventilation curve. The rate of evaporative water loss in some birds, mammals, and lizards is in fact proportional to the ventilation rate (McNab 2002), since the skin is quite impermeable to water vapor in many species.

Birds have no sweat glands and when exposed to high ambient temperatures they rely upon increased evaporation of their respiratory tract as an important avenue for heat dissipation in order to avoid overheating. This is achieved by an increase in the respiratory ventilation, commonly referred to as panting, or by adaptations such as gular flutter. Evaporative water loss in birds has been extensively investigated by several researchers in the past 50 years, and we will discuss this topic with greater detail in chapter 5, section 1.

2.3.5 Heat Balance

The net thermal exchange of resting animals with the environment is the sum of the heat transfers by radiation, conduction, convection, and evaporation:

$$Q_{net} = Q_{rad} + Q_{cond} + Q_{conv} + Q_{evap}$$

This equation assumes that animal body temperature is constant. Otherwise, an additional term must be added for positive or negative heat storage (Q_{stor}) to represent an increasing or decreasing body temperature, respectively. Furthermore, if the animal is active, another term must be added, i.e., the work done against the environment (Q_{work}):

$$Q_{net} = Q_{rad} + Q_{cond} + Q_{conv} + Q_{evap} + (Q_{stor} + Q_{work}) \tag{11}$$

The work term is quite small for most animals.

Many difficulties are associated with the use of the above equation. One is that heat exchange with the environment occurs at the surfaces of a body (integumental and respiratory), and surface temperatures are difficult to measure when compared to measuring core temperature. Replacement of surface temperature (T_s) by body temperature (T_b) is a practical expedient, particularly when internal heat transfer is of importance. However, surface temperatures cannot be ignored when an animal faces a radiant heat load that raises T_s above the value that is expected from heat transfer to the outside at a particular ambient temperature (e.g., Ohmart and Lasiewski 1971).

Another difficulty with this equation is that the area (A) and the coefficients k and h are highly variable. They are the result of changes in posture, ptilomotion, pilomotion (changes in erection or depression of feathers and fur, respectively), peripheral blood circulation, and ventilation rate.

A practical response to the difficulties mentioned above is to combine the four modes of heat exchange into two terms, one for "dry" heat exchange, which depends on a temperature difference and incorporates radiation, conduction, and convection, and another for evaporative water loss. If core temperature is replaced by surface temperature, considering that most of the existing temperature differences between the body and environment are found between the core and body surface, we have:

$$Q_{net} = C' \cdot (T_b - T_a) + LE, \tag{12}$$

where C' is a coefficient (in J/[h.°C]) called *thermal conductance* (or the coefficient of "dry" thermal conductance), and T_a is ambient temperature. This equation was described by Burton (1934) and is applicable over all relevant temperatures of biological importance, but cannot be used if an appreciable external radiant heat load exists, i.e., strong sunlight.

If the net thermal exchange between a body and the environment is studied at cool to cold T_as, conditions in which the evaporation of water makes up only about 10% of heat transfer, the equation above can be simplified to

$$Q_{net} = C \cdot (T_b - T_a) \tag{13}$$

where C is now the "total" thermal conductance, i.e., it includes evaporative as well as non-evaporative routes of heat transfer. Thus, C has been called "wet" thermal conductance. This model of energy expenditure is usually used with data on O_2 consumption as a function of T_a.

2.3.6 Thermoneutrality

At moderate to cool ambient temperatures, heat transfer between a body and its surrounding environment mainly varies with the difference in temperature (ΔT) (see equation 13). For body temperature to remain constant, all heat transfer from the animal must be replaced. In the absence of an external heat source such as sunlight, heat must be produced by metabolism. Therefore,

$$VO_2 = C \cdot (T_b - T_a) \tag{14}$$

where metabolism is measured as the rate of O_2 consumption (VO_2) and C is the thermal conductance, here defined in terms of O_2 flux. The units for rate of O_2 consumption are normally mL O_2/h, and those for thermal conductance are mL O_2/h·°C. Mass-specific units are usually mL O_2/g·h and mL O_2/g·h·°C, for rate of metabolism and thermal conductance, respectively. Of direct relevance to the heat budget, the rate of metabolism can also be expressed as energy used per unit time (i.e., power). This is derived by assuming the appropriate energy equivalent for O_2, which is slightly dependent on the foodstuffs metabolized (≈ 20.09 J/mL O_2).

The above equation indicates that the rate of metabolism is proportional to the difference of temperature between an endotherm and its surrounding environment, at least at low ambient temperatures. The slope of the metabolism-ΔT curve equals C when the curve extrapolates through the origin. The rate of metabolism never equals zero but reaches a minimal value when ΔT falls below a value equal to $T_b - T_l$, where T_l is defined as the lower limit of *thermoneutrality*. This means that rate of metabolism is independent of ΔT over a range in ΔT (called the region or zone of thermoneutrality), and by implication, a corresponding range of T_a's, because all variation in ΔT reflects a variation in T_a when T_b is constant. The zone of thermoneutrality exists because heat transfer to the outside environment remains constant, i.e., heat production is constant but thermal conductance is adjusted to compensate for the change in the $T_b - T_a$ gradient. Conductance changes can be accomplished by postural adjustment of the exposed surface area, changes in ptiloerection in the case of birds, or changes in peripheral vasoconstriction or vasodilation.

The rate of metabolism in the zone of thermoneutrality is the basal rate of metabolism (BMR). Equation 14 holds under conditions when the radiant and air temperatures are similar and when air velocity is low, as long as evaporative cooling is not an important avenue of heat transfer (McNab 2002).

2.4 Body Size

The consequences of body size on form and function in living organisms has been an important area of research in physiology, ecology, and evolutionary biology for many decades (see Huxley 1932; Kleiber 1932; Brody 1945; Thompson 1961; Calder 1984; Schmidt-Nielsen 1984; Brown and West 2000; McNab 2002). Real organisms usually are not isometric (i.e., function does not scale in direct proportion to size), even when organized on similar patterns. Instead, certain proportions change in a regular but non-linear fashion. In biology, such non-isometric scaling is referred to as *allometric* (from the Greek *alloios*, meaning different). Within this context, the impact of body size on the thermal relations and metabolism of birds is critical to our understanding of their biology.

In birds, as in all organisms, the total rate of metabolism increases with body size, but not in an intuitive linear manner. Instead, the relationship is allometric, and is usually expressed as a power function:

$$y = a \cdot m^b, \tag{15}$$

or when transformed into the corresponding logarithmic equation:

$$\log_{10} y = \log_{10} a + b \cdot \log_{10} m, \tag{16}$$

where y is rate of metabolism (e.g., BMR) as a function of body mass (m) raised to the power (b), and a is a coefficient.

McKechnie and Wolf (2004) in their thorough analysis of the allometric relationships of BMR in birds state that "Understanding how animals partition energy resources into maintenance requirements, activity, growth and reproduction, as well as the evolution of patterns of energy allocation, is a central goal of ecological and evolutionary physiology." They go on to observe that "Attempts to identify the sources of selection responsible for variation in the maintenance energy requirements of endotherms have typically focused on the minimum maintenance metabolic rate during normothermy, or basal metabolic rate (BMR; e.g., Elgar and Harvey 1987; Lovegrove 2000; Tieleman and Williams 2000; Lovegrove 2003)." However, it should be pointed out that direct selection on BMR seems unlikely in most circumstances.

According to McKechnie and Wolf (2004), "One common approach to identifying metabolic adaptation involves the comparison of observed BMR with that expected on the basis of allometry (Reynolds and Lee, 1996; Garland and Ives, 2000; Lovegrove, 2000; Tieleman and Williams, 2000).Typically, hypotheses concerning adaptation in BMR are tested by generating conventional and/or phylogenetically independent prediction intervals (Garland and Ives 2000), with data falling outside these prediction intervals considered to differ significantly from the expected values. In the last 2 decades, considerable effort has been invested in developing the statistical procedures necessary to correct for the potentially confounding effects of phylogenetic relatedness when inferring adaptation (Felsenstein 1985; Harvey and Pagel 1991; Garland et al. 1992, 1993; Garland and Ives 2000)." McKechnie and Wolf (2004) observe "Several avian and mammalian metabolic parameters have been found to exhibit strong phylogenetic dependence (Freckleton et al. 2002)."

McKechnie and Wolf (2004) write "Reynolds and Lee (1996) analyzed data for 254 bird species and demonstrated that the BMR of passerines and nonpasserines was not significantly different once the data were corrected for phylogenetic nonindependence." They examined more closely the data analyzed by Reynolds and Lee, and concluded that a significant proportion of the data the latter included in their analysis did not meet the criteria that defines BMR, and sample size was either unspecified or less than three individuals. McKechnie and Wolf observed that "data inclusion criteria have a profound effect on allometric predictions of avian BMR and on the conclusions reached regarding deviations of observed BMR from expected values."

McKechnie and Wolf used independent contrasts (Felsenstein 1985; Garland *et al.* 1992) to compare the BMR of passerines and non-passerines, and to calculate phylogenetic independent allometric intervals for their updated data set. Their careful analyses were consistent with the general conclusion that BMR does not differ between passerine and non-passerine birds after taking phylogeny relatedness into account. But, more importantly, their analyses showed that allometric equations based on the data set used by Reynolds and Lee (1996) overestimate expected BMR in birds, and potentially lead to incorrect conclusions regarding avian physiological adaptation. McKechnie and Wolf (2004) state "Our analyses are consistent with the dichotomy between the results of conventional and phylogenetically independent analyses of avian BMR (Reynolds and Lee 1996; Rezende et al. 2002). Whereas conventional analyses show the BMR of passerines to be significantly higher than that of nonpasserines (Lasiewski and Dawson 1967; Zar 1968; Reynolds and Lee 1996; Rezende et al. 2002; but see also Prinzinger and Hanssler 1980), phylogenetically independent analyses suggest that the higher metabolic rate of passerines reflects phylogenetic relatedness rather than adaptive variation (Reynolds and Lee 1996; Garland and Ives 2000; Rezende et al. 2002)."

McKechnie (2008) recently reviewed the role of phenotypic flexibility in avian BMR. He states "Comparative analyses of avian energetics often involve the implicit assumption that basal metabolic rate (BMR) is a fed, taxon-specific trait. However, in most species that have been investigated, BMR exhibits phenotypic flexibility and can be reversibly adjusted over short time scales. Many non-migrants adjust BMR seasonally, with the winter BMR usually higher than the summer BMR. The data that are currently available do not, however, support the idea that the magnitude and direction of these adjustments varies consistently with body mass. Long-distance migrants often exhibit large intra-annual changes in BMR, reflecting the physiological adjustments associated with different stages of their migratory cycles." According to McKechnie (2008), the emerging view of avian BMR is of a highly flexible physiological trait that is continually adjusted in response to environmental factors.

2.5 Water and Ion Fluxes

As we shall see in the following chapters, birds are found in different and challenging habitats, and lead rather diverse modes of life. Nectarivorous birds may ingest copious amounts of water with their food, and they must eliminate excess water to preserve cell volume. Seabirds drink sea water and they must eliminate the excess of salts like sodium chloride, magnesium, potassium, and others in order to maintain ionic and water balance. Although birds have great mobility, desert birds are non-fossorial, diurnal animals that, potentially, face periods in which they have difficulties in accessing freshwater, which they need in greater quantity when exposed to high ambient temperatures since they actively heat regulate by evaporation.

Before we discuss how birds specifically deal with such challenges, we will first present a few relevant physiological principles which apply not only to birds but to organisms in general.

The basic structural unit of organisms, the cell, can as a very crude approximation be described from the point of view of osmotic condition as a solution contained within a membrane. Surrounding the cells in multicellular animals are solutions that constitute the external environment of the cells. The solvent is always water which is present in very variable proportion, from over 99% to less than 50% of the total mass of solution.

2.5.1 Osmosis

Osmosis is the movement of water through a semipermeable membrane (i.e., permeable to water but impermeable to solutes like ions, carbohydrate molecules, etc.) that separates two solutions with different concentration of solutes. Concentration, therefore, is of primary importance to osmotic exchange in the characterization of a solution.

Osmotic movements can be prevented by applying a physical pressure to the more concentrated solution, counteracting exactly the tendency for the water molecules to move across the membrane into it. The pressure, or more accurately the energy per mole of solute (π, in atm or in J) exerted on a membrane that permits only water to move through is given by

$$\pi = i \cdot C \cdot R \cdot T, \tag{17}$$

where R is the gas constant (0.082 L·atm/K·mole, or 8.31 J/K·mole), T is absolute temperature, and C is the molal concentration of a solute (number of moles of solutes per kilogram of water, at a fixed temperature). The coefficient i indicates the mean number of particles produced from a mole of solute when in solution. Because all biological solutions contain salts, it is convenient to express the concentration of a solution in osmolals ($c = i \cdot C$), which takes the molal concentration of solutes into consideration along with the number of particles produced when the solutes are dissolved. Equation 17 is a slightly modified version of the van't Hoff equation:

$$\pi \cdot V = n_B \cdot R.T$$

where $n_B / V = iC$, and n_B the amount of solute.

Osmotic pressure according to this view is the result of a partial pressure produced by the bombardment of solutes against a membrane, just as a partial pressure is exerted by a particular gas species in a mixture of gases.

If two solutions are separated by a semipermeable membrane, water moves from the solution with the lower concentration to the solution with the higher concentration (Fick's law). The flux of water (w) depends on the difference in osmotic potential (the tendency of a solution to attract water), $\Delta\pi$:

$$w = A \cdot P \cdot \Delta\pi$$

$$w = A \cdot P \cdot (\pi_2 - \pi_1)$$

and using equation 17, we have

$$w = A \cdot P \cdot R \cdot T \cdot (c_2 - c_1),$$

where A is the surface area of the membrane separating the solutions, c is the concentration in osmolals, and P is the coefficient of osmotic permeability.

This model, however, does not fully explain the high rate of many osmotic processes, nor does it explain the action of certain hormones that can increase the rate of water flow across membranes (e.g., antidiuretic hormone in vertebrate kidneys). Recent work shows that osmosis partly and perhaps predominantly occurs via specific membrane water channels, controlled by proteins known as porins or aquaporins, having a central water-filled pore; these are now known to be present in all organisms. Their presence explains the fairly common observation of higher osmotic permeability in one direction for animal cells, usually with inward osmosis happening faster than outward osmotic flux.

Recently, Yang *et al.* (2004) were able to identify and functionally characterize for the first time an avian aquaporin in the kidneys of Japanese Quails (*Coturnix coturnix*). Their study indicates that avian antidiuretic hormone enhances urine concentration in avian kidneys via the quail aquaporin water channel, and this mechanism might be relevant for birds exposed to water deprivation.

Aquaporins have also been found in the nasal glands (Müller *et al.* 2006) of Mallard ducklings (*Anas platyrhynchos*). In activated glands downregulation of aquaporins in capillaries and duct cells may prevent dilution of the initially secreted fluid, enabling the animals to excrete large volumes of a highly concentrated salt solution. Marine birds would certainly benefit from such a mechanism when drinking sea water or eating marine items as we shall see later in this chapter.

Besides the kidneys, the lower intestine of birds also functions in osmoregulation, and aquaporins have been observed in the proximal rectum, distal rectum and in the coprodeum of House Sparrows (*Passer domesticus*), thus allowing water movement across their intestinal cell membranes (Casotti *et al.* 2007).

2.5.2 Bird Regulation of Water and Ion Fluxes

Carnivorous and frugivorous birds routinely acquire all the water they require through their food. Species living in xeric environments can survive on the metabolic water produced from a dry diet even in the absence of drinking water (Krag and Skadhauge 1972; Skadhauge and Bradshaw 1974). For most birds, however, drinking is a necessary route of water intake.

The regulation of water and ion fluxes (commonly referred to as osmoregulation) in birds involves the interacting contribution of various organs and organ systems, i.e., the kidneys, intestinal tract, salt glands (if present or functional), and skin and respiratory tracts (as routes of evaporative water loss) (Goldstein and Skadhauge 2000). Discussions of topics in bird osmoregulation can be found in Skadhauge (1981), Wideman (1988), Hughes and Chadwick (1989), Braun and Duke (1989), Dantzler (1989), Gertsberger and Gray (1993), Skadhauge (1993), Elbrønd *et al.* (1993), Brown *et al.* (1993), and Goldstein and Skadhauge (2000), and it is not the purpose of this volume to discuss the subject with the detail one finds in these reviews.

Most of the principles outlined in this section have been used by several investigators to study avian osmoregulatory organs or organ systems. The kidneys are usually considered the primary organs of osmotic regulation. However, avian kidneys empty their output into the lower intestine, after which the urine may reside in and be modified by the coprodeum, colon, and ceca. Kidney function must therefore be integrated with these organs, and under some circumstances, and in some species, salt glands undertake a primary osmoregulatory role. In many birds, the majority of total water loss may in fact be evaporative, not excretory.

Birds have a pair of kidneys and ureters, which transport urine to the urodeum of the cloaca. Birds do not possess a urinary bladder, though the cloaca may serve as its functional equivalent in some species and circumstances. The functional units of the kidney are the tubular structures called nephrons. Of the total nephron population, between 10 and 30% possess loops of Henle (Goldstein and Braun 1989), and several studies suggest habitat-related patterns in kidney structure, such as smaller kidneys, larger volume of medulla, or small volume of cortex in desert species (Thomas and Robin 1977; Warui 1989; Casotti and Ricardson 1992).

Like the mammalian kidney, the avian kidney filters a large volume of fluid (approximately 11 times the entire body water content each day for a 100 g bird) and then recovers most filtered water by tubular reabsorption. In hydrated birds the percentage of filtered water that is reabsorbed is typically greater than 95%. Variation from this situation can be achieved through regulation of either the rate of filtration or the rate of absorption, both of which may be influenced by the avian antidiuretic hormone arginine vasotocin (AVT), a small (8 amino-acid) peptide hormone released from the neurohypophysis. Tubular reabsorption of water in birds can range from less than 70% to more than 99% of the filtered volume. As a result, the final urine concentration may vary from dilute (approximately 40 mmol/kg) to hyperosmotic (2–3 times plasma osmolality). In general, the maximum concentration ability of bird kidneys is less than that of mammalian kidneys.

Water conservation in birds, therefore, results at least in part from renal mechanisms that include the ability to reduce glomerular filtration rate and to increase tubular reabsorption of water. Excretion of uric acid (uricotelism) also provides a means to excreting nitrogenous waste with a minimum of water loss, because of the insolubility and low toxicity of urate salts, and hence their lack of contribution to urinary osmotic concentration. Beyond this, avian osmoregulation depends on other organs that regulate water and ion fluxes.

Birds lack a urinary bladder, and in most species ureteral urine flows backward by retrograde peristalsis from the urodeum into both coprodeum and colon (rectum). This brings the absorbing epithelia of the lower gut into contact with a mixture of urine and chymus coming from the ileum. Urinary refluxing may serve a variety of functions like recovering of water, electrolytes, and nitrogen that would otherwise be lost (Goldstein and Skadhauge 2000).

In birds, the coprodeum and colon form a common storage chamber for ureteral urine and chyme. However, their transport properties are markedly different. When the absorptive areas and the local transport rates of the two organs are compared, the colon has a much higher total capacity than the coprodeum (Goldstein and Skadhauge 2000). The colon therefore recovers most of the salt and water from ureteral urine and chyme, whereas the coprodeum functions as a final "tuner" of total excretion.

Paired caeca are present in the great majority of birds, and some urine and chyme may enter the caeca. Large caeca are typically associated with herbivory and granivory (McLelland 1989). Not only ions and water, but nutrients, particularly short-chain fatty acids (SCFA), the products of fermentation, are absorbed there. *In vivo* perfusion studies demonstrate that a substantial absorption of NaCl and water and secretion of K^+ occurs in the caeca of the domestic fowl (Rice and Skadhauge 1982; Skadhauge *et al.* 1983).

In the large, flightless ratites (ostriches, emus, rheas, etc.) the mass of the gut and its contents are not critical as in flying species, and special adaptation of the lower gut to act as a fermentation chamber is possible. Having the fermentation vat after the stomach and the small intestine allows the herbivore to assimilate the soluble nutrients of food before microbes have a chance to ferment them. This has two advantages, i.e., fermentation reduces the energy in food by between 10 and 20%, and fermenting microbes modify some of the components that are abundant in plant diets and that are important for the nutrition of the herbivore. For example, gut microbes hydrogenate unsaturated fatty acids, including those that are essential to the host, such as linolenic and linoleic acids. Thus, by having the fermenting chamber in a posterior position, hindgut fermentation permits assimilation of these useful nutrients before microbes transform them (Karasov and Martínez del Rio 2007). This is the case in the Ostrich (*Struthio camelus*; Skadhauge *et al.* 1984) in which the more than 10 m long colon of the adult male has SCFA concentrations up to 120 mmol. This has also been observed in the Rhea (*Rhea americana*) however to a minor extent (Skadhauge *et al.* 1996; Elbrønd *et al.* 2009). The Ostrich has a strong sphincter between coprodeum and terminal colon, and no influx of urine has been observed (Skadhauge 1983). As no feces are stored in the Ostrich coprodeum, it functions as a real bladder. In contrast to these species, the Emu

(*Dromaius novaehollandiae*) has a conventional, small coprodeum/colon where little fermentation occurs. Skadhauge *et al.* (1991) showed that the dehydrated Emu has a low urine-to-plasma ratio osmolality (1.4) and a fairly large urine flow rate. To compensate for this, the absorption capacity of the coprodeum/colon is particularly high in this species, as a consequence of extensive mucosal folding.

2.5.3 Salt Glands

The presence of supraorbital glands in marine birds was known for many years, but only in 1958 were Schmidt-Nielsen and colleagues able to determine their salt secreting function. Functional salt glands exist in 10 orders of birds, including all those with marine representatives and at least two orders of terrestrial birds, i.e, several falconiforms (Cade and Greenwald 1966) and the Roadrunner, (*Geococcyx californianus*; Ohmart 1972). Within related groups, glands are larger in species exposed to a hypersaline diet and water; marine species have larger glands than coastal terrestrial birds (Staaland 1967). The fluid secreted by avian salt glands is typically nearly pure NaCl solution, between 500 and 1,000 mM, and more than 75% of excreted Na^+ is lost through the salt glands under conditions of salt loading or dehydration. The division of water loss between salt glands and kidneys depends on the water load. During a heavy saline load, both salt glands and kidneys may be equally responsible for water excretion, whereas during dehydration the renal losses of water may be much more reduced. Despite the low concentration of K^+ in the salt gland fluid, the salt glands may eliminate more than one-third of excreted K^+ after a salt load (Goldstein and Skadhauge 2000).

As has been mentioned earlier, evaporative water losses from the respiratory tract and the skin are important components of avian water balance. Respiratory water loss is an inevitable consequence of high rates of ventilation when birds are exposed to high ambient temperatures. Despite birds' lack of sweat glands, a substantial fraction (often more than 50% and sometimes above 80%) of total evaporation occurs across the skin (see Dawson 1982). During dehydration, birds greatly decrease excretory water loss, thus evaporation constitutes a greater fraction (up to 80%) of total water loss (Skadhauge 1981).

3

Physiological Bases of Fecundity/ Longevity Tradeoffs

As sexually reproducing animals, birds must balance the competing nutritional needs of growth and self-maintenance with those of reproduction. This creates a fecundity-longevity tradeoff (Kirkwood and Holliday 1979). Extrinsic mortality rates will shape the type of life history that best manages these conflicts and promotes the highest reproductive potential. Because physiological processes are intrinsic to animal health and reproduction, they are a logical focus for gaining a mechanistic understanding of the factors influencing fecundity/longevity tradeoffs. Fisher, one of the outstanding evolutionary theorists of the last century, identified the value of physiological studies in examining reproductive/self-maintenance tradeoffs, but also the need to place them in context with variations in environmental and life history characteristics. He said: "It would be instructive to know not only by what physiological mechanisms a just apportionment is made between the nutriment devoted to gonads and that devoted to the rest of the parental organism, but also what circumstances in the life history and environment would render profitable the diversion of a greater or lesser share of the available resources towards reproduction." (Fisher 1930). Physiological research and life history theory have advanced significantly since that time and it is worthwhile to examine further insights that physiological approaches bring to understanding life history variation. We highlight this by examining the costs associated with three major breeding activities (egg production, male mating costs, and parental costs) and the bases of longevity differences from physiological perspectives. We rely mainly on research involving species with biparental care, as these represent the majority of birds and have been more extensively studied.

3.1 Costs of Egg Laying

3.1.1 Nutrient and Energy Demands

The ability of many birds to readily lay replacement eggs following egg loss or removal from a clutch has prompted many to consider the main constraints on clutch size to relate to the costs of rearing chicks rather than those of egg laying. This is a view that was championed by Lack who stated "... clutch size is considered to be ultimately determined by the average maximum number of chicks for which parents can find enough food" (Lack 1947, p. 331). Lack was one of the most influential ornithologists of his time and did much to advance the idea that evolution rewards individual fitness and not populations or species as believed by many of his contemporaries. While this view did much to advance life history theory as we know it today, his assertion that realized fecundity was the primary determinate of individual fitness ignores the potential effects of current reproductive effort on future breeding success. This point was raised by Williams (1966) who resurrected Fisher's view that fecundity will compromise longevity (Fisher 1930).

One very appealing aspect of Lack's theory of clutch size evolution is that it can be very easily tested. If clutch size is ultimately determined by the maximum number of chicks that parents can successfully rear, then adding an additional egg from the nest of another bird at the same stage of laying should not increase – and in fact may decrease - the fledging rate of the foster parents. This experimental approach has been used to examine clutch size optimality in many marine and terrestrial species, furnishing substantial data for broader analysis. For example, Ydenburg and Bertram (1989) examined the results of 25 clutch enlargement studies on 21 colonial seabird species and found that 15 of these species reared more fledglings when their clutch was supplemented with an additional egg compared to control birds with natural clutch sizes. Vanderwerf (1992) came to the same conclusion after conducting a meta-analysis of 42 independent brood-enlargement studies that included both marine and terrestrial species. This seems contrary to Lack's hypothesis, but investment in one reproductive event may reduce the chance of future reproduction. Because natural selection will favor individuals that have the highest lifetime reproductive success, the tendency of females to lay fewer eggs than they or the pair can rear suggests that other factors besides success in a single breeding event shape clutch size optimisation.

One way to identify constraints on clutch size is through experimental manipulation of egg laying. If females are optimising their current and future reproductive potential by laying a particular number of eggs at a given breeding attempt, then stimulating them to lay additional eggs should have negative consequences. A series of experiments addressing this question were undertaken on Lesser Black-backed Gulls (*Larus fuscus*; Monaghan *et al.* 1998; Nager *et al.* 2001). This species normally lays three-egg clutches, but will lay additional eggs if eggs are removed from their nest. The experimental procedure involved removing the first egg laid in some clutches, with females responding by laying three additional eggs, therefore they laid four eggs in total. Upon clutch completion, eggs were cross-fostered between the manipulated birds and control birds, which had three-egg clutches. This procedure uncoupled issues of egg quality from parental quality regarding differences in egg production on subsequent hatching success and chick rearing.

The effects on reproductive success of laying one extra egg were striking. Three days after clutch completion, females stimulated to lay four eggs had significantly smaller pectoral muscle size and a 5% lower body mass than control birds. This energetic imbalance persisted into the chick rearing stages and was associated with much higher nest failure (42% in four-egg birds versus 12% in controls), a 30% reduction in fledging, and significantly reduced fledgling body mass in birds reared by four-egg females compared to controls (Monaghan *et al.* 1998). This extra demand on four-egg females also affected their long-term reproductive potential, as only 67% of the four-egg females returned to breed during the following two years, compared to 100% of the controls (Nager *et al.* 2001).

While it may seem implausible that laying a single additional egg could have such devastating consequences on reproduction, this represents 33% increased allocation for a species usually laying three-egg clutches. It is insightful to examine the energy and nutrient requirements of egg formation in the context of what a female needs for her daily maintenance. Using egg composition data for birds ranging from 11 to 3400 g, Meijer and Drent (1999) determined the total protein and fat contained in a representative clutch and compared this with estimates of energy and nitrogen requirements of laying females. By reanalysing these data in terms of the relative cost of each egg laid, further insight into the consequences of clutch size variation can be made (Table 3.1). This analysis reveals a striking outcome: each egg laid will impose a substantial burden on daily nitrogen requirements, but relatively little demand on energy intake, particularly for small species. The inverse relation between body size and relative cost of egg lipids does not reflect lower yolk investment by small birds. Quite the contrary, these birds produce relatively large eggs as a proportion of their body mass, but the amount of energy contained in their egg lipids is a much smaller proportion of their mass-specific metabolic rate than that in larger birds (Table 3.1).

Although the daily impact on nitrogen need is ameliorated in those species having greater than one day laying intervals between eggs (Adelie Penguin, *Pygoscelis adeliae*; Willow Grouse, *Lagopus lagopus*; European Kestrel, *Falco tinnunculus*), on average, each bird faces a doubling of daily protein requirement for each egg laid. Because these estimates presume 100% conversion efficiencies of maternal protein and lipids to eggs (Meijer and Drent 1999), the actual requirements will be substantially higher. Nevertheless, the comparison suggests access to protein is far more likely to constrain clutch size than access to energy.

A direct way to test whether limitations of energy or protein are more important in constraining egg size and number is to provide supplementary food to breeding birds. This has been done a number of times and the results are highly variable. What makes such experiments problematic is the large interannual variation in food availability, which strongly influences the relative improvement

Table 3.1 Physical and physiological characteristics of female birds from selected species and estimates of the protein and lipid energy required to produce a single egg. Data are from Meijer and Drent (1999), with permission from Wiley-Blackwell

General Characteristics

	Body mass (g)	No. of eggs	Egg Mass/ body mass (%)	Egg protein as % daily requirement	Egg lipid as % daily energy need	Mass-specific BMR (kJ/g)
Adelie Penguin	3400	2	3.5	246	28	0.23
Lesser Snow Goose	2950	3–6	4.2	312	38	0.24
Eider	2163	4.6	4.8	258	38	0.27
Mallard	1300	8–11	4.0	189	31	0.31
Willow Grouse	615	10	3.6	116	22	0.37
American Coot	500	8.3	6.0	138	17	0.40
Kestrel	300	5	7.0	162	14	0.46
Starling	90	5	7.8	90	5	1.05
Red-billed Quelea	22	3	9.1	96	4	1.55
Pied Flycatcher	16	6	10.4	104	5	1.70
Zebra Finch	15	4	6.5	113	5	1.72
Blue Tit	11	11	10.9	51	4	1.88

offered by food supplementation. Most early studies of this type focused on providing additional energy to breeding birds, generally resulting in a few days advancement of the initial date of laying, but rarely affecting clutch size or egg volume (Nager *et al*. 1997). The results of experiments using protein-rich foods have also been variable, but some have shown significant effects on egg production. More importantly, from a functional perspective, a few of these have shown that protein quality is more important than protein quantity in effecting improvements. Bolton *et al*. (1992) supplemented Lesser Black-backed Gulls with either animal fat, fish, or cooked eggs. The egg volumes of fat-enriched birds did not differ from controls, whereas fish-supplemented birds had significantly larger eggs than controls but significantly smaller eggs than egg-supplemented birds (Fig. 3.1; Bolton *et al*. 1992). Fat-fed Blue Tits (*Parus caeruleus*) similarly showed no increase in egg size over control birds, but those supplemented with cooked eggs showed significant increases in egg volume (Ramsay and Houston 1998). Egg proteins differ from many muscle proteins in having higher relative amounts of particular essential amino acids, especially methionine and cystine (Murphy 1994). Ramsay and Houston (1998) examined the effects of protein quality on egg production in Blue Tits. Their high quality protein contained five essential amino acids in the proportion found in eggs (methionine, lysine, cystine, threonine, and tryptophan), whereas the other food contained an equivalent concentration of aspartate, a non-essential amino acid. Birds receiving the low quality protein gained no improvement in egg or

(a)　　　　　　　　　　　　　　(b)

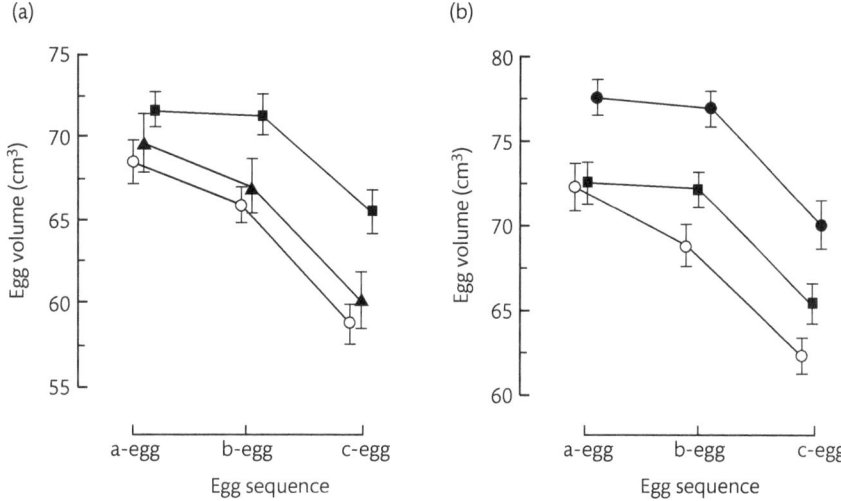

Fig. 3.1 Mean egg volume of eggs laid sequentially (eggs a through c) from 3-egg clutches of Lesser Black-backed Gulls (*Larus fuscus*) eating either: Panel (a): natural diet (○), fish-supplemented diet (■), fat-supplemented diet (▲), or Panel B: natural diet (○), fish-supplemented diet (■), or egg-supplemented diet (●). Each point represents a mean and the error bar indicates 1 standard error. (From Bolton *et al.* 1992, with permission from Wiley-Blackwell).

clutch size over controls, but those getting the high-quality diet had significantly larger clutches (Ramsay and Houston 1998).

The reproductive gains of being able to lay and then rear a larger clutch are obvious, but the reproductive advantages of laying larger eggs are much less clear. In general, chicks hatching from large eggs are predicted to have higher survivorship than their clutchmates (Williams 1994). Functional interpretation of cause and effect in such comparisons is confounded by the possibility that parents laying larger eggs may be better providers after hatching and are able to lay early in the season when conditions are best. Evidence from cross-fostering variably sized eggs from different clutches among parents laying either larger or smaller eggs in Common Terns (*Sterna hirundo*), demonstrates that egg quality is a significant determinant of fledging success. A completely unambiguous demonstration of egg-size effects is shown by the Australian Brush Turkey (*Alectura lathami*), a megapode. Eggs of the Brush Turkey are heated by exogenous sources, usually mounds of fermenting vegetation tended by the male, and receive no parental care after hatching. Consequently, the entire reproductive investment made by the female is in egg production. Egg size varies markedly in this species, and chicks from the largest eggs showed larger hatching mass, greater motor performance, and significant correlations with mass at 10 months of age (Goth and Evans 2004).

3.1.2 Self-Maintenance/Egg Production Tradeoffs

The above studies demonstrate that clutch size and egg quality are constrained by the ability of laying females to allocate sufficient essential amino acids and other key nutrients to egg production. Whether birds rely mainly on endogenous supplies to meet breeding demands (capital breeders) or exogenous acquisition (income breeders), there will be competition between resources for self-maintenance and those for reproduction. Even in income breeders, decreases in locomotor and heart muscle masses can occur during egg laying (Fogden and Fogden 1979; Houston *et al.* 1995a; Williams and Martyniuk 2000). Zebra Finches (*Taeniopygia guttata*), for example, may lose up to 14% of their pectoral muscle protein during the laying period (Houston *et al.* 1995c), with radio-labelling methods confirming the incorporation of muscle protein into eggs (Houston *et al.* 1995b). Because the muscles of adults contain about half the concentration of methionine and cystine as eggs, approximately twice as much muscle protein would have to be mobilized to furnish a given amount of suitable egg protein (Murphy 1994). Examination of Zebra Finch pectoral muscle during the laying period showed no evidence that specific amino acids were being selectively removed from flight muscle, but rather there was depletion from the general protein pool of the sarcoplasm, with the suggestion that this would not compromise flight function (Cottam *et al.* 2002). A study of breeding female Blue Tits, however, found progressive mass-independent decreases in flight take-off speed with each egg laid, revealing a significant decrease in pectoral muscle function as breeding progressed (Kullberg *et al.* 2002). By the time the last egg was laid, females showed up to 20% reduction in escape speed, presumably making them much more vulnerable to predators. Veasey *et al.* (2000) similarly found a significant relation between the extent of post-laying pectoral muscle loss and escape takeoff speed in Zebra Finches.

Somewhat surprisingly, muscle loss during egg laying is a relatively common phenomenon among income-breeding passerine birds. Among 11 species for which muscle volume was estimated during breeding, seven exhibited muscle volume decreases during the laying period (Houston *et al.* 1995c). It is important to recognize, however, that loss of pectoral muscle will not necessarily result in a loss of flight ability. Because flight performance is related more to pectoral muscle size relative to overall wing loading than to their absolute size *per se*, there will be little change in escape speed if there are concurrent and proportionate mass losses in other tissues. Long-distance migrating species typically undergo muscular hypertrophy *en route* to breeding areas and then undergo consequent muscle and body mass reduction following arrival, with some of these nutrients getting incorporated into egg production (Morrison and Hobson 2004). Diet quality, however, will importantly affect the extent of pectoral muscle loss in breeding females. Female Zebra Finches fed diets containing the full complement of essential amino acids two weeks before laying had significantly lower losses of pectoral muscle after egg laying than birds on a low-quality diet (Selman and Houston 1996a). Similarly, Williams and Martyniuk (2000) showed that Zebra Finches provided with nutritionally complete diets had proportionately larger pectoral and heart masses than females given seed alone or seed with protein supplement. Thus compromises in flight performance will depend importantly on diet quality preceding and during the egg laying period and how this affects flight muscle size in the context of variation in all body tissues.

It is very clear that the nutritional state of a female has important consequences in terms of the number and size of eggs she can afford to lay. Laying fewer eggs than can be reared will reduce her current reproductive potential, but laying too many may penalize her later breeding success as well as that of the current clutch. It is important to recognize that a female's physiological status at the time of laying dictates the developmental conditions of her chicks. Even if food abundance and quality improve after laying, chicks may be irreversibly affected by egg composition, which is tightly linked to attributes of the mother at the time of egg formation. For example, female Zebra Finches provided with a high-protein supplement containing egg before they began laying produced males with longer tarsi than females given low-protein supplementation, irrespective of the dietary treatment of the birds that incubated or raised them (Gorman and Nager 2004). Furthermore, females mated to males hatched from high-protein eggs had significantly higher fecundity on their first breeding than females paired with low-protein males. This demonstrates that maternal condition at the time of laying can critically affect the fecundity of progeny, which, in turn, will affect the mother's fitness.

In addition to providing the more obvious nutrient and energy sources, breeding females must also allocate appropriate levels of antibacterial enzymes, antibodies, and antioxidants to each egg formed to ensure production of healthy chicks (Blount et al. 2000). Unlike mammals, birds have a very limited opportunity to pass antibodies to their chicks, with some immunoglobulins (Ig) being incorporated into yolk during follicular development (IgY) and others (IgA) being transferred with albumen during egg formation (Klasing 1998). Immunoglobulin A is important for ensuring that the appropriate microflora colonize the chick's gastrointestinal tract, whereas the IgY immunoglobulins provide maternal antibodies that may protect the chick from parasitic infections while their immune system develops its own capabilities. There is some evidence that antibody allocation is constrained in birds in poorer physical condition (Hargitai et al. 2006; Pihlaja et al. 2006), but there is little consensus regarding what levels of maternal antibodies are optimal for chick health (Grindstaff et al. 2003; Boulinier and Staszewski 2008). In contrast, there is good evidence that egg antioxidant content has important consequences for chick hatchability, growth rate, early post-hatch viability, and immune development (Surai 2002).

The main antioxidants actively transferred to eggs are vitamins A and E and carotenoids (Costantini and Moller 2008). Because these antioxidants are not synthesized by birds, they must be obtained from the diet. Thus, unless laying females obtain appropriate amounts of these in their diet, they will sacrifice their own ability to cope with oxidative stress if they lay more eggs than they can nutritionally afford.

There is a suggestion that females may be particularly vulnerable to oxidative stress during egg formation (Williams 2005). The onset of egg production is preceded by rapid rises in plasma 17β-estradiol that stimulates growth of the ovary and oviduct and synthesis of yolk precursors such as vitellogenin and very low-density lipoproteins (Challenger *et al.* 2001). The associated enlargement of the reproductive tissues, and thus total body mass, not only increases flight costs for the laying female, but also increases the female's maintenance costs, as reflected by a 22% increase in resting metabolic rate (RMR) in ovulating European Starlings (*Sturnus vulgaris*; Vezina and Williams 2002). These added metabolic expenditures will likely increase the rate of reactive O_2 species (ROS) formation by aerobic processes at a time when the female's ability to maintain her antioxidant status is most challenged. The potential for ensuing oxidative stress to lead to long-term oxidative damage may be increased in breeding birds by the abundance of lipid-rich precursors associated with yolk formation (Williams 2005). Polyunsaturated fatty acids are particularly vulnerable to oxidation (Holman 1954) and, once oxidized, they form products that are themselves ROS (Hulbert *et al.* 2007). This creates a volatile chain reaction that could result in substantial damage to membranes, proteins, and other structures that are critical for maintaining health and vitality.

3.1.3 Fitness Consequences

Overall it is obvious that the condition of a female at the time of laying will have far-reaching implications for the quantity and quality of progeny she produces, but also for her own health and future reproductive potential. Given the immediate demands of egg production, it is likely that there will be at least some nutritional strain on laying females. The extent of this strain will depend on the nutritional status of the female at the start of laying, how many eggs she lays, and how well she accesses critical nutrients from the environment during egg production and after clutch completion. Because immune capability is strongly affected by energy and nutrient availability (Houston *et al.* 2007; Klasing 2007), egg laying has a strong potential to reduce immune function in laying birds. Females that initiate laying in relatively poor condition or persist in laying despite their nutritional deficits are also expected to experience higher levels of corticosterone (CORT), the avian glucocorticoid. Consistent with this prediction, Love *et al.* (2008) measured an inverse relation between the body condition of laying female

European Starlings and their plasma levels of CORT. The strongly immunosuppressive effects of glucocorticoids (Sapolsky *et al.* 2000) will impose health risks, but the overall cost of elevated CORT at this time is more far reaching. As a steroid, CORT is highly lipophilic and readily incorporated into the yolk of developing eggs. To examine the consequences of elevated maternal CORT on chick development, Love *et al.* (2005) experimentally manipulated plasma CORT levels in free-living female European Starlings. They found elevated plasma levels of CORT resulted in significantly increased yolk concentration of this hormone and this, in turn, selected against the production of male chicks. This was the result of male progeny having significantly greater embryo mortality, retarded pre-hatching development rates, decreased post-hatching growth rates, and significantly reduced inflammatory response to immune challenge compared to their sisters (Love *et al.* 2005). Similarly, peahens (female *Pavo cristatus*) in poor physical condition produced few male chicks, which the authors interpreted to be a pre-zygotic effect (Pike and Petrie 2005). Condition-dependent skewing of offspring sex ratios is expected to maximize female reproductive output via consequent reproductive success of progeny if offspring quality is correlated with maternal quality. Thus daughters produced by poor-quality mothers will have greater reproductive potential than poor-quality sons, as such males will be less competitive in gaining access to mates.

Most studies of clutch size optimality have been designed with the assumption that the number of eggs reflected optimisation of chick rearing and the laying schedule was timed to reflect feeding potential at the time of chick hatching and not nutrient constraints on the female at the time of laying. As a consequence, brood enlargement has been a standard means of evaluating clutch size optima, rather than eliciting extra egg laying to test whether fecundity constraints occur earlier. The potential costs of egg production outlined here identify the potential risks confronting females that lay more eggs than they can nutritionally afford and suggest that decisions at the time of egg laying can have significant effects on a female's lifetime reproductive success. This is beautifully illustrated by a study of Visser and Lessells (2001) on Great Tits (*Parus major*). They manipulated workloads on females by either: inducing two extra eggs to be laid following egg removal and then returning the eggs taken (full costs); adding two eggs to a bird's clutch at the start of incubation (free eggs); adding two 2-day old chicks to a bird's brood (free chicks) or letting birds rear the number of eggs they laid (controls). The study took place over a two-year period and examined not only the effect of treatment on the number of fledglings that were produced each year, but also the number of fledglings returning the following year and the survival of females. Because each fledgling is genetically related to its mother by 50%, treatment effects on female fitness were calculated as the product of 0.5 times the number of offspring recruited the following year plus the number of surviving females. Although food availability varied between the two years, females induced to lay two additional eggs had significantly reduced fitness compared to control birds in both years (Fig. 3.2). By contrast, females receiving either extra eggs or extra chicks gained fitness benefits compared to control birds (free eggs and free chicks were treated as if they were produced by the female). Surprisingly, off-spring recruitment did not differ between treatments, but female survivorship was substantially lower in birds induced to lay extra eggs compared to all other treatments. The cost of laying extra eggs is most clearly seen by comparing fitness outcomes of the manipulated clutches to that of the controls for the two years combined (Fig. 3.3). Females clearly have the capacity to incubate and rear more chicks than they do, but they suffer significantly reduced fitness when forced to lay more eggs. This strongly suggests that optimal clutch size for a laying female is the one she elects to lay given her physiological status at the time of laying.

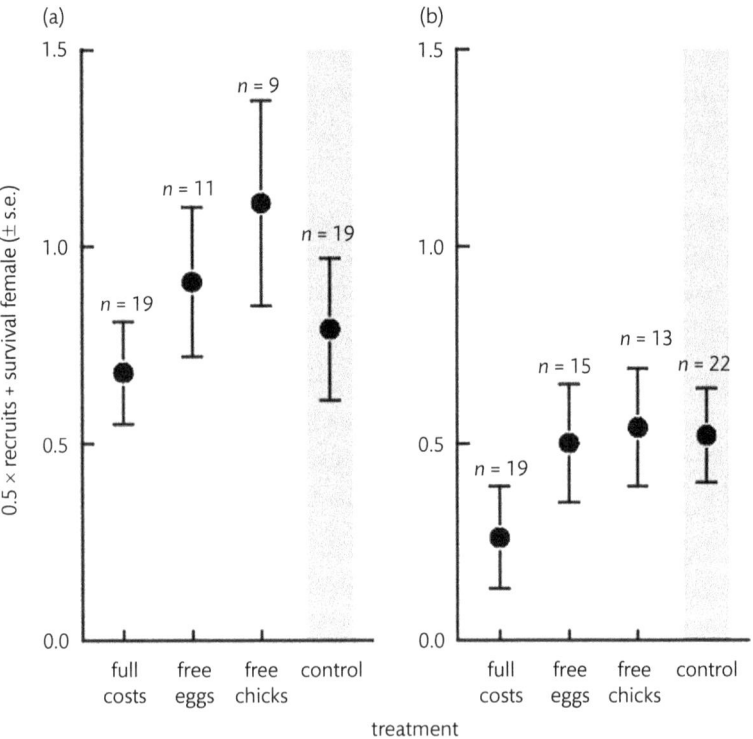

Fig. 3.2 Mean female Great Tit (*Parus major*) fitness values (defined as 0.5 × the number of recruits resulting from first and second broods + female survival) in relation to variation in experimental manipulation of egg production and incubation costs. Panel (a) represents results from 1998 and panel (b) from 1999. Sample sizes (n) for each treatment are indicated above the standard error bar. (From Visser and Lessells 2001, with permission from the Royal Society).

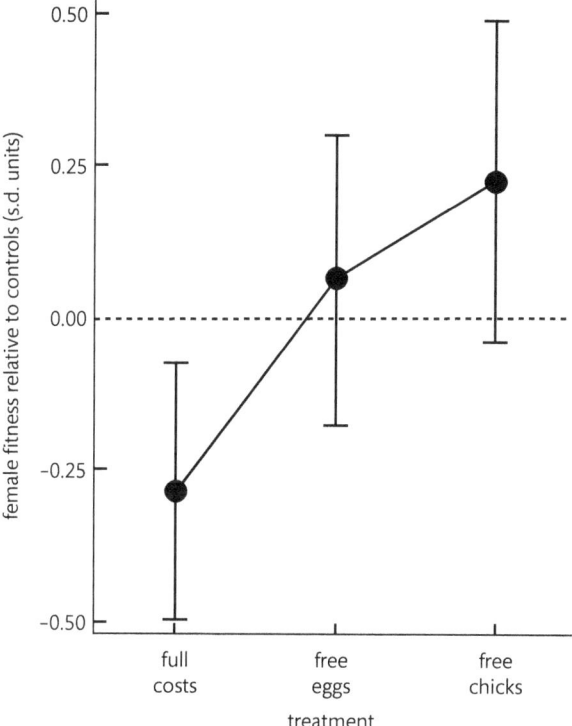

Fig. 3.3 Fitness differences in female Great Tits (*Parus major*) laying 2 additional eggs (full costs), receiving 2 additional eggs during incubation (free eggs), or given 2 extra chicks (free chicks) relative to control females with unsupplemented clutches. Fitness is as defined in Figure 3.2, with the plotted values representing (mean treatment residual − mean control residual)/(S.D. of control residuals) based on an ANOVA of fitness values with year as the only factor. Error bars are the standard error of the difference between the treatment and the mean control value. (From Visser and Lessells 2001, with permission from the Royal Society).

Conditions that require more eggs to be laid (e.g., predation, breakage, infertility, etc.) will likely compromise her future breeding success.

3.2 Male Mating Costs

3.2.1 Testosterone Phenology

In almost all male birds, the transition from prebreeding to breeding life stages is associated with gonadal enlargement and enhanced testosterone (T) secretion. This rise in T level activates a suite of morphological, physiological, and behavioral changes that are critical for successful reproduction. These include development of accessory reproductive organs, regulation of spermatogenesis, and expression of many morphological and behavioral secondary sexual characteristics that directly affect attractiveness to mates (Wingfield *et al.* 2001a). While particular baseline levels of T are needed to attain reproductive condition, species vary tremendously in the levels of T they secrete, the temporal pattern of its secretion over a breeding season, and in their sensitivity to T. The biggest differences are related to breeding system, with species having biparental care showing rapid reductions in T at the time of incubation, whereas males of polygynous species maintain high levels over most of the breeding season (Wingfield *et al.* 1990). Among biparental species, T secretion rises during territory establishment and peaks at about the time of egg laying, when efforts to exclude other males from access to their mate are most intense. This pattern is shown by biparental species with very different life histories and with very different patterns of territorial aggression throughout the year (Fig. 3.4). Gambel's White-crowned Sparrow (*Zonotrichia leucophrys gambelii*) flocks for most of the year and only exhibits territoriality during breeding, coincident with elevated T (Wingfield and Farner 1978). In contrast, the European Robin (*Erithacus rubecula*), Stonechat (*Saxicola torquata rubicola*), and Song Sparrow (*Melospiza melodia*) exhibit territoriality outside the breeding season, but territorial aggression in nonbreeding periods is not associated with T (Wingfield *et al.* 2001b). Spotted Antbirds (*Hylophylax naevioides*), a tropical forest species, are territorial and mate for life and males maintain plasma T at baseline levels throughout the year (Hau *et al.* 2000). Interestingly, the only time that male Spotted Antbirds exhibit elevated T is when they are challenged by another male, with T reaching its highest level when contests take place during the breeding season (Wikelski *et al.* 1999). Substantial rises in T secretion occur dynamically in males of many other biparentally breeding species during male-male aggression, particularly during territory establishment and when females are sexually receptive (Wingfield *et al.* 1990).

The absence of elevated T in territorial male Stonechats, European Robins and Song Sparrows outside the breeding season (Fig. 3.4) and in birds directing aggression towards predators, demonstrates that aggression *per se* is not T dependent (Wingfield and Marler 1988). Instead, T-related aggression is believed to be specifically associated with male-male conflicts in a reproductive context, such as during establishment and maintenance of territories and while mate guarding (Wingfield *et al.* 1990). This means that males of most species will experience elevated T at the onset of breeding, but the duration and extent of this rise depends both on the frequency of male-male conflicts and the type of breeding system. For example, Ball and Wingfield (1987) found that starlings living in a dense colony with high competition for nest boxes had significantly higher T levels than starlings breeding in less dense locations.

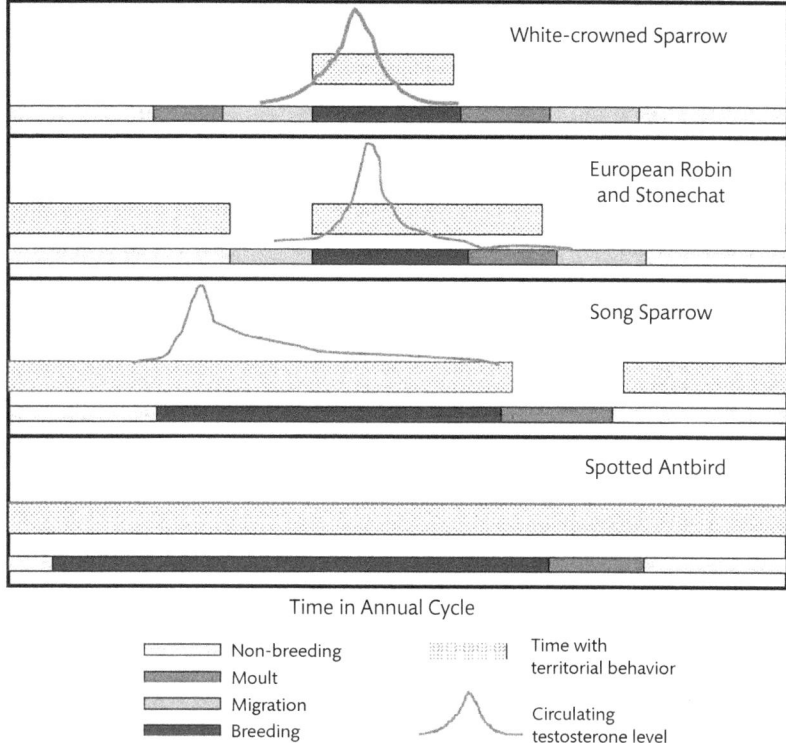

Fig. 3.4 Schematic representation of territorial behavior in males of five passerine species in relation to circulating testosterone (T) and life-history stages. The White-crowned Sparrow (*Zonotrichia leucophrys gambelii*) is territorial only during breeding when T is elevated. The Stonechat (*Saxicola torquata*) and European Robin (*Erithacus rubecula*) are territorial at both breeding and non-breeding grounds. The Song Sparrow (*Melospiza melodia morphna*) is territorial through much of the annual cycle, whereas the tropical Spotted Antbird (*Hylophylax naevioides*) is territorial when accompanying its breeding partner over its breeding life. (Redrawn from Wingfield *et al.* 2001b).

Because more than 80% of bird species have biparental care (Cockburn 2006), mating opportunities for males essentially cease at the conclusion of egg-laying. Consequently, elevated T levels will fall rapidly at the onset of incubation in most male birds and will remain low until the next round of mating opportunities (Hegner and Wingfield 1986). Thus if T-affected changes in behavior and physiology impose added demands on a male, most of these costs will be coincident with those of egg laying in females. What form might these costs take? At the very least, T-related increases in aggression will result in males spending more time in agonistic activities and less in self maintenance, potentially leading to nutritional imbalance. Furthermore, the conspicuous nature of these conflicts will also lead to greater mortality among frequent combatants from predators. There are also reports (see below) that elevated T causes immunosuppression as well as increased metabolic rates. Accordingly, this phase of the annual cycle could be costly to both sexes.

3.2.2 Behavioral Effects of Testosterone

The first experimental evidence to prove that avian secondary sexual characteristics and stereotypic male behaviors were dependent on a secreted substance was obtained by Arnold Berthold in 1849. It was common knowledge that castration of immature male chickens (capons) prevented them from expressing large combs, wattles, crowing, male-male aggression or mating behavior. Berthold discovered that capons could develop these traits if their testes or those from another bird were implanted intraperitoneally following castration. When Berthold dissected the experimental birds, he discovered that the implanted organs had attracted a vascular supply from the host. From this he deduced that the morphological and behavioral traits of roosters depended on a blood-borne testicular substance (Berthold 1849). More than a century and a half later, the influence of testicular hormones on male behavior remains a very active area of research.

At the onset of a breeding period, males undergo an increased rate of T secretion, which corresponds with dramatic changes in behavior. Even males from flocking species begin to show increased territorial behavior and aggression to other males as their T levels rise. There is also evidence that T regulates many of the intrasexual and intersexual behaviors that are associated with breeding males. This conclusion is reached both from field studies comparing endogenous hormone levels of focal birds to the behaviors they are expressing and through experimental manipulation of T under breeding conditions. For example, Johnsen (1998) found a significant correlation between plasma T level in male Red-winged Blackbirds (Agelaius phoeniceus) and the frequencies of aggressive interactions with other males, sexual chases, and flights with their showy epaulets exposed. Similarly, Saino and Moller (1995) found that T levels directly accounted for variation in the rate of mate-guarding by male Barn Swallows (*Hirundo rustica*).

Another approach to studying the influence of T on male behavior is to experimentally manipulate T levels by implanting birds with T-filled silastic tubes and then compare their activities to control birds. In general, experimentally elevated T gives rise to increases in behaviors associated with mating, but decreases in those associated with paternal effort. For example, elevated T in Dark-eyed Juncos (*Junco hyemalis*) gives rise to an increased territory size

(Chandler *et al.* 1994), increased mate-guarding activity (Chandler *et al.* 1997), extended sperm production (Kast *et al.* 1998), enhanced attractiveness to females (Enstrom *et al.* 1997), and an increased number of extra-pair copulations (Raouf *et al.* 1997). These activities, however, come at the expense of reduced fat stores and overall body condition in T-enhanced males (Ketterson *et al.* 1991), and reduced involvement with paternal duties such as feeding young (Ketterson *et al.* 1992).

Testosterone supplementation results in significant increases in singing rates in Dark-Eyed Juncos (Ketterson and Nolan 1992), Pied Flycatchers (*Ficedula hypoleuca*; Silverin 1980), House Sparrows (*Passer domesticus*; Hegner and Wingfield 1987a), European Starlings (De Ridder *et al.* 2000), House Finches (*Carpodacus mexicanus*; Stoehr and Hill 2000), and Great Tits (Van Duyse *et al.* 2002). Increased territory size has been documented in T treated Song Sparrows, as well as inducing polygyny in males (Wingfield 1984a). On the other hand, T elevation inhibits paternal care of eggs or young in Pied Flycatchers (Silverin 1980), House Sparrows (Hegner and Wingfield 1987a), Reed Warblers (*Acrocephalus scirpaceus*; Dittami *et al.* 1991), Barn Swallows (Saino and Moller 1995), Spotless Starling (*Sturnus unicolour*; Moreno *et al.* 1999b), European Starlings (De Ridder *et al.* 2000), House Finches (Stoehr and Hill 2000), and Rufous Whistlers (*Pachycephala rufiventris*; McDonald *et al.* 2001). Thus T generally stimulates mating behavior at the expense of parental activities. There are, however, a number of exceptions to this general pattern. For example, T supplementation does not reduce male care of nestlings in Great Tits (Van Duyse *et al.* 2002), Lapland Longspurs (*Calcarius lapponicus*), or Chestnut-collared Longspurs (*C. ornatus*; Lynn *et al.* 2005). Male care of nestlings is important for successful fledging in Great Tits (Bjorklund and Westman 1986), Lapland Longspurs (Hunt *et al.* 1999), and Chestnut-collared Longspurs (Lynn and Wingfield 2003). Unlike almost all other studies, T-supplemented Blue Tits showed no increase in interactions with other males, aggression to a decoy male during simulated territorial intrusion, or in extent of mate guarding (Foerster and Kempenaers 2005). Similarly, T was correlated solely with courtship activities, but not aggression in Buff-banded Rails (*Gallirallus philippensis*; Wiley and Goldizen 2003).

These deviations from a generalized pattern of T-affected behaviors demonstrate that species can evolve modifications in the extent of T coupling to particular behaviors. It is well known, for example, that T can directly stimulate androgen receptors, or exert its influence indirectly following conversion to estradiol via aromatase enzymes located in specific target tissues (Balthazart *et al.* 2004). Thus, although T is essential as a substrate for estradiol production, the rate of estradiol formation will depend importantly on aromatase activity levels. For example, male Pied Flycatchers arrive at their breeding sites in advance of females and try to secure a territory containing a nest hole. Although birds have enlarged testes at this time and are actively secreting T, they vary tremendously in their level of aggression, as indicated by the rate of attack on a decoy during simulated territorial intrusion (Silverin 1993). Silverin then compared rate of attack by individual male Pied Flycatchers to both their circulating level of T and their brain aromatase activity rates. He discovered that rate of aggression was not correlated with plasma T, but significantly corresponded with aromatase activity in the anterior diencephalon (Silverin *et al.* 2004).

Modification to T sensitivity can also result from temporal and spatial variation in expression of androgen receptors for T-coupled behaviors and estrogen receptors for behaviors involving conversion of T to estradiol. The year-round aggression displayed by the Spotted Antbird, despite having very low levels of T outside the breeding season (Fig. 3.4), is facilitated by upregulation of estrogen receptors in the preoptic area and androgen receptors in the nucleus taeniae in the nonbreeding period (Canoine *et al.* 2007). There is also evidence that sex steroid effects can be further modified by steroid-binding proteins and steroid receptor co-regulators, thus permitting further modification of T-affected behavior among species (Ball and Balthazart 2008). This provides a broad template upon which T-affected behaviors can be tailored according to life history requirements (Hau 2007).

Regardless of interspecific differences in mechanisms of action, any species involved in reproductive activities that decrease time and energy for self-maintenance activities is facing a tradeoff between breeding success and nutritional balance. While this will also be influenced by food availability, there is evidence of decreased body mass and fat stores following prolonged T elevation in breeding male Song Sparrows (Wingfield 1984b) and Dark-eyed Juncos (Ketterson *et al.* 1992). Discovering the extent to which these energy imbalances are due to T-affected changes in behavior versus direct effects of T on metabolic rate requires respirometric measurements of T-treated birds.

3.2.3 Physiological Effects of Testosterone

3.2.3.1 Testosterone Effects on Basal Metabolic Rate

The metabolic expenditures of birds vary tremendously over the course of a day. These expenditures can range over 20-fold, being highest when a bird is taking off for flight, least when sleeping during its normal rest period, and at a range of intermediate levels for other activities (Buttemer *et al.* 1986). It is possible, therefore, for T to increase a bird's daily metabolic expenditures indirectly by modifying its behavior, but also directly by increasing its underlying maintenance metabolism, or basal metabolic rate (BMR). Although some have criticized BMR as being unrepresentative of what an animal experiences metabolically in its daily existence, it is a physiological baseline that defines the minimum cost of living for an endothermic animal (Hulbert and Else 2004). In this context, it is important to distinguish between potential direct and indirect effects of T on metabolic rates as these have very different implications for free-living birds. If T affects metabolic rate directly through increased BMR, then an animal would experience persistently elevated rates of energy expenditure when exposed to temperatures within and above their thermoneutral zone, with added demands on their daily energy needs. If, instead, T affected metabolism indirectly through changes in activity, these would be largely context dependent and thus have the potential to be ameliorated by changes in behavior.

There are persistent claims that T affects BMR in birds. Many of these cite a study by Feuerbacher and Prinzinger (1981), which showed a lower nocturnal metabolic rate in castrated male Japanese Quail (*Coturnix coturnix japonica*) than intact birds. What is ignored, however, is that T treatment failed to restore BMR in these birds or in those studied by Hanssler and Prinzinger (1979). Similar metabolic differences have been shown between castrated and intact male Bramblings (*Fringilla montifringilla*; Rautenberg 1952–53) and Spotted Munia (*Lonchura punctulata*; Gupta and Thapliyal 1984), but in both studies, T treatment to castrates failed to change BMR. The decreased BMR in castrates may be partly due to the cost of supporting gonadal tissue. For example, Chappell *et al.* (1999) found a significant relation between testicular mass of postbreeding House Sparrows and BMR.

A more recent study reported that T significantly increased BMR in House Sparrows (Buchanan *et al.* 2003). They measured metabolic rates in intact males and three groups of castrates; a high-T, low-T, and control group. The castrates do show a significant increase in BMR in relation to T level, which supports their conclusion that T directly affects BMR in these birds. However, the intact group also had T levels as low as the control castrates, yet had a BMR intermediate to the two T treated groups. Because the BMR values for the intact group were more than 50% higher than reported for this species (Hudson and Kimsey 1966), another interpretation is that birds did not fully settle in the limited time allowed for these measurements (three hours), and that those with higher T settled least. This interpretation is strengthened by more recent findings that T-treatment had no effect on BMR of intact House Sparrows under short-day conditions (Buttemer *et al.* 2008a), in castrated males under long-day exposure (Buttemer *et al.* unpubl. data), or in intact White-plumed Honeyeaters (*Lichenstomus penicillatus*) during their breeding period (Buttemer and Astheimer 2000). The conditions used for the overnight metabolic measurements in the latter two studies were conducive for BMR determination, which resulted in control and T treated birds having indistinguishable BMR values from those reported (Hudson and Kimsey 1966) or predicted (Aschoff and Pohl 1970a) for these species.

On balance, there is little or no support for T having a direct effect on metabolic rate in birds. This does not mean that birds with high T levels will not experience higher daily energy costs, but if they do it will be a result of T-affected changes in behavior and not changes in underlying metabolic processes. From this we would expect breeding males with a higher probability of male-male conflict to have substantially higher daily energy needs than those from more stable social environments.

3.2.3.2 Testosterone-Affected Immunosuppression

The idea that T may be immunosuppressive stems from observations that reproductively active vertebrate males generally show lower immune responsiveness to antigenic challenge and higher parasite burdens than females (Grossman 1984; Alexander and Stimson 1988). This interpretation underpins a mate-selection model proposed by Folstad and Karter (1992) as the immunocompetence handicap hypothesis (ICHH). The ICHH proposes that males require increased levels of T for expression of secondary sexual characters, but these elevated T levels simultaneously place them at greater risk of parasitic infection. Importantly, because parasitic infection also influences T secretion, T levels will be inversely related to parasite burden. Thus, males with superior genetic resistance ("good genes") to parasitic infection will have higher T levels than poorer quality males, enabling them to exhibit more effective T-dependent sexual signals. Thus, because T-dependent traits are more costly to lower-quality individuals, females can gain an honest appraisal of the bearer's health, and genetic quality by inference, from the quality of his signals.

A core assumption of The ICHH is that the biochemical substance (T in this case) both promotes secondary sexual trait expression and simultaneously compromises the immune system (Folstad and Karter 1992). Not all secondary sexual traits in birds are T-dependent, however. There are examples of male plumage characteristics that are either independent of T (Owens and Short 1995; Kimball and Ligon 1999), or expressed outside the breeding season when T levels are very low (Buchanan *et al.* 2003). With these discrepancies aside, if females rely mostly on T-dependent secondary sexual signals, then T-induced immunosuppression must be a general phenomenon in males for the ICHH to be valid. This central tenet of The ICHH has prompted much experimental examination of the relation between T and male immunity.

One of the first studies to test the ICHH experimentally manipulated the plasma T levels in free-living male Barn Swallows and then measured their ectoparasite load weeks later (Saino *et al.* 1995). Testosterone treated birds had a significantly greater infestation of feather lice than control birds, which was interpreted to be supportive of the ICHH. Eens *et al.* (2000) performed a similar experiment in captive Moorhens (*Gallinula chloropus*) and likewise found T treated birds to have significantly greater ectoparasite burdens than control birds. Endoparasite infection levels have also been shown to be increased by T. Male Dark-eyed Juncos given exogenous T to attain breeding-levels had a significantly greater infection level of *Leucocytozoon fringilliarum*, a haemoparasite, than control birds (Deviche and Parris 2006). While the results from these studies are in complete agreement with outcomes predicted by the ICHH, there is no way to judge if they are a direct result of T-induced immunosuppression. In the case of field studies, there are too many uncontrolled variables to distinguish direct from indirect effects of T. For example, the activity patterns and use of habitat by T treated males may vary from control birds, and may result in different levels of exposure to the parasites or their vectors. In the case of ectoparasites, increases could result from T-affected differences in preening and, therefore, have little to do with immune capability.

The uncertainties of inferring compromised immune function from increased parasite burdens has prompted many researchers to evaluate immune responses using more direct measures. These typically involve quantifying the responses of T treated and control birds to one or more immune challenges and then inferring an overall effect of T on immune capability. The immune challenges that are chosen are based as much for their ease of quantification as for functional considerations. The most frequently used measure is the extent of swelling following intradermal injection of phytohaemagglutinin (PHA) into the patagium (wing web). This has been referred to as a delayed-type hypersensitivity test, owing to the occurrence of peak swelling 24 to 72 hours later, and involving T-cell-mediated infiltration by granulocytes, macrophages, and lymphocytes (Martin *et al.* 2006). Another commonly used immune challenge is to inject sheep red blood cells (SRBC) intraperitoneally and then determine humoral antibody response indirectly from haemagglutination measures of serially diluted plasma (e.g., Lochmiller *et al.* 1993). There has been increasing use of ELISA-based assays to determine antigen-specific antibody response to novel antigens such as keyhole limpet haemocyanin (KLH), as they afford much greater resolution of B-cell proliferation than agglutination assays (e.g., Hasselquist *et al.* 1999).

The varied responses displayed by birds when immune challenged demonstrate that T effects on immunity are not as clear-cut as originally proposed. For example, T treatment had no effect on humoral antibody response to KLH immunisation in Red-winged Blackbirds (*Agelaius phoeniceus*; Hasselquist *et al.* 1999), but it significantly reduced European Starling antibody response to KLH immunisation (Duffy *et al.* 2000). While such differences in response to the same immune challenge might represent species differences in T-affected immunity, this assumes that the immune challenge that was measured was a valid surrogate of overall immunity. It is typically assumed that T-induced immunosuppression is occurring whenever T treated birds show lower responsiveness to a particular immune challenge than control birds, but such results are rarely validated by further functional testing. For example, it is not clear to what extent a bird is immunocompromised to specific naturally-experienced pathogens if it shows a 30% reduction in PHA sensitivity, or a 20% reduction in antibody production to SRBC following prolonged T treatment.

When more than one immune challenge is used, results remain unclear; sometimes a similar directional effect of T occurs in both tests (e.g., Duffy *et al.* 2000), but sometimes a significant T-affected decrease in one measurement (e.g., SRBC agglutination) is not mirrored in another measure (e.g., PHA challenge; Casto *et al.* 2001). The lack of consistent T effects across multiple immune measures in the same individual has been found in a number of studies, but there is also inconsistency in how T affects a given immune response in the same species. For example, T treatment significantly reduced House Sparrow antibody formation to SRBC in one study (Buchanan *et al.* 2003), but showed no effect in another (Buttemer *et al.* unpubl. data). Surprisingly, Evans *et al.* (2000) reported that T treated House Sparrows had higher antibody production than control birds after accounting for T-related increases in CORT secretion. The latter result raises a critical issue: to what extent might CORT be responsible for reported T-related reductions in immune responsiveness?

Owen-Ashley *et al.* (2004) designed an elegant experiment to test the possibility that the reported immunosuppressive effects of T might have a correlative rather than a causative basis. They experimentally manipulated the amount of Three endogenously produced androgens in Song Sparrows and then quantified the birds' response to immune challenges as well as measuring CORT levels. The steroids were dehydroepiandrosterone (DHEA), 5α-dihydrotestosterone (DHT), and T. Dehydroepiandrosterone is an androgen precursor (Fig. 3.5), whereas DHT and T are produced by the testes during territory establishment and pair formation (Owen-Ashley *et al.* 2004). Importantly, DHT and T bind to androgen receptors in target tissues, but only T can be converted to estrogenic products by aromatase (cyp19, Fig. 3.5). Thus, if DHT and T have similar effects on immune performance, then a direct immunomodulation effect by T will be supported. By contrast, if T shows immunosuppressive effects but DHT does not, then indirect mechanisms must be responsible for this outcome. The two most obvious candidates would be estrogens formed from aromatisation of T or T-stimulated rises in CORT secretion. The results were unambiguous in showing that T reduced the extent of patagial inflammation following PHA injection and inhibited antibody formation to both tetanus and diptheria antigens. However, because the DHT treated birds did not differ from controls (Fig. 3.6), there was no evidence that T exerted its effects by direct stimulation of androgen receptors. At the same time, T treated birds showed a rise in CORT (Fig. 3.7), suggesting T-induced immunosupression results from CORT elevation, estrogen production following aromatisation of T, or some combination of the two. Because these birds were held individually in cages and given free access to food, the rise in CORT among T treated birds represents a direct effect of T on CORT rather than an indirect effect due to nutritional or other stressors.

There are a number of other studies that have identified rises in CORT following experimental manipulation of plasma T levels. Increases in basal levels of

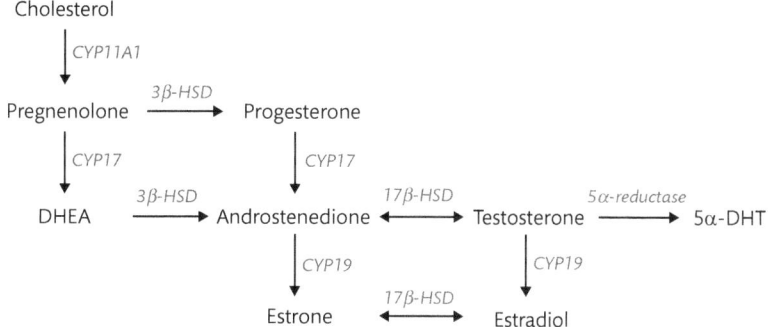

Fig. 3.5 Schematic representation of sex steroid synthesis. Hormone abbreviations are: DHEA, dehydroepiandrosterone; DHT, dihydrotestosterone. Enzymes are shown in blue-colored, italicized abbreviations as: *CYP11A1*, cytochrome p450 side-chain cleavage; *CYP17*, cytochrome p450 17α-hydroxylase/C17,20lyase; *CYP19*, P450 aromatase; 3β-HSD, 3β-hydroxysteroid dehydrogenase/Δ5-Δ4isomerase; and *17β-HSD*, 17β-hydroxysteroid dehydrogenase. (Redrawn from Hau 2007).

CORT have been reported in European Starlings (Duffy *et al.* 2000), house Sparrows (Evans *et al.* 2000; Buchanan *et al.* 2003), and Dark-eyed Juncos (Ketterson *et al.* 1991; Klukowski *et al.* 1997; Schoech *et al.* 1999; Casto *et al.* 2001). Klukowski *et al.* (1997) reported that experimentally elevated T also increased the levels of corticosterone-binding globulins (CBGs). This was confirmed by Deviche *et al.* (2001) who showed that CBG levels in male Dark-eyed Juncos were increased by exposure to long-day photoperiods, but were also directly influenced by their T levels. Although birds lack sex-steroid-specific binding globulins, Dark-eyed Junco T shows a moderate affinity to CBG, resulting in most of the T being bound to CBG when CORT levels are near baseline. However, because CORT has higher affinity for CBG (Breuner and Orchinik 2002), a change in CORT secretion is expected to dynamically alter plasma levels of unbound T. Deviche *et al.* (2001) estimate that the amount of CORT secreted in a typical acute episode of stress would rapidly displace CBG-bound T, with a consequent six-fold rise in free-T levels in plasma.

Corticosterone is the principal glucocorticoid secreted by birds and it is known to be immunosuppressive, particularly when plasma levels are elevated for extended periods (Sapolsky *et al.* 2000). Because CORT secretion typically increases when animals are socially or physiologically perturbed, there is ample opportunity for T-related changes in behavior to indirectly increase CORT secretion. A good example of this comes from a study of male Gouldian Finches (*Erythrura gouldiae*; Pryke *et al.* 2007). Gouldian Finches

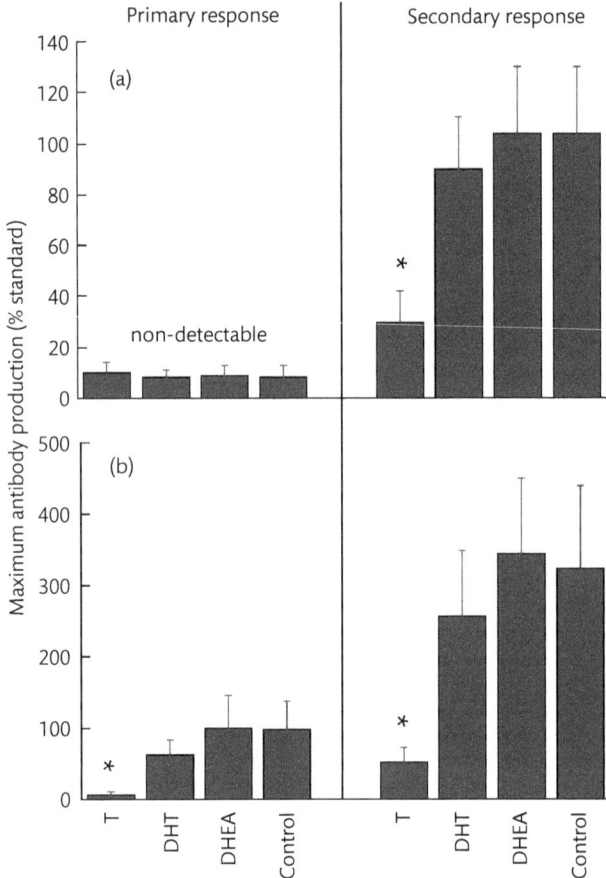

Fig. 3.6 Mean peak primary and secondary antibody production levels in Song Sparrows (*Melospiza melodia*) given implants filled with testosterone (T), dihydrotestosterone (DHT), dehydroepiandrosterone (DHEA) or empty (controls) following vaccination with diptheria (Panel A) or tetanus (Panel B) antigens. Primary responses to diptheria were nondetectable (N.D.). Asterisks indicate significant differences (P<0.05) compared to other groups and the error bars indicate 1 standard error. (Redrawn from Owen-Ashley *et al.* 2004).

have two distinct color morphs: red-headed males that are conspicuously aggressive and dominant, and behaviorally submissive black-headed males (Pryke and Griffith 2006). To examine the physiological consequences of exposure to different frequencies of the same or different colored morphs, birds that had previously been socially isolated were placed six per cage in various ratios of black-head:red-head morphs after first obtaining a control

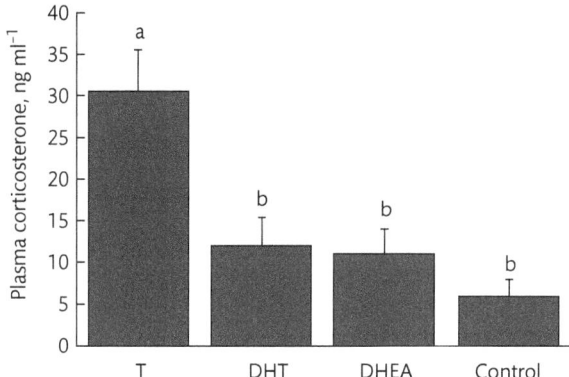

Fig. 3.7 Baseline mean plasma corticosterone levels in Song Sparrows (*Melospiza melodia*) 19 days after receiving empty implants (control) or implants filled with testosterone (T), dihydrotestosterone (DHT), or dehydroepiandrosterone (DHEA). Error bars represent 1 standard error and shared letters among treatments indicate no significant difference in corticosterone levels. (Redrawn from Owen-Ashley *et al*. 2004).

blood sample for hormone characterisation. Each bird's immune response to an intradermal injection of phaetohaemagglutinin (PHA) as well as their plasma levels of CORT and T were evaluated after six days of social interactions. When birds were socially isolated, the two color morphs had indistinguishable plasma levels of T and CORT and showed the same level of wing-web swelling following PHA injection (Fig. 3.8). But when caged together, an increased frequency of red-headed morphs resulted in higher levels of T among red-headed birds, but reduced T in black-headed males. Interestingly, CORT levels increased to a similar extent as T in the red-headed males, but remained at the same levels among black-headed birds regardless of the color-morph composition. It is intriguing that the immune response to PHA in relation to red-head frequency corresponds better to the variation in CORT in both morphs than it does to T.

The potential for T to increase CORT levels, whether directly or indirectly, complicates the requirement of The ICHH to have a hormone that directly stimulates secondary sexual characteristics while suppressing immune functions. However, unless rises in CORT are directly scaled to T secretion, reconciliation with the ICHH is problematic. Such scalings are unlikely to be straightforward, as T-related rises in CORT will depend importantly on social and nutritional contexts. From a functional viewpoint, there are more fundamental issues that need to be considered before accepting or dismissing T as an immunological liability.

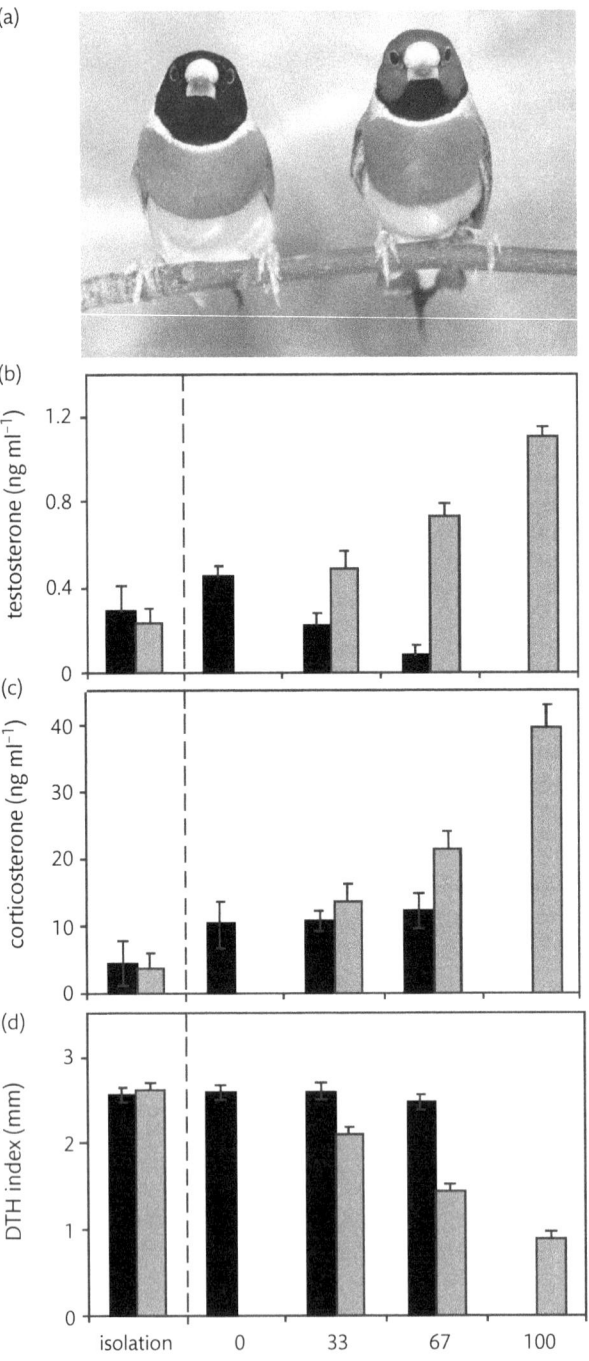

Fig. 3.8 Panel (a) Black-headed (left) and Red-headed male Gouldian Finch (*Erythrura gouldiae*) morphs. Socially naïve birds were maintained in visual isolation from opposite morphs prior to the experiment (isolation period) and had blood samples collected during this period. They were then transferred to cages at densities of 6 birds/cage at ratios of red:black of 0:6, 2:4, 4:2, or 6:0. Blood samples collected 6 days later were used to determine plasma concentrations of testosterone (panel b), corticosterone (panel c) and delayed-type hypersensitivity to phytohaemagglutinin injection as measured by wing web swelling (panel d). Black and grey bars depict measurements from black- and red-headed morphs, respectively, and all values are shown as least-square means ± one standard deviation. (From Pryke *et al.* 2007, with permission from the Royal Society).

3.2.3.3 Immunosuppressed or Immuno-Optimized?

A common assumption in all studies of T-affected immunity is that the immune measure being studied is a valid surrogate of the animal's overall immune capabilities. If this were so, then different measures of immune function would be expected to show the same response to T treatment. While this occasionally occurs (Owen-Ashley *et al.* 2004; Duffy *et al.* 2000), most studies using multiple immune assays have confounding results, with evidence of T-affected immune depression in one assay, but lack of an effect in others (Buchanan *et al.* 2003; Casto *et al.* 2001). In a study of the correlation between immune competence and quality of a secondary sexual signal, Faivre *et al.* (2003) quantified bill color in relation to humoral immunity to SRBC and a delayed-type hypersensitivity test to PHA. Bill color in European Blackbirds *(Turdus merula)* is T-dependent and ranges from pale yellow to deep orange. They found that the PHA response correlated directly with increasing orange color, but SRBC antibody production was inversely related. Thus, one result supports the ICHH and the other one refutes it. The extent of inconsistencies in experimental studies of The ICHH is revealed by a meta-analysis showing that the weak effect of T on direct measures of immunity becomes insignificant after controlling for multiple studies on the same species (Roberts *et al.* 2004).

One interpretation of the variable nature of T-affected immune responses is that it represents a redistribution of immune cells rather than immunosuppression (Braude *et al.* 1999). Thus, T induces birds to allocate immune resources in anticipation of where they are most needed. Such immunoredistribution has been demonstrated for glucocorticoids, whereby acute secretion of glucocorticoids provokes migration of leucocytes from peripheral circulation to lymph nodes, skin, and other peripheral locations (Dhabhar and McEwen 1997). There is no experimental evidence, however, that has validated T-dependent immunoredistribution in birds. Another interpretation for the variation in specific immune measures is that it represents selective downregulation of immune components during times of resource limitation (Viney *et al.* 2005). In this view, immunity is dynamically and adaptively adjusted to optimize protection against the most relevant pathogens using the immune resources afforded by the animal at the time.

The reasoning of Viney *et al.* (2005) is evolutionarily appealing, as it defines optimality of immune function in terms of fitness outcomes. Assuming that immunity and reproductive activities compete for the same resources, the ensuing unavoidable tradeoffs should be managed in ways that optimize lifetime reproductive success (Zuk and Stoehr 2002). Thus, optimal allocation of resources towards immunity and reproduction would be expected to vary among species or populations with different life expectancies (Lee 2006), or with different intensities of parasite exposure (Sheldon and Verhulst 1996). In this context, animals would benefit by responding to a range of cues, including T, to dynamically optimize their immune capability in relation to their current or expected nutritional status, but the optimal response to these cues would differ among species and populations. Thus, the effects of T and other hormones on immunity would be expected to vary across species and time.

A further complication in interpreting tradeoffs between male reproductive condition and immune function is that most studies rely on exogenous supply of T to mimic reproductive hormonal status. Natural elevations of T are a result of sequential stimulation along the hypothalamic-pituitary-gonadal axis. Thus, following receipt of appropriate stimuli, the hypothalamus secretes gonadotrophin-releasing hormone (GnRH), which, in turn, stimulates cells within the anterior pituitary to secrete follicle stimulating and luteinizing hormones (FSH and LH, respectively). Upon reaching the gonads, FSH and LH promote testicular development and production of T (Adkins-Regan 2008). There are feedback mechanisms, whereby rising T levels inhibit GnRH, FSH and LH secretion (Ottinger *et al.* 2002). Thus, under natural conditions, elevated T would be a consequence of concurrent rises in GnRH and LH, but when T is experimentally elevated, GnRH and the gonadotrophins FSH and LH will be artificially reduced. Thus, T treated birds will have a different hormonal milieu than birds experiencing natural rises in T. This will have important consequences for interpreting immune tradeoffs associated with breeding if the hypothalamic and pituitary hormones inhibited by T are themselves immunomodulatory.

There is evidence that T-induced immunosuppression of B-cell proliferation in mice is mediated by GnRH. By comparing the effects of various steroid treatments on B-cell production in GnRH-deficient and GnRH-sufficient littermates, it was determined that T-induced immunosuppression was due to its inhibitory effects on GnRH, which itself stimulates B-cell production (Jacobson and Ansari 2004). If this mechanism applies to birds, T-treated animals will have a pharmacologically depressed GnRH, which will distort the immunosuppressive potential of T. In another study examining the effects of gonadotrophins on immune and other performance measures in Side-blotched Lizards (*Uta stansburiana*), Mills *et al.* (2008) discovered that LH, FSH and T exerted independent and often contrasting effects on innate and adaptive immune responses. Furthermore, the interactive effects of these hormones on immune functions varied between the color morphs they studied, emphasizing the evolutionary flexibility in immune function modulation.

3.2.3.4 Overview

Males undergo dramatic behavioral and physiological changes when they prepare for breeding. The notion that T is a major effector during these changes is not refuted, but there has been too little appreciation of the evolutionary plasticity in how species respond to these signals in accord with their different life history needs. Similarly, too much attention has been paid to how T directly affects immune performance at the expense of understanding how other associated hormones and nutritional status interact in this process. In measuring immune function, the assumption that single or even multiple immune measures reflect overall protection against pathogens and infections must be validated, and the fitness consequences of presumed deficiencies explored.

The redirection of behavior from maintenance-dominated towards breeding activities is likely to compromise nutritional balance, but the extent will depend on food availability, social context, and life history characteristics. An important corollary of nutritional balance is the effect it will have on both immune performance and oxidative stress. There is ample evidence that nutritional deficits compromise immune capacity (Klasing 2007; Lochmiller *et al* 1993), but certain critical antioxidants are only obtained from the diet. With the recent suggestion that T itself might invoke oxidative stress (Alonso-Alvarez *et al.* 2006), the influence of variations in nutritional plane on fitness takes on greater emphasis.

3.3 Costs of Parenting

The class Aves is unusual amongst vertebrates in the extent of male involvement with post-fertilisation reproductive activities. Male birds can (and do) contribute extensively to all phases of egg, chick, and post-fledgling care, and benefit from the fitness gains that result from biparental involvement in these activities. In a thorough review of parental care data for over 5000 bird species, Cockburn (2006) validated biparental care in 75% of the species, but inferred that it was likely to occur in 81% of those surveyed. By contrast, female only care occurred in 8% and male only care in 1% of the species. Single parent care was highly biased towards species producing precocial nidifugous young, a process that requires relatively little post-hatching parental effort. In the case of tropical frugivores and nectarivores, breeding occurs when food is highly abundant and provisioning young is not difficult, placing higher value on females selecting males for good genes rather than for direct or indirect care of progeny (Gowaty 1996). Cooperative breeding, where more than two individuals assist with a brood, occurs in 9% of avian species. The absence of parental care is rare among birds, involving 1% of all species and comprising mainly brood parasites that lay eggs in the nests of others and megapodes, which display parental care after laying but not after hatching (Cockburn 2006).

The involvement of one or both parents in reproductive activities varies tremendously between species, but we will focus on two of the most studied categories, incubation and chick rearing, to identify the costs they impose on parents.

3.3.1 Costs of Incubation

3.3.1.1 Incubation Temperature Considerations

Embryonic development requires egg temperature (T_{egg}) to attain levels above a threshold value, termed the physiological zero temperature (Lundy 1969), before embryo growth is initiated (Fig. 3.9). For most species, this temperature threshold is sufficiently above ambient conditions to require direct skin contact (brooding) by an incubating bird before development progresses. This allows a high level of control as to whether or not eggs in a multiple-egg clutch hatch synchronously. Thus birds initiating incubation prior to clutch completion will have a spread of hatching dates amongst the clutch (asynchronous hatching), whereas those deferring incubation until the last egg is laid will likely have synchronous hatching.

In addition to the timing of incubation, the temperature stability during incubation has a strong effect on incubation time and hatching. For many species, there is a limited range of egg temperatures that promote optimal growth (Fig. 3.9). Although most species can tolerate departures lower than this optimal range for variable periods, lower temperatures will retard growth rate which leads to delayed hatching time and inefficient transfer of egg nutrients to growth (Olson *et al.* 2006). Such delays will not only prolong the time the eggs and chicks are exposed to predators, but also produce chicks of smaller size and lower quality. Even at moderate temperatures, female chicks hatching from eggs having more consistent incubation were found to have much higher fecundity when they bred (Gorman and Nager 2004). Thus, incubating birds must weigh the fitness

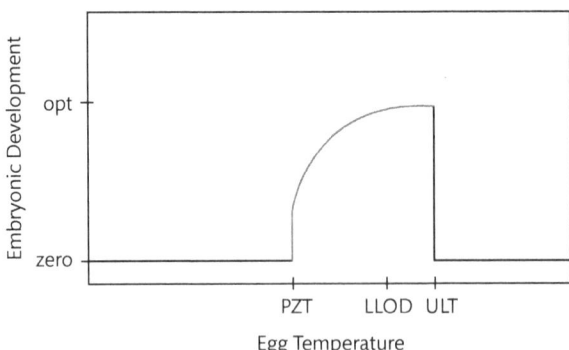

Fig. 3.9 General relationship between avian embryonic development and egg temperature. The temperature range compatible with development (highlighted by the gray line) is bounded by the PZT, physiological zero temperature and the ULT, upper lethal temperature. The lower limit of optimal development (LLOD) is a few degrees Celsius from the ULT. (Redrawn from Lundy 1969).

consequences of allocating time and energy towards incubation duties against the costs it imposes on their future fecundity and survival.

Ambient Temperature Considerations   Birds as endothermic animals can be exposed to a range of ambient temperatures (T_a), designated the thermoneutral zone (TNZ), over which they do not require metabolic expenditures above resting levels (RMR) to maintain stable body temperatures (T_b). When sitting on eggs at ambient temperatures within their TNZ, no extra heat production is needed to maintain optimal T_{egg} (reviewed by Williams 1996b). However, the incubating bird-egg complex has different heat-transfer properties than the incubating bird by itself, resulting in the onset of thermogenesis occurring at a slightly warmer temperature and the rate of heat production rising more rapidly as T_a is further reduced (Fig. 3.10). The extent of this rise will be affected by the insulative quality of the nest (Buttemer *et al.* 1987), but also by the size of the bird. Both the lower boundary of the TNZ and the relative insulation of birds are size dependent, resulting in large birds tolerating much lower T_a before they need to increase heat production and showing smaller proportionate rises in metabolic rate for a given drop in T_a below this. As a consequence, a 42 kg Emu, fully

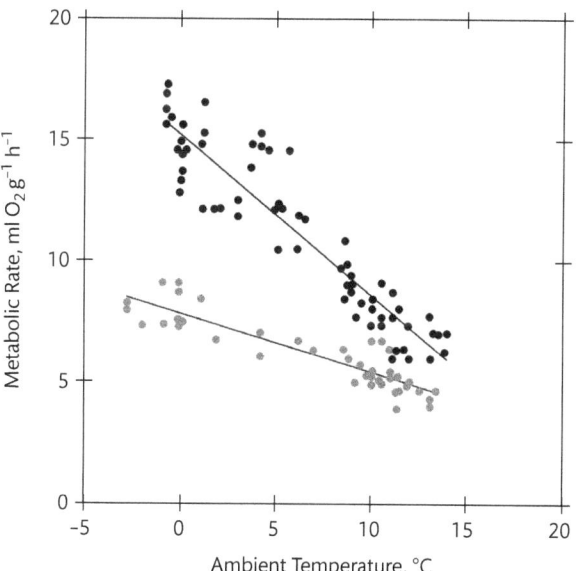

Fig. 3.10 The relationship between rate of oxygen consumption and ambient temperature in female Blue Tits (*Parus caeruleus*) when standing above eggs in a nest (black-filled circles), or when incubating 13 eggs (gray-filled circles). (Redrawn from Haftorn and Reinertsen 1985).

exposed to the cold winter sky, can incubate eggs at T_a close to freezing without elevating its metabolic rate above RMR (Buttemer and Dawson 1989). Another factor that affects the heat required for incubation is whether incubation is continuous or intermittent. If an incubating bird leaves the nest, the eggs will cool and require an additional allocation of heat for rewarming. The amount of heat required, of course, will depend on the extent of egg cooling, which depends on nest properties, egg size and number, and ambient thermal conditions.

3.3.1.3 Incubation Behavior Considerations

There are many patterns of incubation among birds (see review by Deeming 2002), ranging from continuous by either sex, continuous, but shared by both sexes, and many different combinations of discontinuous incubation by one or both parents or helpers. Continuous incubation by one sex occurs in mainly in capital breeders, whereby energy stores are accrued in advance of egg laying and incubation and thereby permit eggs to be maintained under very stable thermal conditions (e.g., Emperor Penguins, Emus, Eider Ducks). In some species, the female incubates continuously and relies upon the male for all of her food (e.g., hornbills). In many seabirds, one bird waits on the eggs until nest relief by its partner, which protects the eggs from predators and ensures stable developmental temperatures.

Most species, whether single sex or biparental in incubation duties, are income breeders and intersperse incubation duties with feeding. Thus feeding is either deferred in capital breeders, postponed for discrete periods in paired-continuous incubators, or sporadically scheduled in single sex or biparental intermittent incubators. Furthermore, because the potential for egg cooling increases at lower T_a, birds having to feed during the incubation period typically shorten their foraging periods and thus their time off the nest as T_a decreases (Conway and Martin 2000). Consequently, apart from the energy demanded by incubation thermal requirements, the disruption to foraging further challenges the ability of breeding birds to maintain their nutritional and energy balances. Thus, clutch size may be constrained by both the ability of parents to provide the amount and quality of heat needed for each developing chick, but also by the fitness costs if clutches above a particular size compromise current and future reproductive efforts.

3.3.1.4 Evidence of Incubation Costs

Physiologists usually express costs in units of energy, whereas evolutionary biologists refer to costs in terms of reproductive or fitness consequences. These units can be interrelated, as animals need to invest energy in the materials and activities of reproduction, and energy costs in excess of energy gains can lead to compromised health, reproductive performance and, consequently, lifetime reproductive success. Field measurements of daily energy expenditure (DEE) using stable isotopes consistently show costs during the incubation stage to be somewhat lower than those during chick rearing (Williams 1996b), but this difference disappears in species where the female is solely responsible for incubation (Tinbergen and Williams 2002). In fact the highest incubation-stage costs recorded, four to six times basal metabolic rate (BMR), were measured in the Orange-breasted Sunbird (*Nectarina violacea*; Williams 1993). Because females are the sole incubators in this species and lay eggs in winter, they must forage intensively to find sufficient food for themselves while off the nest, but return frequently to rewarm their rapidly cooling eggs. For the most part, though, incubation costs are in the range of two to three times BMR (Tinbergen and Williams 2002).

Energy costs for incubation and chick rearing must be placed in the context of Time available for feeding, food abundance, and the "commute distance" to the food source. For example, if temperatures become warmer between the start of incubation and chick hatching, parents will be able to spend more time away from the nest after chick hatching with less developmental consequences from temperature declines. On the other hand, all the food they gather while incubating is for themselves, whereas that gathered during chick-rearing must support themselves and the energy and nutrient demands of rapidly growing chicks with combined masses that may be far in excess of their own. Thus differences in energy balance, not energy expenditure, may be more representative of immediate costs to incubating adults. These kinds of imbalances would be expected to manifest themselves in health-related measurements and also subsequent chick-rearing abilities.

The most direct way to examine the potential constraints of clutch size on incubation costs is to manipulate clutch size and determine the consequences. This has been done in a sufficient number of species to identify a general trend in the outcomes. Thomson *et al.* (1998) examined data from clutch manipulation studies and concluded that clutch enlargement above normal levels resulted in reduced hatching success, protracted incubation periods, and higher parental energy expenditure. A few subsequent studies have taken care to modify clutch size during incubation and then restore it to pre-manipulation size at the time of hatching to separate incubation effects on chick rearing from chick provisioning effects. Two of these support the general findings of Thomson *et al.* (1998) in showing clutch enlargement leading to extended incubation periods in Barn Swallows (Engstrand and Bryant 2002) and House Wrens (*Troglodytes aedon*; Dobbs *et al.* 2006). Some of the effects of clutch enlargement may not be evident until long after hatching. For example, Cichon (2000) and Ilmonen *et al.* (2002) reported that clutch enlargement in Collared (*Ficedula albicollis*) and Pied Flycatchers affected fledgling success, but not hatching success, female body condition, or female ability to mount a humoral immune response to a foreign antigen. Similarly, clutch enlargement in Great Tits had no effect on the fitness of Their current clutch, but significantly reduced fledgling success of their second clutch in the same season. Furthermore, females producing a second clutch after experiencing an enlarged first clutch had a reduced survivorship compared to control birds or those with reduced clutches (de Heij *et al.* 2006).

These studies reveal the importance of viewing fitness consequences beyond hatching success and also establish incubation costs as another reproductive phase placing constraints on clutch size evolution. The consistent protraction in incubation time with clutch enlargement has obvious fitness consequences in prolonging nestling exposure to predators. The reduction in fledgling success accompanying the higher incubation demands of enlarged clutches and reduced survivorship of incubating adults provides further evidence that impacts on fitness occur through the entire reproductive process.

3.3.2 Costs of Chick Rearing

The degree of development of chicks at hatching varies tremendously across species, ranging from those fully independent of their parents, such as the extremely precocial Australian megapodes, to those fully dependent on parents for food and heat, including many highly altricial passerine species. There is a continuum of hatchling developmental states between these two extremes (O'Connor 1984), with a matching spread of post-hatching parental effort needed to achieve fully independent young. Precocial species produce larger eggs with substantially higher energy content than altricial species (Carey 1996), but expend far less energy looking after chicks than altricial parents and, in consequence, many of these species show uniparental care. Thus, their reproduction/maintenance tradeoffs mainly involve egg production and incubation for females and territoriality/mate attraction activities and occasionally incubation for males. By contrast, altricial species spread their tradeoffs over more activities and both sexes tend to share in the effort, particularly after hatching.

Once chicks hatch, they initiate begging behaviors that elicit feeding responses from attending adults (Kilner 2002). Breeding adults are very sensitive to chick begging, a trait that brood parasites exploit when they lay their eggs into nests of unrelated hosts (Lotem 1998). This demand for food will increase as the growth period progresses, resulting in very high demands just prior to fledging. For example, studies using doubly labelled water have shown that the DEE of adult Kittiwakes (*Rissa tridactyla*) rises from about 2.6 times BMR the first week of chick feeding to 4.3 BMR towards the end of nesting (Fyhn *et al.* 2001). The foraging demand on birds feeding chicks will be directly affected by chick number, but also by food availability and the foraging skills of the adults. This places a constraint on clutch size in altricial species, as inadequate provisioning after hatching may risk losing the entire clutch, but it also creates evolutionary incentives for the use of asynchronous hatching and even siblicidal brood reduction strategies to improve fitness among many species.

Because it is relatively cheaper to produce an altricial egg than a precocial one, altricial females can afford to take more risks in anticipating an optimal clutch size. For example, if food were abundant during egg synthesis, they could lay a large clutch without much penalty. However, if food decreased at the time of hatching, or their male partner was an inferior provisioner, they might jeopardize the survivorship of the entire clutch by having too many chicks to feed. In such situations, parents might selectively feed particular chicks in the clutch, but if they do not select the right ones there may be fitness penalties. Many species avoid this complication by establishing a nestling hierarchy before hatching. This is done most simply by initiating incubation before the last egg is laid, but another strategy is to sequentially decrease the amount of maternally derived androgens transferred to eggs in a clutch (Sockman *et al.* 2006). Because yolk androgens are known to promote growth in developing chicks (Schwabl 1996; Lipar and Ketterson 2000), this would reinforce hatching asynchrony among the clutch. Hatching asynchrony, as first described by Lack (1947), is believed to be advantageous when food availability is below a particular threshold value. Under such conditions, earlier-hatched chicks obtain a greater share of available food at the expense of their smaller, less developed siblings, thus ensuring fledging of at least some of the clutch. In years of more abundant resources, all the chicks can obtain adequate food and parental output is maximized.

Experimental studies that have manipulated hatching synchrony among clutchmates are inconsistent in supporting Lack's interpretation, but they have shown that fledging success alone ignores other fitness measures. Amundsen and Slagsvold (1991) showed that chicks from asynchronous broods had higher fledging body masses than those from synchronous broods, and that this resulted in a greater probability of recruitment of these offspring as subsequent breeders (Amundsen and Slagsvold 1998). Asynchronous hatching also facilitates active brood reduction by chicks via enhanced competitive or aggressive inequalities among offspring, resulting in last-hatched chicks being killed by earlier siblings, particularly in poor years. Experiments examining the effect of hatching synchrony in Cattle Egrets (*Bubulcus ibis*), a siblicidal species, found that fledging rates did not differ among nests, but the amount of food required to rear chicks was least in broods with normal asynchronous hatching and highest in those hatching synchronously (Mock and Ploger 1987). Irrespective of hatching mode, however, chick number and stage of development will provoke high workloads in feeding adults, which may affect their future breeding and survivorship.

3.3.2.1 Fecundity/Longevity Tradeoffs

Clutch and brood manipulation are very effective experimental procedures to examine fecundity-related tradeoffs. This is usually done by removing eggs or chicks from some clutches, adding the same number to others, and allowing a control group to rear their original clutch size. Most of these studies reveal that chicks from enlarged clutches fledge at a reduced body mass and suffer higher mortality (summarized in Dijkstra *et al.* 1990). Although most studies do not monitor longer-term effects on adults, those that do show that birds provisioning enlarged broods often experience reduced fecundity in later attempts (e.g., Tinbergen 1987; Hegner and Wingfield 1987b; Parejo and Danchin 2006), as well as reduced survivorship (e.g., Askenmo 1979; Nur 1984; Reid 1987; Saino *et al.* 1999). These are classic cases of the "cost of reproduction" expected from life-history theory.

An excellent example is unambiguously revealed by long-term studies of the European Kestrel (Daan *et al.* 1996). Unlike most studies that infer mortality by failure of adults to return to a given breeding location in subsequent years, an uncertain assumption given the possibility of dispersal, Daan *et al.* (1996) had knowledge of the date and location of death from recoveries of ringed birds. Kestrel parents with enlarged broods had 60% mortality, those with reduced broods had 22%, and control birds had intermediate survivorship (Fig. 3.11). These mortality patterns correspond to differences in DEE among these groups, reflecting the direct increase in foraging costs with clutch size (Daan *et al.* 1996).

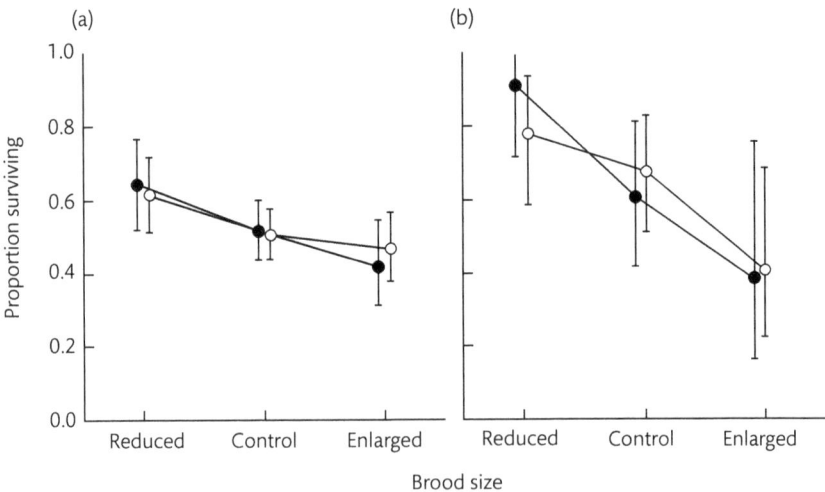

Fig. 3.11 First-year survival indices for European Kestrel (*Falco tinnunculus*) parents raising experimentally enlarged, experimentally reduced, or natural broods. Survivorship in panel (a) is based on local recoveries at the study site (local survival), whereas survivorship in panel (b) is based on the fraction of birds surviving 1 year among all recoveries reported freshly dead over their annual range (global survival). Open symbols are based on multiple sampling over the 3-year period, whereas closed symbols are restricted to the first recording of birds surviving at least one year. (From Daan *et al.* 1996, with permission from Wiley-Blackwell).

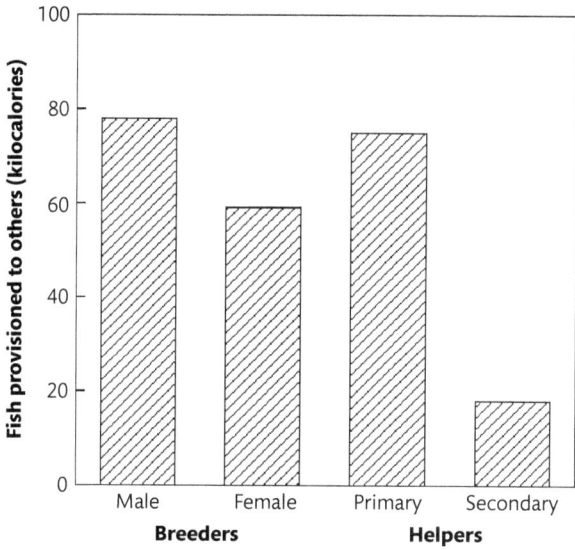

Fig. 3.12 Non-breeding primary helper Pied Kingfishers (*Ceryle rudis*) feed much more fish to the mother and her chicks than do secondary helpers. The primary helpers are related to the breeders they assist, whereas the secondary helpers are not. (Redrawn from Alcock 1993, after Reyer 1984).

A similar relation between foraging workload and mortality occurred in Pied Kingfishers (*Ceryle rudis*; Reyer 1984). This is a cooperatively breeding species, where males from earlier broods (primary helpers) achieve indirect fitness benefits by helping to provision their mother and her chicks at some cost to survivorship, while unrelated non-breeding males (secondary helpers) have minor involvement in feeding, but improve their chances of securing mates in later years. The significantly higher feeding rate of primary compared to secondary helpers (Fig. 3.12), is associated with a 54% survivorship among primary helpers compared to 74% for secondary helpers (Reyer 1984). The implied relation between workload and mortality in Kestrels and Pied Kingfishers may be exacerbated by the motivation of breeding birds to provision chicks at their own expense. This could lead to immediate energy and nutrient shortfalls that in turn, may result in medium to long-term immune and heath compromises.

3.3.2.2 Immunity Tradeoffs

Brood manipulation experiments have also been used to see if immune function is compromised by the demands of breeding. Again, these experiments typically involve increasing the modal clutch size (n) by adding two eggs for enlarged clutches (n+2), removing two eggs for reduced clutches (n−2), and comparing the responses of birds in these groups to control birds (clutch size = n). In one such study, food provisioning rates of Pied Flycatcher females were more than two-fold higher in enlarged compared to reduced clutches when chicks were three days old, with control females showing intermediate provisioning rates. At this time, female cell-mediated response to PHA immunisation was inversely related to feeding effort (Moreno *et al.* 1999a). A similar correspondence between decreased cell-mediated response and brood-manipulated increases in feeding rate was demonstrated in Barn Swallows (Pap and Markus 2003).

If workload and immunity were causally linked, one would expect immune performance to be negatively correlated with rate of energy expenditure among individuals. Moreno *et al.* (2001) tested this idea in brood-manipulated Pied Flycatchers by measuring DEE isotopically and quantifying provisioning rates of adults. They found that female DEE corresponded directly with provisioning rate, but was inversely related to their cell-mediated immune response to PHA challenge. These results collectively suggest that provisioning young does come at the expense of parental self-maintenance. Increases in CORT plasma levels in birds with higher brood sizes have been reported in several studies (Ilmonen *et al.* 2003; Hegner and Wingfield 1987b; Silverin 1982), indicating energy imbalance in these birds. Furthermore, female Blue Tits raising enlarged broods expressed higher levels of heat-shock proteins (HSP60) than birds with control or reduced clutch sizes, but significantly less immunoglobulin (Merino *et al.* 2006). Because heat-shock proteins are expressed during situations perturbing cellular homeostasis (Sorenson *et al.* 2003), this further indicates that breeding places high demands on adults.

It seems intuitive that the tradeoffs to parents will depend both on the demands of offspring and the ability of parents to satisfy them. Thus, brood-size related tradeoffs should be reduced in times of food abundance, but increased when food is more difficult to obtain. These interactions were examined in Tree Swallows (*Tachycineta bicolor*), whose clutches ranged naturally from three to six eggs (Lifjeld *et al.* 2002). When examining cell-mediated immunity in relation to clutch size, there was a significant inverse relation between clutch size and extent of swelling following PHA challenge, but only when foraging opportunities were good. During cold weather and low insect availability, all birds displayed depressed cell-mediated responses, irrespective of brood size. Thus, cell-mediated immunity was maintained when the birds could afford it, but progressively depressed when the demands of chicks exceeded the ability of parents to deliver without penalising themselves.

Within a breeding population, there will be a wide range of foraging abilities and physical condition among adult birds, although these two traits may be highly correlated. Because of this, it is likely that individuals that are better able to meet their own nutritional needs will also be better placed to provide for offspring without compromising their health than less effective individuals. This was demonstrated in a three year breeding study of Gould's Petrels (*Pterodroma leucoptera*) examining the physical properties of adults in relation to chick growth and quality at fledging (O'Dwyer *et al.* 2007). These seabirds undergo a series of prolonged incubation bouts interspersed with at-sea feeding to replace energy reserves depleted during incubation. Parents differ significantly in their body condition at the onset of incubation, but those in best condition at this time remain in good condition at each subsequent stage of breeding. Furthermore, individuals scoring highest in body condition throughout incubation also have higher prolactin levels (O'Dwyer *et al.* 2006) and chick provisioning rates, produce chicks with higher growth rates, and their young fledge at a younger age and in better condition (O'Dwyer *et al.* 2007). Such asymmetries are likely to be common among breeding birds and should result in similar asymmetries in health and fecundity compromises associated with increased reproductive demands.

The interaction between individual quality and the immune consequences of increased reproductive demand was examined in Tree Swallows by Ardia *et al.* (2003). Nests were manipulated to create clutches that were 50% smaller or larger than the original clutch size, whereas controls retained their starting clutch size. All females were then given primary and secondary injections of SRBC to assess their immune performance. Although females with enlarged clutches showed reduced secondary antibody production compared to birds with smaller clutches overall, clutch initiation date was the strongest predictor of secondary immunity. Thus, early breeders, which are dominated by high-quality females, displayed superior ability to mount immune defence irrespective of chick demand. Significantly, post-breeding survivorship for the next three years was significantly higher in birds displaying strong secondary immune responses, which means reproductive fitness is strongly condition-dependent. A similar conclusion was reached by Hasselquist *et al.* (2001), showing a significant inverse relation between date of egg laying in Tree Swallows and humoral antibody production. Furthermore, they manipulated the flight costs of females by clipping four primary flight feathers on each wing. Females with higher workloads due to increased flight costs displayed significantly reduced capacity to mount an immune response to the novel antigen.

3.3.2.3 Breeding-Related Oxidative Stress

The suggestion that increased workload might underlie increased mortality in breeding birds (Reyer 1984; Daan *et al.* 1996) gains some support from demonstrations that increased workload from enlarged clutches or trimmed flight feathers compromises immune functions. There is also experimental evidence that increased reproductive demands may impose greater oxidative stress on breeding Zebra Finches (Alonso-Alvarez *et al.* 2004; Wiersma *et al.* 2004). Both studies manipulated brood size two days after hatching to create two chick and six chick broods. After 12 days of chick rearing, the red blood cells (RBC) of birds rearing larger broods had significantly greater sensitivity to peroxidation damage than those with broods of two, or non-breeding birds (Alonso-Alvarez *et al.* 2004). This assay reveals the ability of antioxidants within and immediately outside the RBC to quench experimentally controlled free radical delivery. The greater peroxidation damage to RBCs from large-brood parents signifies they have a reduced antioxidant capacity.

Wiersma *et al.* (2004) examined antioxidant capacity more directly, by measuring the activity of two major antioxidant enzymes, superoxide dismutase (SOD) and glutathione peroxidase (GPx) from pectoral muscle samples taken 19-20 days after chick rearing. They found brood-size related decreased antioxidant capacity, with birds rearing large broods having about 25% less enzyme activity than those with small broods. Because oxidative stress is considered to have a major influence on the rate of ageing and overall health (Halliwell and Gutteridge 1999), these findings are important. This suggests that immunocompromise and oxidative stress may be mechanisms underlying fecundity/longevity tradeoffs.

3.4 Oxidative Stress and Longevity

Extrinsic mortality rates profoundly influence life history characteristics. Species having low rates of adult survivorship will benefit from high rates of fecundity, which, as we have seen, competes with self-maintenance. Such species will also begin breeding as soon as mature, which further compromises life span (Alonso-Alvarez *et al.* 2006). Longevity traits are not likely to evolve in such species, because individuals with genes promoting longer life spans will not realize greater fitness due to the high levels of extrinsic mortality. By contrast, species with low rates of extrinsic mortality will benefit from traits promoting long-term survivorship, if lifetime reproductive success is improved by living longer. This, in turn, will select for a complementary suite of biochemical, physiological, morphological, and behavioral traits that collectively promote the attainment of an increased maximum lifespan potential (MLSP). Although there are many biochemical and physiological processes that influence rates of ageing, those associated with oxidative stress, particularly from mitochondrial free-radical production, are proposed to be especially important (Harman 1956; Sohal and Weindruch 1996; Beckman and Ames 1998).

3.4.1 Oxidative Stress

The oxidative stress theory is derived from Harman's free radical theory, which proposed a mechanistic link between metabolic rate and ageing (Harman 1956).

According to this theory, O_2 free radicals are an inevitable byproduct of aerobic metabolism, which, in turn, leads to cumulative damage of vital biological molecules and eventual death. This idea developed into the oxidative stress theory, which acknowledges that i) aerobic metabolism produces other ROS in addition to free radicals; ii) organisms use antioxidant defences to diminish ROS; and iii) mechanisms that repair or remove ROS-damaged molecules are available to organisms. Thus oxidative stress will arise when ROS production exceeds the capacity of antioxidant defences. Because mitochondria are the main producers of ROS, it is logical to assume that increased rates of O_2 consumption lead directly to increased rates of ROS formation. From this interpretation one would predict that animals with high rates of metabolism should have correspondingly elevated ROS production and faster rates of ageing. This "live fast, die young" belief gains some credibility from the fact that small animals have higher mass-specific metabolic rates than larger ones and usually have shorter lives.

There are, however, many species that defy this simplistic interpretation. For example, the MLSP of birds, which is the greatest lifespan recorded for a species, is two-fold greater than in mammals of similar size (Fig. 3.13). Paradoxically, the metabolic rate of birds averages 50% higher (Fig. 3.13), resulting in birds consuming an average of three-times as much O_2 in their lifetime as comparably O_2 sized mammals. It is also obvious that there is far greater difference in lifetime energy expenditure among mammals and birds in a given size class than between the two classes (Fig. 3.13). This is clearly shown by comparing the BMR and MLSP of selected galliformes ranging in size from 48 g King Quail (*Coturnix chinensis*) to 3.4 kg Wild Turkey (*Meleagris gallopavo*) to selected species of procellariiformes from 48 g Leach's Storm Petrel (*Oceanodroma leucorhoa*) to 8.1 kg Wandering Albatross (*Diomedea exulans*) (Fig. 3.14; Buttemer *et al.* 2008b). These two groups have statistically indistinguishable metabolic rates, yet the MLSP values of the seabirds average 4.6 times those of the fowl. This clearly demonstrates that aerobic metabolic rate alone does not explain longevity differences between species. Thus there must be differences between species in i) the amount of ROS production per unit of O_2 consumed; ii) their antioxidant defense capacity; and iii) their sensitivity to ROS production, or combinations of these.

3.4.2 Reactive O_2 Species Production

Reactive O_2 species include both radicals, molecules containing one or more unpaired electrons that are capable of independent existence, as well as nonradical derivatives of O_2. There are three main ROS formed by the reduction of univalent O_2: superoxide and hydroxyl radicals, and hydrogen peroxide (H_2O_2). Of these, the hydroxyl radical is the most reactive, but H_2O_2, while not a free radical, can diffuse from its site of production and form hydroxyl and other radicals at distant cellular locations, thus propagating oxidative damage (Hulbert *et al.*

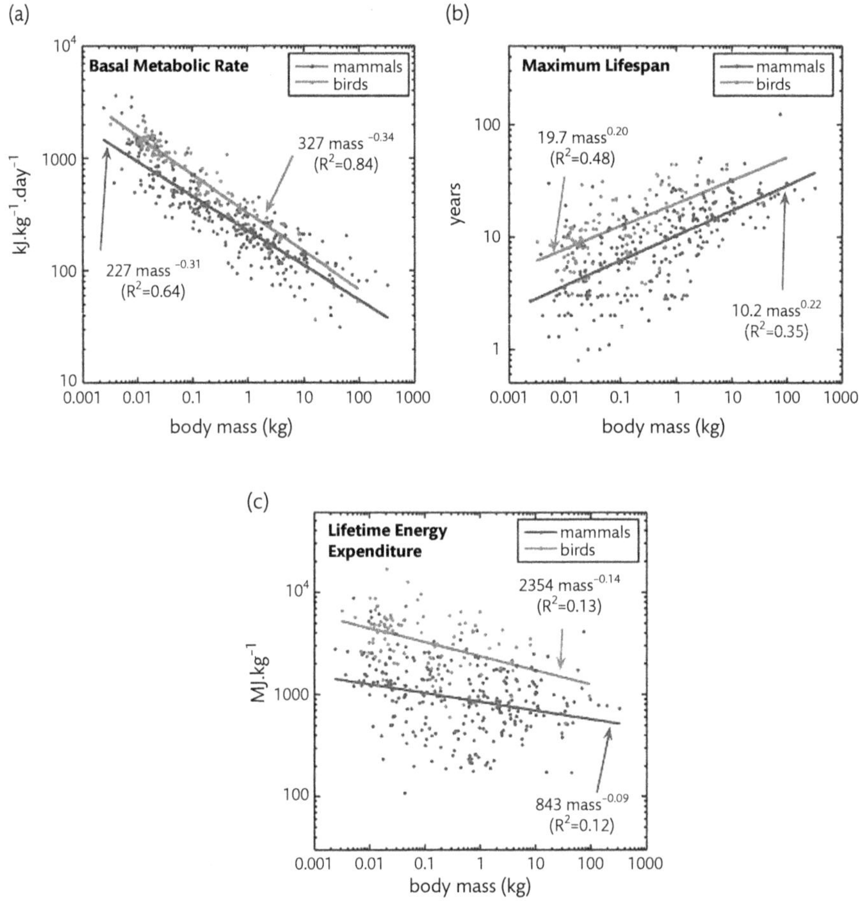

(a)

Basal Metabolic Rate

— mammals
— birds

$327 \, mass^{-0.34}$
$(R^2=0.84)$

$227 \, mass^{-0.31}$
$(R^2=0.64)$

kJ.kg^{-1}.day^{-1}

body mass (kg)

(b)

Maximum Lifespan

— mammals
— birds

$19.7 \, mass^{0.20}$
$(R^2=0.48)$

$10.2 \, mass^{0.22}$
$(R^2=0.35)$

years

body mass (kg)

(c)

Lifetime Energy Expenditure

— mammals
— birds

$2354 \, mass^{-0.14}$
$(R^2=0.13)$

$843 \, mass^{-0.09}$
$(R^2=0.12)$

MJ.kg^{-1}

body mass (kg)

Fig. 3.13 The relationship between body mass of birds and mammals and mass-specific metabolic rate (panel A), maximum lifespan (panel B), and lifetime energy expenditure (panel C). The lighter-toned dots and lines represent data from birds and the darker-toned dots and lines represent data from mammals. Metabolic rate and body mass data for birds are from C. White (pers. comm.) and for mammals from White and Seymour (2003). The maximum life span data are from Carey and Judge (2000). (From Hulbert *et al.* 2007).

2007). While there are a number of endogenous sources of ROS in animals, the mitochondrial respiratory chain is the dominant site of ROS production in healthy tissues under most conditions (Sanz *et al.* 2006), particularly at mitochondrial complexes I and III during state 4 respiration (Fig. 3.15; Pamplona *et al.* 2002). The possibility that differences in MLSP between birds and mammals might be reflected by differences in mitochondrial ROS formation was examined in pigeons (MLSP ~35 years) and rats (MLSP ~three years). Importantly,

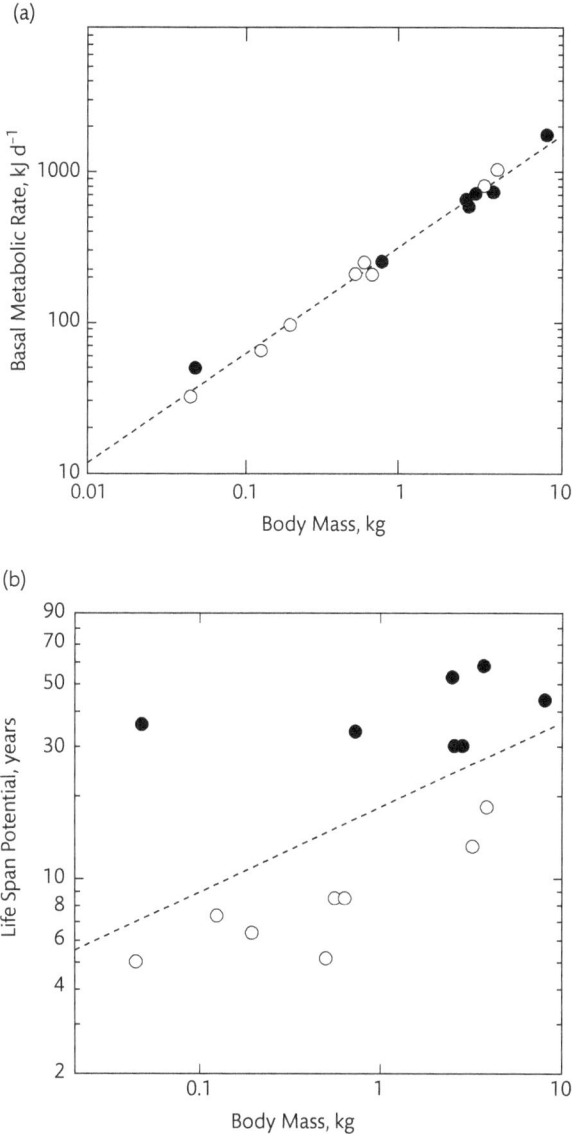

Fig. 3.14 The relationship between basal metabolic rate and body mass in Procellariiformes (7 species – filled symbols) and Galliformes (8 species – open symbols) (panel A) and maximum life span potential (panel B). The lines in both plots are best-fit least-squares regression through respective data. (From Buttemer *et al.* 2008b).

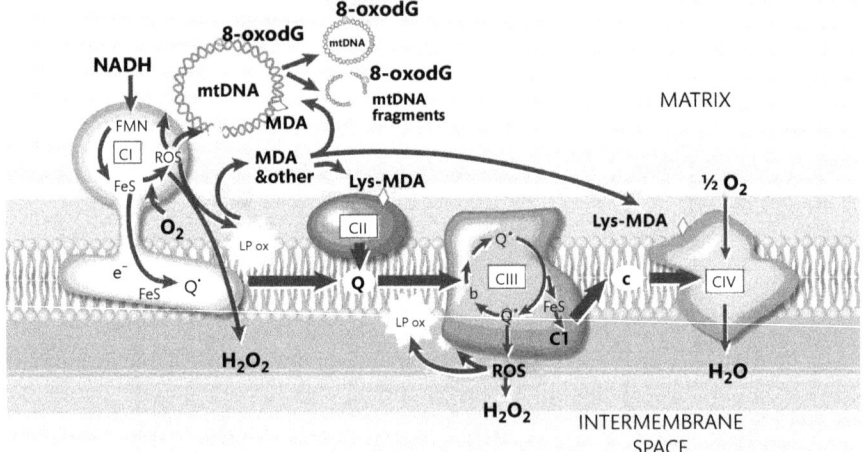

Fig. 3.15 Schematic representation of mitochondrial processes pertinent to aging. The diagram identifies mitochondrial complexes I and III as the main sites of mitochondrial reactive oxygen species (ROS) generation. Oxygen radicals attack lipids, carbohydrates, proteins and DNA. The products of lipid peroxidation include highly reactive molecules that can case lipoxidation damage to mitochondrial DNA and proteins. (From Hulbert *et al.* 2007).

for reconciling the ability of birds to have both a higher MLSP and a higher BMR than mammals, the rate of mitochondrial ROS production per unit of O_2 consumed was considerably lower in pigeons than in rats (Barja *et al.* 1994). Further studies comparing mitochondrial ROS production of other long-lived bird species to a short-lived mammal confirmed this result (Herrero and Barja 1998).

The mechanisms responsible for these differences in ROS production have not been identified, but it is believed that differences between species in rate of mitochondrial ROS production per unit of O_2 consumed are confined to complex I (Barja 2004). Of The number of physicochemical factors contributing to variation in ROS production, proton motive force is particularly important (Lambert and Brand 2004). A change as little as 10 mV in mitochondrial membrane potential, for example, results in a 70% decrease in superoxide production (Miwa *et al.* 2003). The decrease in ROS production in isolated mitochondria between states 4 and 3 may account for the paradox that exercise does not result in decreased longevity in rodents (Herrero and Barja 1997).

One mechanism believed to enable dynamic regulation of excessive ROS production in some mammalian mitochondria involves the activation of uncoupling proteins (UCP2 and UCP3) by ROS-generated aldehydes (Echtay 2007). The UCPs are so named because when activated they rapidly dissipate the mitochondrial proton motive force, thus uncoupling substrate oxidation from phosphorylation. These UCPs can be activated by 4-hydroxynonenal (HNE) and hydroxyhexenal, peroxidation products of n-6 and n-3 fatty acyl chains, respectively. In mammals, high rates of mitochondrial superoxide production will provoke peroxidation of mitochondrial membrane lipids, HNE and hydroxyhexenal levels will rise and activate UCPs. Once activated, the UCPs will increase mitochondrial proton flux, which, in turn, will decrease membrane potential and reduce ROS production. Whether similar mechanisms underlie the low ROS production rates in birds remains to be verified.

Unlike mammals, birds possess only a single UCP homolog, called avUCP. Comparisons of avUCP from Domestic Fowl (*Gallus gallus*) and Swallow-tailed Hummingbirds (*Eupetomena macroura*) show it has about 70% homology to mammalian UCP2 and UCP3, but only about 50% homology to UCP1 (Raimbault *et al.* 2001; Vianna *et al.* 2001). There is experimental evidence of upregulation of avUCP following cold exposure in Muscovy Ducks (*Cairina moschata*; Raimbault *et al.* 2001), chickens (Collin *et al.* 2003; Toyomizu *et al.* 2002), King Penguins (*Aptenodytes patagonicus*; Talbot *et al.* 2004), and during recovery from torpor in hummingbirds (Vianna *et al.* 2001). During heat stress, avUCP is downregulated in three-month old cockerels (*Gallus gallus*; Mujahid *et al.* 2006). These responses imply a thermogenic, rather than a ROS-protective role for avUCP. However, Talbot *et al.* (2004) demonstrated that mitochondria containing up-regulated avUCP had superoxide-induced uncoupling. This result, however, does not unambiguously prove a ROS-protective mechanism, as avian adenine nucleotide translocase (avANT) was also upregulated in cold-exposed King Penguins and ANT is known to induce mitochondrial uncoupling in the presence of lipid peroxidation products (Echtay 2007). Furthermore, Criscuolo *et al.* (2005) expressed avUCP in yeast mitochondria and these failed to show uncoupled respiration when exposed to superoxides. However, the lack of effect may be due to differences in the types of peroxidized fatty acid products of yeast compared to avian mitochondrial membranes. A further consideration regarding the potential for avUCP to reduce whole body ROS production is that its distribution varies markedly between species and with age. It is found mostly in skeletal muscle of eight week old chickens and at moderate levels in spleen and brain (Evock-Clover *et al.* 2002), whereas it is confined to skeletal muscle in adult chickens (Raimbault *et al.* 2001). By contrast, adult Swallow-tailed Hummingbirds displayed high levels of avUCP mRNA expression in pectoral muscle and slightly lower expression in heart, and liver (Vianna *et al.* 2001).

Thus, the involvement of avUCP in ROS-protection of avian mitochondria is unresolved. Determination of whether species with very different MLSPs have corresponding differences in rates of mitochondrial ROS production under different physicochemical states must first be established before assigning functional significance to differences in avUPC or avANT expression. Furthermore, there may be modification to other mitochondrial constituents such as cytochrome b (Rottenburg 2007) or differential expression of subunits in complex I (Sanz *et al.* 2006) that contribute significantly to differences in ROS production among species.

3.4.3 Antioxidant Defenses

The pervasive threat of ROS to living systems is evidenced by the ubiquitous appearance of SOD among bacteria (Hassan 1989), plants (Alscher *et al.* 2002), and animals (Fridovich 1989). These antioxidant enzymes convert superoxide radicals to O_2 and H_2O_2. There are many antioxidants in animals that collectively contribute to reducing, and sometimes preventing, oxidation of oxidizable substrates (Halliwell and Gutteridge 1999). The principle antioxidants in birds fall into two main categories: endogenous enzymatic antioxidants and low-molecular weight antioxidants. There are variants of SOD in the different cellular compartments, with MnSOD located in the mitochondrial matrix, Cu/ZnSOD in the cytosol and intermembrane space, and a different Cu/ZnSOD in extracellular locations. Because SOD produces H_2O_2, it is not a complete antioxidant, but there are two functionally complementary enzymes, catalase (CAT) and GPx, that eliminate H_2O_2. The smaller molecular size of the non-enzymatic antioxidants permits them to remove ROS at sites that are inaccessible to larger enzymatic antioxidants. These molecules scavenge ROS and are oxidized in the process and then must be reduced to regain their antioxidant capacity. The main lipophilic nonenzymatic ROS scavengers are the tocopherols and carotenoids, whereas the main hydrophilic ones are glutathione (GSH), ascorbate, and uric acid.

Because oxidative stress occurs whenever rates of ROS production exceed those of free-radical scavenging, species with high MLSP might be expected to have greater antioxidant capacity than species with reduced maximum life spans. Perhaps surprisingly, most studies comparing endogenous levels of tissue antioxidants with MLSP in vertebrates show a negative relation, or lack of significance, between MLSP and antioxidant levels (summarized in Hulbert *et al.* 2007). If we presume that a species antioxidant defenses are optimized to the levels of oxidative stress they experience, then variations in rates of ROS production rather than antioxidant capacity are more likely to underlie interspecific differences in MLSP. This interpretation gains further support from studies attempting to extend MLSP in species through dietary or pharmacological increases of their antioxidant levels. Although these experiments sometimes show increased average lifespans of treated animals, they invariably have no effect on MLSP (Hulbert *et al.* 2007; Sanz *et al.* 2006).

The relation between antioxidant capacity and MLSP in birds has received much less attention than in mammals, but a study characterizing circulating antioxidant levels among 95 species provides useful insight (Cohen *et al.* 2008). Birds were sampled among tropical and north-temperate assemblages, and represented species with very different life histories and sizes. Species showed substantial variation in both total antioxidant capacity (TAC) and in the constituent antioxidants that comprised TAC. Among all birds, uric acid, the dominant circulating antioxidant, corresponded significantly with TAC (Cohen *et al.* 2007). When analysed in a multidimensional context of morphological, physiological, and life history traits, several patterns emerged. Birds with a "live fast, die young" life history strategy (larger clutch size, shorter incubation period, shorter life span) had higher circulating antioxidant levels than long-living species (Cohen *et al.* 2008). This reinforces the view that the potential for oxidative stress varies inversely with MLSP. Furthermore, the higher circulating levels of uric acid found in shorter-living species challenges the view that uric acid levels are directly linked to longevity in birds (Simoyi *et al.* 2002).

The other broad pattern to emerge from Cohen *et al.*'s (2008) comparative study was that shorter-lived species displayed a reduction in TAC after experiencing a one hour period of capture stress, whereas longer-living species increased their TAC levels following a stress episode. Although chronic stress is associated with rises in lipid peroxidation (Lin *et al.* 2004a), some species can dynamically modify antioxidant defenses during acute stress episodes. Chickens injected with CORT responded by increasing levels of uric acid and TAC as well as showing an increased enzymatic antioxidant activity of SOD in cardiac muscle (Lin *et al.* 2004b). As a result of these adjustments, there was no evidence of a change in lipid peroxidation rate. The capacity of long-lived species to dynamically increase their TAC during acute stress should offer similar advantages during short-term homeostatic challenge. It is important to emphasize, however, that the dependence of birds on dietary intakes of carotenoids and tocopherols as well as their need to maintain uric acid levels through adequate protein supply (Machin *et al.* 2004) means that food availability and its quality will affect antioxidant capacities.

3.4.4 Membrane Lipids and Peroxidation Resistance

The earlier mentioned comparisons of long-living birds and shorter-living mammals identified two traits that were proposed to account for differences in MLSP: birds had lower rates of mitochondrial ROS formation per unit of O_2 consumed (Herrero and Barja 1998; Barja *et al.* 1994), and had mitochondrial and cell membranes that were much less susceptible to lipoxidation damage (Pamplona *et al.* 1999a; Pamplona *et al.* 1999b). These differences can be explained by the types of fatty acids these species incorporate into their membranes and the fact the fatty acids differ substantially in their potential for peroxidation damage by ROS. Among acyl chains, single-bonded carbon atoms situated between two double-bonded carbons are those most prone to oxidative attack. Thus, polyunsaturated fatty acids (PUFA), especially those with multiple double bonds, are far more likely to suffer peroxidation damage than monounsaturated fatty acids (MUFA) or saturated fatty acids (SFA). This was determined empirically by Holman (1954), who showed substantial differences in O_2 uptake rates among fatty acids. For example, the polyunsaturated docosahexaenoic acid (DHA) with six double bonds (22:6 n-3) is 320 times more susceptible to peroxidative damage than the monounsaturated oleic acid (18:1, n-9). By combining the relative susceptibility of different fatty acids (based on their number of double bonds; Fig. 3.16) with fatty acid characterization of membrane lipids, it is possible to determine a peroxidation index (PI) for any membrane. Membranes having high PI values will be more prone to attack by ROS than membranes with low PI scores. Importantly, because the products of membrane lipid peroxidation are themselves ROS (Halliwell and Gutteridge 1999), the process is autocatalytic and can propagate irreversible damage to the membrane and nearby cellular structures, including proteins and DNA. Thus, species having a greater proportion of SFA and MUFA in their membranes (lower PI) will be more resistant to oxidative damage than species having proportionately more PUFA (higher PI) and will, therefore, be less prone to ageing effects associated with oxidative damage (Hulbert et al. 2007). This is the basis of the membrane pacemaker hypothesis

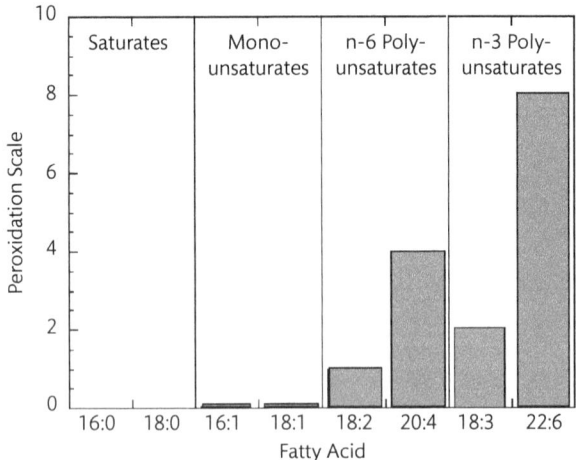

Fig. 3.16 The relative susceptibilities of selected fatty acids to peroxidation. Values are from Holman (1954), and were empirically determined as rates of oxygen consumption. They are expressed relative to the rate for linoleic acid 18:2 n-6 which is arbitrarily given a value of 1. The omega-6 polyunsaturated fatty acids are designated n-6 and the omega-3 polyunsaturated fatty acids are designated n-3. The number for each fatty acid refers to acyl chain length:number of double bonds. (Redrawn from Hulbert et al. 2007).

of ageing, which predicts that MLSP will correlate with membrane lipid composition (Hulbert 2005).

Membranes, of course, serve many functions, and variation in their fatty acid composition will obviously be influenced by these varied requirements. One important consequence of membrane lipid composition is its influence on membrane fluidity. Saturated fatty acids are highly viscous, but the insertion of a single double bond in an acyl chain markedly decreases its viscosity, with the addition of more double bonds having little additive effect (Brenner 1984). Thus, reducing the number of double bonds to achieve reduced peroxidation sensitivity should not be constrained severely by membrane fluidity requirements. Membrane composition of birds and mammals, however, does show significant size-related differences in particular constituents (Hulbert *et al.* 2002a; Hulbert *et al.* 2002b). The membranes of both animal groups show a significant allometric increase in their proportion of MUFA, but significant allometric decreases in proportion of n-3 PUFA, including DHA (22:6, n-3), which is particularly susceptible to lipoxidation (Fig. 3.16).

The size-related differences in membrane PUFA content, particularly DHA, are believed to underlie allometric differences in metabolic intensity among birds and mammals (Hulbert and Else 1999). This is based on theoretical grounds and empirical evidence that the molecular activity of embedded enzymes is directly related to the degree of membrane polyunsaturation (Else and Wu 1999; Wu *et al.* 2001). Thus, smaller species are confronted with a higher mass-specific O_2 usage (i.e., higher mass-specific metabolic rate) and potential for ROS formation, while simultaneously having more peroxidation-susceptible membranes than larger ones (Fig. 3.17). While these size-related differences in membrane composition may partly account for the inverse relation between body size and longevity (Fig. 3.13), the consistent difference in MLSP between particular bird taxa (Fig. 3.14) provides an opportunity to test the membrane pacemaker hypothesis and determine whether MLSP and membrane PI covary.

A comparison of cardiac muscle phospholipid composition of long-lived petrels and albatrosses (collected as bycatch from commercial fisheries) to published values for short-life span galliformes (Szabo *et al.* 2006) showed significant differences in membrane fatty acids (Buttemer *et al.* 2008b). The myocardial phospholipids of procellariiformes had significantly more MUFA and less PUFA than those from galliformes (Fig. 3.18a). These differences were largely responsible for the lower PI among petrels compared with the fowl (Fig. 3.18b), a result entirely consistent with predictions of the membrane pacemaker hypothesis of ageing (Hulbert 2005). Even more striking, however, is that the magnitude of the differences in PI accord well with the differences in MLSP between these groups. Based on evaluations of PI of skeletal muscle phospholipids in relation to MLSP for birds and mammals, a decrease of only 19% in PI is associated with each doubling of MLSP (Hulbert *et al.* 2007). Applying this relation to the 36% lower

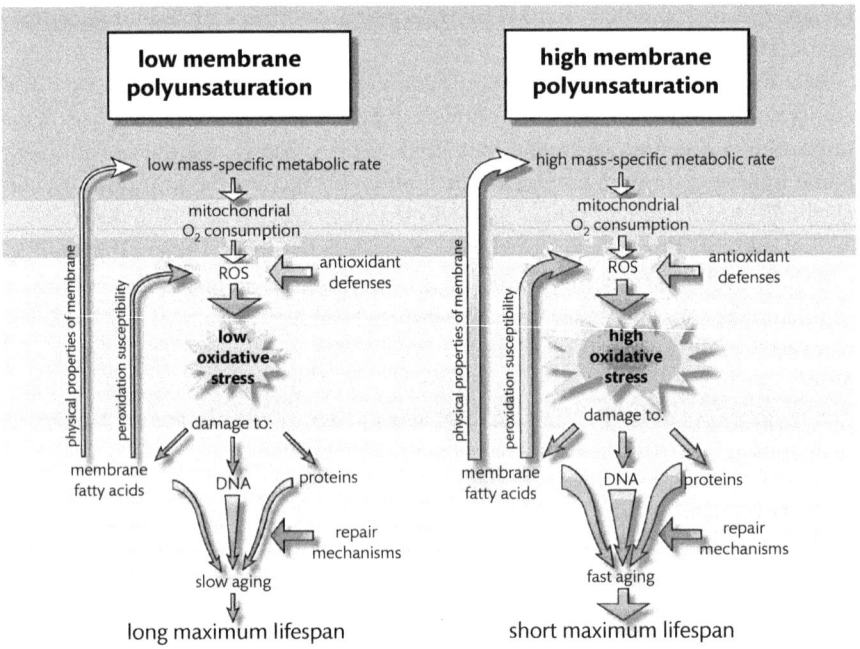

Fig. 3.17 Schematic diagram emphasizing the importance of membrane fatty acid composition on the rate of reactive oxygen species (ROS) formation and consequent oxidative stress. Two different examples are presented: a low-level and a high-level of membrane polyunsaturation. The thickness of the arrows represents the relative intensity of the depicted processes. (From Hulbert *et al.* 2007).

PI in petrels predicts a 4.5-fold greater MLSP for petrels compared to fowl. This value is indistinguishable from the 4.6-fold average MLSP difference noted earlier (Fig. 3.13), but somewhat higher than the size-corrected 3.2-fold difference in MLSP (Buttemer *et al.* 2008b).

There is not sufficient characterisation of avian membrane lipids to test the robustness of the predicted relation between PI and MLSP across a range of species, or even within tissues of the same species. It is well known, for example, that brain phospholipids in mammals contain a very high content of n-3 PUFA, particularly DHA. Furthermore, unlike membrane lipids from other organs, there was no allometric change in n-3 PUFA or DHA content across species ranging from shrews to cattle (Hulbert *et al.* 2002b). While this was suggested to represent the importance of maintaining high-rates of neuronal activity irrespective of brain size (Hulbert 2008), it also means that all mammalian brain mem-

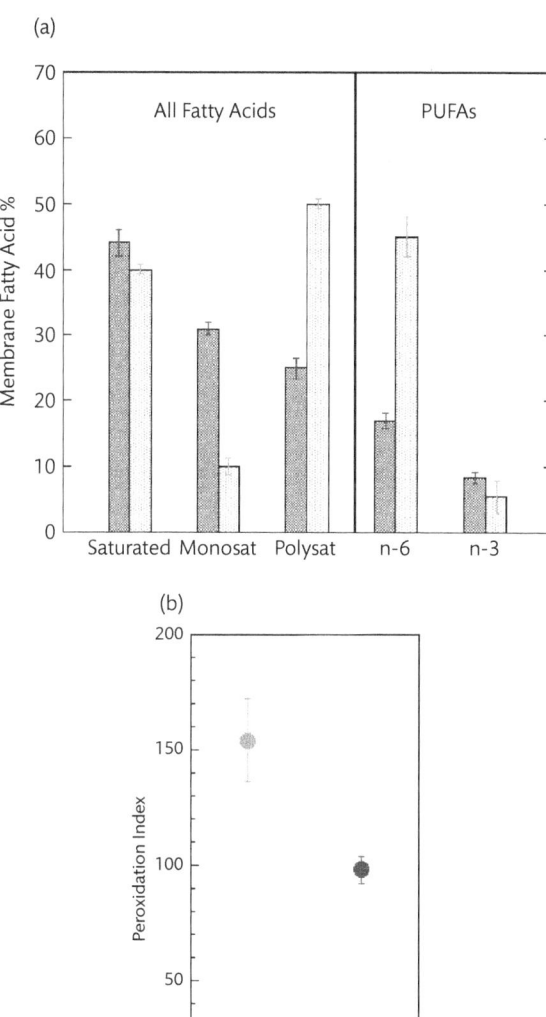

Fig. 3.18 Panel A: Fatty acid composition of cardiac muscle phospholipids in Procellariiformes (darker bars) and Galliformes (lighter bars; fowl data from Szabo *et al.* 2006). PUFAs refer to polyunsaturated fatty acids. Panel B: The average peroxidation indices, measures of lipid vulnerability to oxidative damage, of the cardiac muscle phospholipids in the same birds. Means and standard errors are presented, with the darker symbols representing Procellariiformes and the lighter ones representing Galliformes. (From Buttemer *et al.* 2008b).

branes will share a high PI value. This should place neuronal tissues at much greater risk of peroxidation damage, but there is ample evidence that astrocytes intimately surrounding neurons provide protection from oxidative damage (Desagher *et al.* 1996; Wilson 1997; Watts *et al.* 2005). It is very likely that trade-offs between the advantages of having peroxidation-resistant membranes and the need to have specific membrane characteristics for particular functional requirements will together shape optimal membrane lipid composition. Whether such compromises are offset by increases in intracellular antioxidant capacity to different extent in long-living versus short-living species remains to be determined.

4
Adaptations: Obtaining and Processing Food

In this chapter, we will emphasize the unique physiological features related to feeding and nutrition found in birds that have permitted them to evolve as a very successful taxon. Basic physiological principles are presented within the context of the taxon as well as in the context of adaptations to specific environments. Comparisons are made whenever they are pertinent and serve to better explain a specific function.

Organisms obtain the energy and nutrient supplies they need to sustain life from the surrounding environment. Many species can be viewed as "ecosystems architects", which, according to Carlos Martínez del Rio, "are organisms that not only change the function of an ecosystem, but that also shape the spatial configuration of the elements of the landscape". Many bird species fit this description, in large part because of their abundance, high energy turnover, and great mobility. The intricate interactions between birds and their environments are probably one of the reasons birds have evolved the greatest species diversity among terrestrial vertebrates.

As one example of these interactions, nectar-feeding birds are members of a tightly-linked mutualistic partnership with flowering plants, and because of this, the birds' physiological and behavioral traits can have profound ecological and evolutionary consequences for their plant "partners" (Martínez del Rio et al. 1992). Birds differ in their sugar preferences which, in turn, are deeply influenced by digestive traits; and in turn, sugar preferences affect how flowers "behave" in nectar production. However, there are interesting nuances behind this idea that demonstrate the complexities involving food choices in birds (Martínez del Rio et al. 2001; Fleming et al. 2004; Karasov and Martínez del Rio 2007). In Africa, for instance, many flowers pollinated by Lesser Double-collared Sunbirds (*Nectarinia chalybea*) secrete sucrose (Nicolson and Van Wick 1998). However, sunbirds, which are the African counterparts of hummingbirds, do not show a distinct preference for sucrose (Jackson et al. 1998; Lotz and Nicolson 1996; Lotz and Schondube 2006) as hummingbirds do. More recently, it has been found that sugar preferences may be concentration dependent. Schondube and Martínez del Rio (2003) discovered that the sugar preferences of the Magnificent Hummingbird (*Eugenes fulgens*) and a nectar-feeding passerine, the Cinnamon-bellied Flowerpiercer (*Diglosa baritula*), are distinctly concentration dependent. At low nectar concentrations the birds preferred hexoses, but at high concentrations they preferred sucrose. The physiological reasons for such concentration dependence in sugar preferences are still unknown, but illustrate how complex this situation is. Nicolson (2002) has documented a positive correlation between sucrose content and total sugar concentration in the nectar secreted by some bird-pollinated plant genera. Therefore, digestive traits can not only affect how animals behave, but also influence their interaction with their living and evolving resources, such as the plants that they pollinate and whose seeds they disperse (Karasov and Martínez del Rio 2007).

How energy flows in ecosystems has been a matter of extensive study and it is not the purpose of this book to explore it in depth. However, a few important points should be addressed here for a better understanding of the role of birds in a broader environmental and ecological context.

The energy currency of most organisms is ATP, and birds are no different in this regard. As mentioned earlier in this volume, flapping flight requires very high power levels and hence a high rate of aerobic metabolism. Flapping flight, in turn, allows birds to have great flexibility in terms of their movement and distances covered. However, flight may comprise a large fraction of their total energy budget as it constitutes one of the most power intensive (energy transformation per unit time) types of locomotion among animals (Tucker 1973). Flying birds, therefore, must make a large investment of energy to find sufficient food in their environment to power their extensive movements, especially during the dramatic non-stop migration flights which many species perform every year.

Flight provides a great ability to travel long distances rapidly, but geophysical barriers - oceans, mountains, and deserts, for example - may pose constraints for many migrating birds. Surprisingly, the constraints may be more apparent during stopovers for rest and refueling than during the migration flights themselves. In this regard, the timing (tempo) of the appearance and disappearance of available sources of energy, as well as nutrients, is very important to guarantee survival. This kind of challenge is also faced by some non-flying species such as Emperor Penguins (*Aptenodytes forsteri*), which must often walk for several days across sea ice to reach their breeding site during the severe winter in Antarctica. These are just a few examples, which we will further explore in this chapter.

Flying poses an additional problem, i.e., how much weight can be carried. At various times in their lives, flying birds must lift not only their "basic" body mass, but also developing eggs, gut contents, food for offspring, or extensive fuel reserves. The limits to how much they can carry are tightly constrained by wing area and shape, and available muscle power. These limitations may be particularly relevant in the case of long-distance migrants. In this case, however, it was shown by Kvist *et al.* (2001) that total metabolic power input increased with fuel load in Red Knots (*Calidris canutus*), but proportionally less than the predicted mechanical power output from the flight muscles. The most likely explanation is that the efficiency with which metabolic power input (fuel use) is converted into mechanical power output by the flight muscles increases with fuel load. This may explain why some shorebirds, despite the high metabolic power input required to fly, routinely make nonstop flights of 4,000 km or longer.

In digestive physiology and nutritional ecology, as in other aspects of organismal performance, there will often be trade-offs. For example, if energy supply were limited by the digestion and assimilation of food, total availability of energy could be increased by adding tissue devoted to food processing. However, such reallocation of mass would leave less tissue available to support foraging, growth, or reproduction (Ricklefs 1996), especially given the weight constraints imposed by flight.

These kinds of questions emphasize the importance of using an integrative approach through which one can better understand how constraints in one part of the system may affect other parts of the same system, or how a particular need at some point in time may create constraints in other parts of the system. It is relevant to know whether such limitations may or may not pose survival risks. Birds have been objects of extensive investigation over the past years and perhaps more so than any other taxa because they constitute good model systems to address some of these important questions.

We will review the diversity of feeding habits among birds and the consequences for their nutritional ecology. We will also emphasize different digestion strategies. Digestion is important to ecologists for many reasons, but according to Karasov and Martínez de Rio (2007) there are three particularly relevant ones. "First, digestive processes determine how efficiently animals extract energy and nutrients from their foods. A low digestive efficiency translates into more time and effort spent collecting food to meet a fixed required net input. Hence, a low digestive efficiency translates into less matter and energy available for maintenance, growth, and reproduction. Second, digestive efficiency determines the flux (amount per unit time) and makeup of materials transferred to other trophic levels in the ecosystem. Third, interactions of animals with plants, such as pollination and seed dispersal, sometimes hinge on particulars of the animals' digestive systems."

Food selection is often treated as a challenge of acquisition, but it is also a problem of utilization, processing, and digestion of food. What animals eat and excrete shapes their role in ecological communities and determines their contribution to the flux of energy and materials in ecosystems. The balance between energy acquisition and expenditure is critical for survival and reproductive success. Because energy intake and expenditure integrate all aspects of an individual's life, changes in energy management are closely tied to all aspects of its life history, including diet quality, nutritional requirements, allocation of time, and body plan.

4.1 Energy and Nutrition

According to Ricklefs (1996), the functioning of organisms "may be characterized in energetic terms, but energy fluxes resulting from food intake, excretion, and evaporative water loss are also intrinsically tied to fluxes of materials. These material fluxes either are adjusted to meet the demands of the organism or result fortuitously from the composition of the diet. An organism eats to satisfy its requirements for energy and nutrients." Diets, in general, are not balanced. That is, one or more components of the diet are present in excess of need compared with other components (King and Murphy, 1985). Under such circumstance, all excess material must be either excreted or accumulated, or the composition of the diet must be brought into line with the organism requirements. For instance, the diet of most carnivores contains nitrogen in excess, and therefore an adequate energy intake results in more nitrogen than the individual actually needs. Birds excrete this excess nitrogen as uric acid and other nitrogenous waste products. On the other hand, when the diet contains high proportions of carbohydrates and lipids, a nutritionally adequate diet may contain an excess of energy. Therefore, "to bring diet and requirement into line, individuals with energy-rich diets must either accumulate energy as lipid deposits, increase their energy expenditures, reduce their nutrient requirements relative to energy needs, or selectively forage for the required nutrient" (Ricklefs, 1996). Energy imbalance and nutrient requirements can be a major issue when nutrient demands are high, as during egg formation (King, 1974), or the growth period (Ricklefs, 1974).

How Oilbird (*Steatornis caripensis*) chicks manage energy constitutes an interesting case study highlighting the problem of nutrient-energy imbalance in birds (Ricklefs, 1996). According to Snow (1962), Oilbirds feed almost exclusively on the oily fruits of oil palms and lauraceous trees. These food items have a high energy density but provide low concentrations of nutrients (such as proteins, minerals, or vitamins) (Ricklefs, 1996). The diet of the Oilbird would be nutritionally inadequate to support the rapid nestling growth of most bird species, but Oilbirds have adapted in several ways (Thomas et al., 1993). First, chicks grow slowly compared to other altricial species of similar size (Ricklefs, 1976); slow growth greatly reduces the daily requirement for nutrients while increasing the cumulative requirement for energy. Second, during the early phase of the growth period, nutrients accumulate rapidly in the growing tissues, and large amounts of fat are stored (Snow, 1962). These fat deposits act as an energy sink (Snow, 1962), "taking up the excess energy provided by the high dietary intake needed to satisfy nutrient requirements" (Ricklefs, 1996). Oilbird chicks accumulate lipid at a faster rate during the first half of the nestling period (Thomas et al, 1993), when fat content of the body increases one to two times faster than protein in tissues (White, 1974). Much of this fat is subsequently metabolized after the rate of tissue growth has slowed down and the ratio of required nutrients to energy has decreased. In this manner, according to Ricklefs (1996), "energy stored early in development may subsidize the later requirements of larger chicks, as it evidently does in many petrels whose body mass peaks shortly after the middle of the growth period." Oilbird parents may also produce a kind of protein-rich "crop milk" for very small chicks, similar to that produced by Emperor Penguins (*Aptenodytes forsteri*), fla- mingoes, doves, and pigeons (McLelland, 1979), whose diets are also deficient in protein, in order to support the rapid growth of their chicks (White, 1974).

Energy is often closely tied to water because many species obtain water exclu- sively or primarily from their diets, as do the dependent chicks of most species that perform parental feeding. In this case, the ratio of water to energy in the diet gov- erns the individual's water intake (Goldstein and Nagy, 1985) and, in hot and arid environments, may restrict microhabitat use and activity (e.g., Ricklefs and Hainsworth, 1968). For example, about 80% of the energy in the diet of the chick, in procellariiform seabirds, is provided by lipids (Obst and Nagy, 1993), and the meals delivered to the chicks can contain little free water relative to energy (Ricklefs, 1996). In addition, during long incubation bouts the water available to adult birds is limited to that contained in the gut, in the extra-cellular spaces in the body, and, of course, metabolic water. For example, during their 3-day incubation bouts, adult Leach's Storm-petrels (*Oceanodroma leucorhoa*) lose slightly more than 2 g of water from the stomach contents and roughly the same amount from the body itself (Ricklefs et al., 1986). "During this period, approximately 8 g of water are formed from the metabolism of 6.7 g of fat and 1.8 g of protein" (Ricklefs, 1996). Average body water content during this period is 21 g, and therefore the rate of body water turnover (12 g in 3 days) is about 0.19 (19%) per day (Ricklefs et al., 1986). Because such low availability of water is incompatible with heat stress (which would be accompanied by extensive evaporative water loss), small petrels are restricted to cool, humid nest microhabitats. The low water-to-energy ratios in the diets of pet- rels apparently are related to cost-efficient transport of energy over long distances (Ashmole, 1971; Place et al., 1989), since fat has by far the highest energy content per gram – and the lowest water content per gram - of any food type. Therefore, according to Ricklefs (1996), "adaptation to feed over vast areas of the ocean has secondarily restricted the range of thermal environments suitable for nesting through constraints on nestling water economy." Such trade-offs may set "absolute boundaries on permissible conditions for existence with a given diet."

Nectar-feeding birds occupy the other extreme of the spectrum with regards to water economy. Broad-tailed hummingbirds (*Selasphorus platycercus*) weigh ~3 g and take in food containing 290 mM L^{-1} sucrose (~10% w vol^{-1}) ingesting from 4 to 6 times their body mass in water in a 16 hr day (McWhorter and

Martínez del Rio 1999). Water over-ingestion can lead to plasma dilution, hyponatraemia (low plasma sodium), and rupture of red-blood cells due to osmotic swelling. McWorther and Martínez del Rio (1999) applied a mass balance model to broad-tailed hummingbirds and found that 80% of ingested water was absorbed in the gastrointestinal tract and therefore must be processed by the kidneys. Broad-tailed hummingbirds and probably other nectar-feeding birds (Rooke *et al.* 1983; Lloyd 1991) can face water fluxes typical of aquatic and amphibious organisms. Hummingbird gastrointestinal and renal morphology support the conclusion that dietary water is more or less completely absorbed (reviewed by Martínez del Rio *et al.* 2001). Hummingbird kidneys filter large amounts of plasma and recover valuable solutes contained in it instead of concentrating urine (Goldstein and Skadhauge 2000). Preest and Beuchat (1997) documented an increase in the excretion of ammonia, relative to uric acid, in response to increased in water loads in Anna's hummingbirds (*Calypte anna*). Although it is still unclear whether this shift occurs in all species of nectar-feeding birds, the hypothesis is functionally plausible (Martínez del Rio *et al.* 2001). The potential detrimental effects of reducing renal excretion may be ameliorated by exceedingly low rates of nitrogenated end-product formation and excretion, due to a very low-protein diet (Brice and Grau 1991).

The existence of small safety factors (i.e., the ratio of capacity to load, Diamond and Hammond 1992, where load is the amount of nutrient ingested, Weiss *et al.* 1998) is clearly shown in certain species of hummingbirds (McWhorter and Matínez del Rio 2000). Even when exposed to the mild conditions of the laboratory (with unlimited food and steady temperature), they need to enter torpor (a state of profoundly reduced metabolism and body temperature) to maintain energy balance (Hiebert 1993). In nature, diurnal energy demands are unpredictable and can sometimes increase as a result of low ambient temperatures and/or low flower densities that force hummingbirds to increase flight distances (Calder 1994). Under these conditions, if digestive characteristics limit the supply of energy, balancing the energy budget requires reduced energy outputs at night. Thus, the metabolic strategies of hummingbirds appear to be inextricably linked to their digestive abilities (Martínez del Rio *et al.* 2001).

In contrast to the observations made on hummingbirds, Whitebellied Sunbirds (*Nectarinia talatala*), under conditions of high energy requirements (high thermogenic demands at an ambient temperature of 10°C) and when subjected to a midday fast, increased their rate of afternoon feeding following the fasting period, maintaining evening body mass at control levels (Nicolson *et al.* 2005). The difference between these two nectar-feeding birds is likely related to differences in body mass (three-fold): smaller birds such as hummingbirds must feed more frequently and are less able to endure periods of food scarcity, particularly at low ambient temperatures (Brown *et al.* 1978). In addition, sunbirds do not have the metabolic expense of hovering to feed and do not loose valuable time arousing from torpor.

4.2 Energy and Structure and Function

In the growing chick, tissue function must be divided between embryogenesis and physiological tasks such as muscle contraction (Ricklefs, 1979). Ricklefs (1996) states that "according to one set of hypotheses, growth rate is inversely related to the functional maturity of tissues, in which case increasing precocity of development (the growing chick's approach to completion of development) leads to a slower rate of increase in body mass." Precocity of development evolves in response to food resource and probably also the safety of the chick. It profoundly influences parental care and postnatal growth rate. Therefore, according to Ricklefs (1996), during development, "allocation of tissue between embryonic (growth) and mature functions directly influences the energy management of the chick and the family unit although its adaptive significance may primarily reflect non-energetic considerations."

Time is fixed and fully allocated. In other words, time given to one activity is taken from another, unless temporal overlap is possible (i.e. digestion can occur during activity, etc.). The use of time is also constrained by the environment (Ricklefs, 1996). For example, birds can undertake many activities, such as foraging, territorial defense, etc., only during daylight (or night, in nocturnal species). Time and energy are, according to Ricklefs (1996), "closely linked because every activity or function requires time and also creates an energy demand, which must be satisfied by time devoted to foraging and processing food" (King 1974; Mugaas and King, 1981; Masman et al., 1988; Weathers and Sullivan, 1989; Bunnell and Harestad, 1990; Weathers and Sullivan, 1993).

Energy management must take into account the supply of energy, the demand for energy, the time needed to acquire energy, and the risks associated with energy-supplying and energy-demanding activities. The relationship between the allocation of time and requirements for energy was formalized by Clark and Ricklefs (1988), who expressed energy costs in terms of foraging-time equivalents (FTE). For activity i

$$FTE_i = E_i / F[h] \tag{1}$$

Where E_i is the total energetic cost [in Joules, J] of the activity and F is the net rate at which metabolizable energy is acquired during foraging [Joules per hour, or $J\ h^{-1}$]. The energetic cost of an activity is calculated as

$$E_i - D_i M_i\ [J] \tag{2}$$

where D_i is the duration of the activity i [h], and Mi is the rate of energy consumption for the activity [$J\ h^{-1}$]. For the individual to maintain a positive balance,

$$\Sigma D_i + \Sigma FTE_i \leq 24\ h. \tag{3}$$

Many routine activities, such as foraging and digestion, for example, can be accomplished simultaneously. The relationship between time and energy can be understood most readily when time is unambiguously partitioned between different activities (Ricklefs, 1996). This is the case, for example, for pelagic seabirds during the nesting period when foraging and care of offsprings are spatially separated, and time is partitioned between foraging, transit, and care of eggs and chicks (Ricklefs, 1983).

The time allocated to brooding and foraging is influenced by a variety of factors and can be adjusted by a number of adaptations. Body size, for instance, is an important variable because the smaller surface-to-volume ratios of larger individuals are more favorable for energy conservation. Therefore, in shorebirds (Charadriidae, Scolopacidae), the neonates of larger species depend less on their parents for maintaining body temperature than those of smaller species (Visser and Ricklefs, 1993). In addition, the neonates of larger species have better developed capacities for thermogenesis. In smaller species it has been more evolutionary advantageous for parents to transfer heat early in the growth period, as altricial species do, than for the chick to generate its own energy for thermoregulation (Ricklefs, 1996). Some species, however, have intermediate developmental patterns and the chicks are able to use some energy for thermoregulation.

Patterns of energy acquisition and expenditure follow in part from the distinctive ecological relations of birds. Such patterns, however, reflect also various strategies of managing energy, nutrients, time, and risk to cope with extreme ecological circumstances. Studies of energetics provide information on an important component of life history. However, energy must be integrated with other factors that contribute to fitness to arrive at an understanding of how the environment and adaptive responses to it shape the lives of birds (Ricklefs, 1996).

4.3 Digestive Adaptation

There is considerable accumulated evidence that digestive features are influenced by factors such as diet quality and quantity in species from many avian orders including Anseriformes, Galliformes, and Passeriformes. The ecological importance of digestive adjustments is suggested by a number of theoretical arguments and empirical observations (Karasov 1996).

First, models from or "chemical reactor theory" (see below) predict that enzyme activity, nutrient absorption rates, and digesta retention time are positively related to the efficiency of food utilization, which, in turn, influences feeding rate and therefore foraging distance and time (Karasov 1990). Second, digestion rate may limit metabolizable energy intake and thus rates of growth, storage, or reproduction. This may occur if the volume of the gut (or one of its compartments) and/or its turnover time set an upper limit in the volume (or mass) of digesta that can be processed per unit time (Belovsky 1984) or if small intestinal rates of hydrolysis and absorption are rate limiting. Third, the design and adaptability of the gut may constrain diet selection and thus influence niche width (Barnes and Thomas

1987; Kehoe and Ankney 1985; Moss 1974; Piersma *et al.* 1993). The example of passerines in the Sturnidae-Muscicapidae line that lack the enzyme sucrase and behaviorally avoid sucrose illustrates how a subtle digestive characteristic can constrain dietary diversity.

4.3.1 Chemical Reactor Theory

According to Penry and Jumars (1987), the digestive tactics open to an animal constrain its foraging, and the need to identify digestion parameters has become increasingly apparent. They then suggested that principles of chemical-reactor theory (CRT) can be used to formulate optimization constraints in a general theory of digestion.

Penry and Jumars (1987), state that "to design a process of chemical conversion, whether it be industrial ammonia synthesis or digestion of food materials by an animal, is to ask what type and size of reactor and what operating conditions are required to achieve a given extent of a desired reaction. The answer, the process design of a reactor, is obtained as follows:

1. Reactions of interest are identified, and models of reaction kinetics are developed from which rate equations for the reaction are derived.
2. With reaction kinetics identified, the ideal reactor configuration for accomplishing the given reaction is determined.
3. Operations of real reactors are compared with operations of corresponding ideal models, and deviations of real reactor performance from the ideal are analyzed.
4. There are then two options: modify the real reactor to better approximate the ideal model; or modify the model to account for deviations from ideal operations and use the modified model to achieve improved predictions of real reactor performance."

In general terms, reactions are described as either homogenous or heterogeneous with respect to phases (gaseous, liquid, solid): homogenous reactions occur in only one phase, but heterogeneous reactions involve more phases. However, it is important to note that one phase systems can be heterogeneous and vice-versa, depending on the system dynamics involved.

Digestion is the process that transforms ingested food materials, made largely of complex polymers such as starches and proteins into assimilable components (simple monomers), and proceeds through various reactions catalyzed by enzymes and often mediated by microbes. Such reactions, which involve proteins of high molecular weight and other macromolecules, are intermediate between homogenous and heterogeneous reactions (Penry and Jumars 1987). In modeling gut function it is assumed that digestive reactions are homogenous with rates limited only by chemical kinetics.

The basic form proposed for reactions catalyzed by enzymes is

$$k_{+1} \quad k_{+2}$$
$$A + E \Leftrightarrow EA \rightarrow P + E$$
$$k_{-2}$$

Food component A and enzyme E combine reversibly to form complex EA, which then dissociates irreversibly in to product(s) P and free enzyme E. The rate equation (after Briggs and Haldane 1925) is of the form

$$-r_A = k_{+2} C_E C_A / [(k_{-1} + k_{+2})/k_{+1}] + C_A, \tag{4}$$

where C_E and C_A are, respectively, concentrations of enzyme and of food components A, and k_{-1}, k_{+1}, and k_{+2} are the rate constants. This equation simplifies to the more familiar Michaelis-Menten form when C_E is constant:

$$-r_A = v_{max} C_A / (K_m + C_A), \tag{5}$$

where $v_{max} = (k + 2C_E)$ and $K_m = (k_{-1} + k_{+2})/k_{+1}$. Digestive reactions catalyzed by an animal's own enzymes are described with this rate equation (Penry and Jumars 1987).

According to (Penry and Jumars 1987), "digestive reactions involving microbial fermentation are autocatalytic: a reaction product (microbes) acts as catalyst, and the reaction rate is a function of the concentrations both of microbes and of food materials. The modified Michaelis-Menten rate equation often used to describe such autocatalytic biological reactions (Bishcoff, 1966) is of the form

$$-r_A = v_{max} C_A C_M / (K_m + C_A), \tag{6}$$

where C_M is the concentration of microbes."

The basic Michaelis-Menten equation and the modified form for autocatalytic, fermentation reactions relate digestive reaction rates to concentrations (of food substrates, enzymes, and microbes). Changes in temperature, pH, and the composition of the microbial community may also affect the rates of digestive reactions in the gut. At any given point within the gut, these variables are assumed constant over time a steady-state model. "The need to ensure that these steady-state assumptions are met then becomes an important consideration when designing experimental tests of model predictions," according to Penry and Jumars (1987).

The principal objective in reactor design is to identify the gut (reactor) configuration and digestive (operating) strategy that maximize an animal's net rate of production of energy and nutrients from its ingested food (Penry and Jumars 1987).

One can make use of conservation principles in order to predict performance in reactors. Physicochemical reactions and transfers of mass, energy, and momentum are the phenomena that normally occur in chemical reactors. In contrast to many industrial situations, both temperature and pressure variations within an animal's gut are generally negligible relative to changes in composition. Penry and Jumars (1987) emphasize that "reaction and mass transfer are the dominant processes, and reactor-specific equations for conservation of mass suffice." Thus, according to these authors, "formulation of reactor-specific equations begins with a general mass-balance expression written for any component, reactant, or product, over an arbitrary element of reactor volume:

$$I = E + R + W, \tag{7}$$

where I is input of component in moles per unit of time; E, output in moles per unit of time; R, disappearance in moles per unit of time; and W, accumulation in moles per unit of time." There are three kinds of ideal reactors, which differ in the way the reactants are processed (Penry and Jumars 1987), i.e., (1) the "Batch-Reactor," which is filled with reactants, continuously stirred during the reaction, and then emptied of products after a given reaction period; (2) the "Plug-Flow Reactor (PFR)," characterized by a continuous, orderly flow of material through the (usually tubular) reaction vessel, in which reactants continuously enter and products continuously exit with no linear mixing along the flow path. However, perfect radial mixing is assumed, and reactants are uniform in any cross sectional area; and (3) the "Continuous-Flow, Stirred Tank Reactor (CSTR)," characterized by a continuous flow of material through, and perfect mixing of material within, the reaction vessel.

Space-time (in units of time), the ratio of reactor volume to volumetric flow rate, is the time required to process one reactor volume of material input. In reactor models of animal guts, space-time is throughput time, the time for one gut volume of food to be processed. Throughput time is therefore the ratio of gut volume to throughput rate (volumetric flow rate).

It is important to note that "gut-clearance time" is not the synonym for space-time or throughput time. It refers specifically to the time required for an animal entering starvation to evacuate its gut completely. Gut-clearance time is thus throughput time measured under unsteady conditions that are both unnatural and difficult to model, especially for animals ordinarily feeding in a more or less continuous manner. Note also that neither "gut-residence time" nor "gut-passage time" is equivalent to throughput time or batch-reactor holding time. Residence time is the time an individual material element spends in a reactor or a gut. Gut-passage time is mean gut-residence time. Since mean residence time can be precisely and unambigouously defined, its use is preferable (Penry and Jumars 1987).

Penry and Jumars (1987) design objectives for digestive reactions is to identify gut (reactor) configurations and (operating) strategies that maximize the production rate of energy and nutrients from ingested food. Application of chemical-reactor analysis to the study of digestion results in two very general predictions. Under the assumptions given in the study by Penry and Jumars, "an animal's net rate of energy and nutrients transfer from food via catalytic digestive reactions (eq. 5) is maximized when its gut is an ideal PFR. If, in addition to catalytic digestive reactions, fermentation reactions (eq. 6) are important digestive components, fermentative production rate is maximized when a portion of the gut is a CSTR."

The design objective devised by Penry and Jumars (1987), i.e., "of maximizing production rate in minima of volume and time implicitly assumes that digestive costs (e.g., energy and nutrients required to maintain digestive tissues and energy to propel digesta) are increasing functions of reactor volume and time." Their "engineering approach to process design as applied to feeding emphasizes the interdependence of foraging and digestion. Foraging theory suggests what food will be available at reasonable energetic expense. Digestion theory, in turn, provides process constraints that must be incorporated into foraging models."

4.4 Digestive Efficiency

Models derived from chemical reactor theory (Silby 1981; Penry and Jumars 1987; Martínez del Rio and Karasov 1990) delineate important gastrointestinal (GI) features that influence whole-animal feeding rate and extraction efficiency (nutrient absorption/nutrient ingestion): (i) the volume of the GI tract; (ii) hydrolysis or microbial digestion rates due to levels of pancreatic and intestinal enzymes or microbial activity; (iii) nutrient absorption rate; and (iv) digesta retention time. Within this context, reaction rate (ii) and (iii) interacts positively with retention time to influence efficiency, and there is a tradeoff between the rate of digesta processing, which is inversely related to retention time and the thoroughness of digestion:

Digestive efficiency $\approx$ (Retention time $\times$ Reaction rate)/(Concentration $\times$ Digesta volume)

Concentration is the concentration of substrate per unit volume of digesta and is related to diet richness. The parameter of retention time might be modulated by changes in gut motility that alter digesta flow rate or by changes in GI tract volume that affect digesta volume:

Retention time (min) = Digesta volume (mL)/Digesta flow (mL/min)

Digestive efficiencies are typically measured in balance trials in which rates of energy or material input and output of captive animals are carefully measured. The basic calculation for any substance i uses the fluxes (or rates) of intake and elimination, which are either rates of food intake and feces production or the product of those dry mass fluxes and their respective concentrations of i per gram (g) of dry matter:

$D{}^{\cdot}i$ = (intake rate of i – fecal elimination rate of i)/intake rate of i

These digestibilities are called apparent, and therefore receive an asterisk (*), because the feces sometimes contain, in addition to the undigested food residue, so-called endogenous losses such as sloughed cells from the digestive tract, unabsorbed digestive secretions, or microbial material. The apparent digestibility of energy is slightly lower than the true digestive efficiency. Its use in ecology, however, is entirely appropriate because the endogenous losses represent real losses that must be replaced (Karasov and Martínez del Rio 2007).

Another utilization efficiency that can be calculated from a balance trial is the apparent metabolizable energy coefficient (MEC*), which includes the amounts lost in the urine and as gas (e.g., methane):

MEC* = [energy intake rate − (fecal energy loss rate + urinary and gaseous energy loss rate)]/ energy intake rate

MEC* is also "apparent" for the same reason that the digestibility coefficient is best considered "apparent" – it may reflect endogenous losses.

Many animals, including birds, marsupials, monotremes, as well as other non-mammalian vertebrates and invertebrates, produce excreta that are a combination of urine and feces. If urine and feces cannot be separated, then only energy metabolizability can be determined for these animals, not apparent digestibility. For animals that do not produce urine separate from feces, many researchers use the MEC*, sometimes also called an assimilation coefficient or efficiency, as an index for digestive efficiency (Karasov and Martínez del Rio 2007).

An implicit assumption in these models is that there is no large excess in one feature in relation to the others. For instance, if the concentration of the primary nutrient in a food increases, the model assumes that extraction efficiency will decrease unless compensatory changes occur in retention time or hydrolysis and absorption time. The alternative viewpoint might be that digesta are already retained so long in relation to reaction rates that even with a change in load no adjustment is needed. However, this alternative viewpoint does not seem to hold in birds as shown by changes in utilization efficiency that occur when concentration (or load) changes following diet switches (Karasov 1996).

For example, when American Robins (*Turdus migratorius*) and European Starlings (*Sturnus vulgaris*) were switched from a high carbohydrate fruit diet to an insect diet containing high levels of protein and fat, digestive efficiency of the new diet progressively increased over the course of three days (Levey and Karasov 1989). Efficiency was computed as MEC* = 1 − [excreta energy/food energy] (Karasov 1990). Thus, when load is increased, efficiency is initially low until digestive capacity is increased by adjustments in retention time and/or reaction rates.

The mean values for efficiency do not reflect the tremendous variation among species eating a specific diet type. For example, digestion of the disaccharide sucrose among bird species varies from essentially complete (0.98 + 0.01 in Rufous Hummingbird, *Selaphorus rufous*; Karasov *et al.* 1986) to intermediate (0.61 + 0.01

in Cedar Waxwings, *Bombycilla cedrorum*; Martínez del Rio *et al.* 1989) to nil (0 in American Robins; Karasov and Levey 1990). These differences are due to variation in digestive system structure and function. Thus, although the composition of a food is in many cases more important than the species of consumer in determining digestive efficiency, the importance of structural and functional digestive characteristics of animals themselves cannot be overlooked (Karasov and Martínez del Rio 2007).

Soft foods such as fruits and insects require less mechanical processing than hard seeds and mollusks (Coleman 1974; Custer and Pitelka 1975). A number of studies show that bird species that routinely consume soft, easily macerated food have smaller gizzards than species that eat harder foods, and these shifts also occur within species as diet changes (reviewed by Piersma *et al.* 1993). Some birds (e.g. hawks, penguins) essentially lack a specialized muscular gizzard, and rely exclusively on the stomach (proventriculus). In a number of bird species, captive individuals acclimated to soft food initially do not eat their natural hard-shelled food until after a period that corresponds to the time needed for enlargement of the gizzard to a size necessary for cracking the harder food. One may infer that gastrointestinal function is matched to the prevailing diet, and that when load is increased the bird's ability to eat a new food is initially low until digestive capacity is increased – here, through adjustment of gizzard mass.

Red Knots feed on soft-bodied prey (arthropods and spiders) in the summer and hard-shelled mollusks during migration and on their wintering grounds. The mass of the gizzard changes seasonally in Red Knots. It atrophies after they arrive at their breeding grounds in the tundra, when the birds shift their diets from shellfish to soft-bodied prey. After they leave the tundra, Red Knots resume feeding on mollusks and their gizzards grow (Piersma *et al.* 1999a; Piersma *et al.* 1999b). Seasonal changes observed in the field can be replicated in the laboratory. Red Knots fed on soft food reduced their gizzard masses, but when fed mollusks, they increased their gizzard masses by 147% within six days (Dekinga *et al.* 2001). Red Knots can process hard-to-break-down mollusks when their gizzards are enlarged, and this change is crucial for them to meet their daily energy budget when foraging on mollusks (van Gils *et al.* 2003). These adjustments are fascinating examples of "phenotypic flexibility" (see below) in internal structure related to mechanical processing of food. A recent study (van Gils *et al.* 2005) showed that variation in gizzard mass is, however, small in migrating Red Knots. They appear to time their stopovers so that they hit local peaks in prey quality, which occur during the reproductive seasons of the intertidal benthic invertebrates. By selecting stopovers containing high-quality prey, the metabolic rates of Red Knots are kept to a minimum, fuelling rates are maximized, and the overall speed of migration is increased. Together, these effects may reduce their spring migratory period by a full week, a benefit achieved by minimizing size changes in their gizzards.

4.4.1 Phenotypic Flexibility

Physiological adaptation, here defined as genetic change, can be deduced from comparative inter-specific analyses of traits (e.g., organ size and function, nutrient transport rate, and metabolic rate) with appropriate control for phylogeny. "Hypotheses of adaptation and their tests are fundamentally comparative. To propose that a particular character is adaptive implies that the character confers an advantage which promotes the survival or reproductive success of its carriers relative to organisms lacking the trait. The character is to be compared specifically to phylogenetically antecedent conditions as alternative variants within populations or in related evolutionary lineages" (Larson and Losos 1996).

A central theme of evolutionary physiology involves comparative studies of physiological traits in taxa with different life styles (Feder *et al.* 1987; Wainwright and Reilly 1994). Many such physiological traits exhibit considerable variability in ecological time both within and among individuals which may make it difficult to detect important adapted patterns. To be an adaptation a trait must have resulted from direct natural selection for that trait (Lauder 1996), and without appropriate inherited (genetic) variation adaptation is halted (Kirkpatrick 1996).

However, not all variation is due to genetic differences. The form of phenotypic variation that involves a single genotype producing different phenotypes in response to variation in some environmental variable is called phenotypic plasticity (Travis 1994; Piersma and Lindstrom 1997). Some phenotypic plasticity is irreversible. Reversible changes in body composition, organ size, and digestive processes are examples of phenotypic flexibility: norms of reaction (Stearns 1989; Travis 1994) that allow adjustable responses to changes in the environment (Piersma and Lindstrom 1997). Phenotypic flexibility in physiological traits may be a crucial component of the adaptive repertoire of animals that may influence diet diversity, niche width, feeding rate, and thus the acquisition of energy and essential nutrients (Karasov 1996; Kersten and Visser 1996; Pigliucci 1996; McWilliams *et al.* 1997; Piersma and Lindstrom 1997). The patterns and consequences of phenotypic plasticity and flexibility have important implications for animal ecology (McWilliams and Karasov 2001). Defining these implications requires mechanistically linking the study of the trait in its various forms with some ecologically relevant performance criteria (Arnold 1983; Wainwright and Reilly 1994).

4.5 Modulation of Digestive Enzymes

Modulation of digestive enzymes is an important feature of digestive flexibility in animals, and is tightly linked to changes in nutrient absorption rate or digesta retention (Karasov and Hume 1997). At the whole-animal level, such modulation is important in allowing or constraining diet switching or high feeding rates. It has been argued that animals modulate, instead of maintaining high constitutive levels of specific enzymes, because the metabolic expense of synthesizing large amounts of digestive enzymes would be wasted by animals feeding on diets with low levels of the substrates for those enzymes. Thus, the *a priori* expectation for animals (within-individual variation) with biochemical lability is that, for dietary components such as carbohydrates, protein, and lipid, there would be a positive relationship between their level in the diet and the presence or amount of gut or pancreatic enzymes necessary for their breakdown (Caviedes-Vidal *et al.* 2000).

Not all vertebrates modulate intestinal carbohydrate enzymes (Karasov and Hume 1997). One suggested explanation is that the ability to modulate these enzymes has been selected for in omnivores that switch among diets with varying carbohydrate levels, but not in carnivores that always consume diets with little or no carbohydrates (Buddington *et al.* 1991; Afik *et al.* 1995). Studies with avian species, however, are not consistent with this simple hypothesis. Primarily granivorous chickens (Biviano *et al.* 1993) and turkeys (Sell *et al.* 1989) exhibit increased maltase activity when fed diets high in carbohydrate, whether starch, sucrose, or glucose. In contrast to these galliformes, two passerine birds, European Starlings (Sturnus vulgaris; Martínez del Rio, 1990) and Yellow-rumped Warblers (Dendroica coronata; Afik *et al.* 1995), which in the wild consume both insects, fruits, and (in starlings) seeds, have no change in maltase activity when fed diets high, low, or lacking in carbohydrate. Thus, the question is whether the difference between galliformes on the one hand, and passerines on the other, reflects a phylogenetic pattern (maltase modulated in galliformes but not in passerines), a dietary pattern (maltase modulated in granivores but not in insectivore/frugivores), or chance. Caviedes-Vidal *et al.* (2000) tackled this question using the House Sparrow (*Passer domesticus*) as subject because House sparrows are naturally omnivorous passerines, ingesting starchy seeds with large amounts of glucose, as well as high protein/moderate fat insects and high fat/moderate protein seeds (Martin *et al.* 1951). House sparrows fed on a high-protein diet (60.3% protein) show an upward modulation of aminopeptidase-N activity when compared to subjects fed on a lower protein diet. However, although many species more than double carbohydrase activity on high carbohydrate diets (Karasov and Hume 1997), House Sparrows do not (Caviedes-Vidal *et al.* 2000). Interestingly, the same pattern was also found with regards to their pancreatic enzymes (Caviedes-Vidal and Karasov 1995). The granivorous Chingolo (Caped Sparrow, *Zonotrichia capensis*) whose consumption of insects in the wild varies seasonally from 11 to 37% of diet, also do not exhibit a higher carbohydrase activity when eating a higher carbohydrate diet in both field and laboratory (Sabat *et al.* 1998). Yellow-rumped Warblers are insectivorous during their breeding season in northern regions of North America, whereas during migration and while wintering their diet changes to include a large proportion of fruit (51–78%; Yarbrough and Johnston 1965) and even nectar (Gryj *et al.* 1990) in some locales. Despite these natural diet switches, up-modulation for proteases in birds switched from fruit diets to insect diets, and down-modulation for amylase for birds switched from fruit diets to lower carbohydrate insect and seed diets were not observed (Ciminari *et al.* 2001). On the other hand, congeneric Pine Warblers (Dendroica pinus) exhibit some modulation. Specific amylase activity per gram pancreas has been shown to be a third higher in Pine Warblers eating fruit and seed diets than insect diets, and trypsin and chymotrypsin specific activities a quarter to a third higher in Pine Warblers eating insect diets compared with the other two diets (Levey *et al.* 1999). More generally, with the exception of Pine Warbler, dietary modulation of intestinal carbohydrase activity has not been observed in any passerine species tested so far (n = 5 species; Caviedes-Vidal *et al.* 2000).

Whether the differences among species relate to domestication, phylogeny, or some other factor is still unknown. Perhaps there has been increased selection in the wild compared to domesticated omnivorous species for high constitutive enzyme levels because the foods of the former contain plant secondary metabolites capable of interfering with enzymatic digestion (Ciminari *et al.* 2001). Another possibility is that intestinal carbohydrases exist in considerable excess of need, and the limiting step in starch utilization lies elsewhere (Caviedes-Vidal *et al.* 2000). Or even that the expense of maintaining the capacity to produce high levels of enzymes may not be high.

4.6 Modulation of Absorption Rate

It has been proposed that natural selection matches the absorption capacity of the vertebrate small intestine to nutrient intake (Diamond 1991). If there were no such match, then valuable food energy might be wasted in excreta when feeding on diets with high substrate loads that are not absorbed, and/or the metabolic expenses of synthesizing and maintaining the molecular machinery to absorb substrate would be wasted when feeding on diets with very low levels of substrate (Karasov 1996). Hence, according to the adaptive modulation hypothesis, the transport of nutrients tends to be positively related to dietary levels of those nutrients.

An implicit assumption underlying the adaptive modulation hypothesis is that mediated absorption is the primary pathway of glucose absorption and thus its regulation is subjected to natural selection. However, evidence is increasing that passive absorption can be substantial in birds. For example, 80% of the glucose absorbed by Rainbow Lorikeets (*Trichoglossus haematodus*) drinking artificial nectar (0.4 M D-glucose) was absorbed passively (Karasov and Cork 1994). Pappenheimer (1993) suggested that passive absorption is selectively advantageous because it requires little energy and provides a mechanism that automatically matches the rate of absorption to the rate of hydrolysis or luminal concentration.

Determining the relative contribution of active versus passive transport of sugars thus has important implications for understanding and predicting features of avian feeding ecology. For example, it has been argued that, because foods may contain toxins, there would have been selection against reliance on passive absorption in the intestinal brush border (Diamond 1991).

In several avian species low active D-glucose transport has been measured *in vitro*, but a higher ability to absorb D-glucose has been measured *in vivo*. Yellow-rumped Warblers, which switch seasonally between insects and fruit, have *in vitro* active uptake rates of D-glucose only a few percent of what they achieve at the whole-animal level, and active uptake is not up-regulated by high dietary D-glucose levels (Afik *et al.* 1997a). In an elegant study, Afik *et al.* (1997b) showed that based on the Yellow-rumped Warbler's feeding rate (22 mg wet mass min^{-1}), D-glucose absorptive efficiency (89.8%), and diet D-glucose content (10% of wet mass), the calculated rate of D-glucose absorption is 2 mg min^{-1}, or 11 μmol min^{-1}. Their *in vitro*-based estimate of The maximal absorptive capacity of the Yellow-rumped Warbler's small intestine for D-glucose by active transport is no more than 0.18 μmol min^{-1} (Afik *et al.* 1997a), only 16% of the whole-animal level. Their finding (Afik *et al.* 1997b) that L-glucose, the stereoisomer that does not interact with the intestinal brush border Na^+-coupled glucose cotransporter (SGLT1), is almost entirely absorbed confirms initial predictions that passive absorption must be important. Near complete absorption of ingested L-glucose is also found in nectarivorous Rainbow Lorikeets (80%; Karasov and Cork 1994), granivorous Northern Bobwhite Quail (*Colinus virginianus*; 92%; Levey and Cipollini 1996), House Sparrows (82%; Caviedes-Vidal and Karasov 1996), and frugivorous Cedar Waxwings (79%; Karasov and Levey 1990). The similarity of this finding in birds with diverse diet and taxonomic associations suggests that passive absorption is a general phenomenon in birds.

Previous studies assumed that active, mediated transport accounts for essentially all glucose absorption in hummingbirds (Karasov and Diamond 1983). Considering the significant passive non-mediated absorption of water-soluble nutrients in avian species with diverse diets and taxonomic affiliations (reviewed by McWhorter 2005), and the mismatch between maximal glucose uptake capacity extrapolated from *in vivo* measurements and the metabolic demands of hummingbirds in the field (Powers and Nagy 1988), McWhorter *et al.* (2006) reexamined passive non-mediated glucose uptake in hummingbirds. Glucose uptake rates in Anna's Hummingbirds measured *in vitro* are approximately fourfold lower than glucose absorption rates observed *in vivo*. In their study, McWhorter and colleagues assumed that absorption occurs at the saturating rate along the entire length of the intestine, in which case the maximal rate of active glucose absorption by the small intestine was 1.46 μmoles min^{-1} (Karasov *et al.* 1986). This results in a glucose assimilation rate equivalent to 4.33 J min^{-1}.

Hummingbird diets are generally sucrose-rich (Pyke and Waser 1981) and sugar assimilation efficiencies are high (≥95%; McWhorter and Martínez del Rio 2000; Schondube and Martínez del Rio 2003). McWhorter and colleagues also assumed that fructose is assimilated with equal efficiency (McWhorter and Martínez del Rio 2000), and this results in a maximal energy assimilation rate of ca 8.66 J min^{-1}, or 8.3 kJ day^{-1}, assuming that 16 h of foraging time are available to the bird. The field metabolic rate of the Anna's Hummingbird is ca 32 kJ day^{-1} measured using doubly labeled water (Powers and Nagy 1988). McWhorter and colleagues then concluded that the levels of active transport reported by Diamond *et al.* (1986) and Karasov *et al.* (1986) could not have accounted for all glucose uptake. They performed two experiments using Broad-tailed Hummingbirds in which they: (i) measured the fractional absorption of L-glucose (an *in vivo* test of passive non-mediated glucose transport) in individuals feeding on a sucrose solution close to the median concentration found in the wild (approximately 20% by mass); and (ii) probed the effect of food energy density (indirect measurement of digesta retention time) on passive non-mediated glucose absorption. Even though hummingbirds exhibit the highest capacity for active, mediated glucose uptake among vertebrates, the results obtained by McWhorter *et al.* (2006) demonstrate that they must still rely partially on non-mediated passive uptake in order to meet their high mass-specific metabolic demands.

If some birds rely largely on passive glucose absorption, then they may have experienced little natural selection for high glucose active transport capacity or for the ability to modulate glucose active transport, as has been suggested for active vitamin transport (Stein and Diamond 1989). The opportunistic omnivorous feeding behavior of many birds may require a quickly responding digestive system, and a passive absorptive ability is energetically inexpensive to maintain and responds automatically to the concentration of nutrients in a meal. While passive absorption provides these advantages, it has costs. A high intestinal permeability, allowing passive absorption, might be less selective than a carrier-mediated system. Another important issue associated with passive absorption is that valuable nutrients cannot be recovered from the intestinal lumen unless they are at higher concentration in the gut than in the intestinal cells. However, evidence is accumulating that in some vertebrates other small to medium-sized hydrophilic solutes permeate the small intestine in substantial amounts by passive routes (Afik *et al.* 1997b).

In re-circulating intestinal perfusions in anesthetized pigeons (*Columba livia* L.) and rats (*Rattus norvegicus* L.) of similar size, and in intact animals (Lavin *et al.* 2007), the extent of paracellular absorption of small water soluble nutrients such as monossacharides and amino acids in the pigeon is at least double that of the rat. The greater absorption in the pigeon is not explained by greater small intestine surface area or by more tight junctions in the pigeon, and therefore implies greater paracellular permeability in the pigeon.

Gut morphometric analysis shows that birds have significantly shorter small intestine lengths and lower small intestine nominal surface area than non-flying mammals (Caviedes-Vidal *et al.* 2007). In contrast to non-flying mammals, birds have significantly greater villous amplification of small intestine surface area, with no effect of body size. Greater villous amplification is not likely to counterbalance differences in intestine size (nominal surface area and length) in the case of mediated nutrient uptake, but it may provide one mechanistic explanation for the higher paracellular carbohydrate absorption in birds, and it is reasonable to predict that paracellular absorption of peptides and amino acids might be similarly important (Caviedes-Vidal *et al.* 2007).

Less indiscriminant absorption of lumen contents might permit toxins to be absorbed from food. This vulnerability to hydrophilic toxins could be an important ecological factor constraining food exploratory behavior, limiting the breadth of the dietary niche, and selecting for compensatory behaviors such as ingesting substances that inhibit hydrophilic toxin absorption (Diamond *et al.* 1999).

4.7 Modulation of Retention Time

It is possible to make an *a priori* prediction about the "optimal" retention time in relation to diet composition using concepts from the chemical reactor engineering approach (Penry and Jumars 1987). A gut such as that of fruit- and nectar-eating birds can be modeled as two reactors in series, for example. First, there is a CSTR (the crop plus the stomach). Coupled to it is a PFR (the tubular intestine). Because all the digestion and absorption of sugars take place in the intestine, Martínez del Rio and Karasov (1990) modeled the kinetics of the intestinal digestive process of fruit- and nectar-eating animals as those of a PFR. One interesting prediction of the model is that the net rate of energy absorption from free glucose in fruit and nectar, for example, is maximized by expelling some of the digesta prior to complete absorption of sugar and refilling with a higher concentration of food; i.e., extraction efficiency is considerably less than 100%. This is because when glucose absorption is partly passive, its absorption is fastest at high initial luminal concentration and slows as sugar is absorbed and luminal concentration falls. Further predictions are that optimal retention time and extraction efficiency are inversely related to sugar concentration but positively related to foraging costs (Martínez del Rio and Karasov 1990). Thus, increasing the cost of feeding increases optimal throughput times. As food becomes more expensive to acquire, it should be retained in the intestine longer and be absorbed more thoroughly.

Some of these predictions were tested using Rainbow Lorikeets drinking artificial nectar (Karasov and Cork 1996) and Cedar Waxwings eating artificial fruit. Surprisingly, all predictions were rejected. In lorikeets extraction efficiency is uniformly very high (90%) and is not influenced by glucose concentration. Retention time increases significantly with increasing sugar concentration in waxwings. The trend in lorikeets is that stomach emptying is slowed, which would increase instead of decrease retention time. Therefore, these results are not consistent with the premise of maximization of net energy gain. Instead, they indicate that extraction efficiency is regulated by slowing digesta flow when luminal concentration is high, perhaps by enterogastric negative feedback arising from intestinal receptors (Karasov 1996).

Because the birds in these experiments were not growing, storing fat, or reproducing, perhaps the assumption about energy maximization is not appropriate. Physiological studies of the control of digesta flow have generally been performed with non-reproducing adults (Karasov 1996). These studies indicate the existence of control mechanisms that result in longer retention and higher extraction than might be predicted by rate maximization. For example, high concentrations of amino acids and fats in the duodenum inhibit gastric motility in poultry (Duke 1989) through hormonal (possibly cholecystokinin and pancreatic polypeptide) and neural reflexes. Such mechanisms may be present in wild birds because retention times of markers are longer in Yellow-rumped Warblers for high fat diets (Afik and Karasov 1995) and in American Robins fed insects versus fruits (Levey and Karasov 1992). In American Robins switched from fruits to insects, it took at least three days to increase the initially low digestive efficiency on insects. American Robins usually process insects more slowly than fruits, and Levey and Karasov (1992) speculated that initially, the robins processed insects at the same fast rate as they processed fruits and consequently they were less efficient.

Increased gut volume or digesta mixing apparently enables cold acclimated Cedar Waxwings (McWilliams and Karasov 1998a) and hyperphagic Yellow-rumped Warblers (McWilliams and Karasov 1998b) to maintain constant retention time and hence extraction efficiency, despite short-term increases in feeding rates of 25% and 50%, respectively.

Based on recent evidence, it appears that birds minimize feeding time by maximizing extraction efficiency instead of maximizing net rates of energy gain by reducing extraction efficiency in favor of eating more. Modulation of food intake, retention time, and digesta mixing may be the primary ways that high and constant extraction efficiency is maintained during short-term changes in food quality or costs of feeding. From an ecological perspective, the lack of an effect on food quality or costs of feeding on extraction efficiency suggests that estimates of digestive efficiency are robust. However, this conclusion is relevant to situations in which food type is constant (e.g., diets with constant high or low glucose concentrations; McWilliams and Karasov 1998b). When birds switch between diets that differ in primary nutrients (e.g., between diets high in lipid and diets high in sugar), changes in extraction efficiency can be significant (Afik and Karasov 1995).

4.7.1 Geometric Analysis

Many nutritional questions of ecological importance can also be addressed using a modeling framework called geometric analysis (Simpson and Raubenheimer 1995). Geometric analysis of nutrition relies on two constructs: utilization plots and nutrient spaces. A nutritional budget is a model that quantifies the relationship between the intake, utilization (biochemical transformation), and loss of nutrients (in this case not used for energy production). The budget for a given nutrient i can be described as

$$I_i = (A + L)_i \qquad (8)$$

where I is the amount of nutrient ingested per unit time, A is the amount retained in the body, and L is the amount lost per unit time. The terms in A and L can be expanded to identify the fate of The retained nutrient into somatic (s), reproductive (r), and storage tissue (st), for example,

$$A = A_s + A_r + A_{st.} \qquad (9)$$

L can be expanded into losses in feces (f), urine (u), and gases (e.g., methane) exhaled (e),

$$L = L_f + L_u + L_e. \qquad (10)$$

Utilization plots examine nutritional efficiencies through the relationship between uptake in the x axis and the components of the budget in the y axis. To study digestion, the component of the budget plotted in the y axis is L_f. For the many animals that produce excreta that are a combination of feces and urine, including birds, the utilization plots necessarily plot $L_f + L_u$ versus I. For a detailed and clear account of this research framework, refer to the book by Karasov and Martínez del Rio (2007).

4.8 Digestion Rate and Digestive Efficiency

Budgets are quite useful, but they give a static view of the organism. They fail to capture the dynamical interplay between rates and efficiencies (Karasov and Martínez del Rio 2007). For example, the amount of nutrient transferred to the rest of the body increases as a function of the time during which food is processed in the gut. As time increases, so does the amount of nutrient transferred, up to the total amount of nutrient available. The transfer rate function (Raubenheimer and Simpson, 1998) can be represented by a sigmoid line (Fig. 4.1). The average transfer rate (in units of mass per unit time) is represented by the slope of a line from the origin to the transfer function. For a sigmoidal curve, as depicted in this example, there is a maximal rate of transfer, which occurs at time t1. The

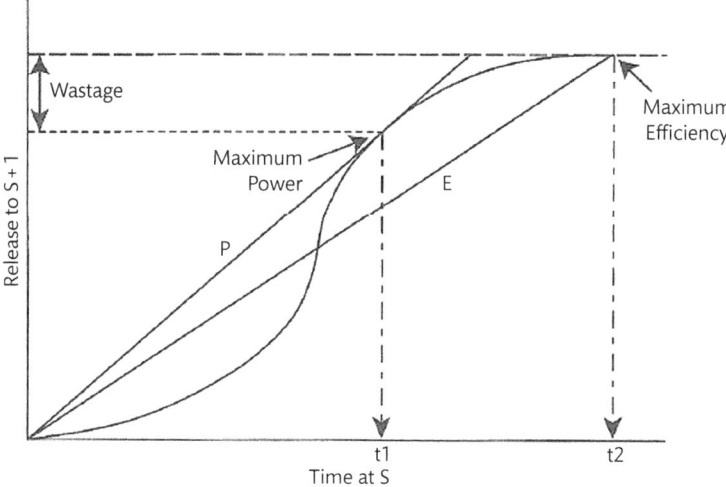

Fig. 4.1 Transfer of energy or material from one system (the gut) to another (the rest of the body). (From Karasov and Martínez del Rio 2007; with permission from Princeton University Press). See text for details.

maximal rate does not maximize the efficiency of transfer, which is maximized at time t2. Thus, if an animal has evolved to maximize its rate of transfer, this is achieved by ending the process at t1, in which case some energy or material is wasted. Here, efficiency of transfer is below what would be possible had the process been extended to time t2. If on the other hand, the animal evolved to maximize transfer efficiency (as might be expected if food is scarce), this is achieved at the expense of the highest rate. Such trade-offs between rate and efficiency may apply to a variety of processes as, for instance, the harvesting of food. The example shown in Fig. 4.1 applies to a single nutritional event (e.g. a meal). However, the concept can also be extended to longer periods and other kinds of situations (Raubenheimer and Simpson 1998).

The rate-efficiency trade-off becomes apparent in comparisons among marine and terrestrial avian predators. Among species in both groups there are small but significant differences in digestive efficiency, even when feeding on the same diets (Hilton *et al.* 1999). In both groups, species that retain food in their gut for less time are less efficient digesters. Other factors besides differences in retention time may explain the differences in efficiency, because lower efficiency is also associated with shorter intestine length (Hilton *et al.* 1999; 2000), and as a consequence of reduced surface area and fewer intestine-bound enzymes for absorption and digestion. Hilton and his colleagues note that raptor species that have relatively less gut mass and shorter retention time tend to be species (e.g., falcons) that pursue active prey. They suggest that natural selection favored these traits because low gut mass and short retention time reduce the mass of tissue and digesta carried, and consequently flight performance is improved.

The transfer functions depicted in Fig. 4.1 are not necessarily fixed but there are likely limits to their adjustability. For example, Karasov and McWilliams (2005) took White-throated Sparrows (*Zonotrichia albicollis*) acclimated to room temperatures (21°C) and switched them to cold conditions (−20°C) either quickly or gradually over several days. In the cold, sparrows, like all homeothermic endotherms, must eat more to counterbalance higher heat losses or they will catabolize their body tissues to supply the extra energy. Sparrows that were switched quickly increased their feeding and digestion rate only 45% and lost body mass, whereas sparrows that were acclimated slowly increased their rates by 83% and maintained body mass. Sparrows that were switched quickly experienced an energy deficit and should have been motivated to eat more, but did not.

These results define two types of limits to energy flow. First, an immediate limit that was only 45% above what sparrows needed routinely for energy balance at 21°C (Fig. 4.2). The difference between that upper limit and the routine demand can be defined as the sparrow's "immediate reserve capacity" (Diamond

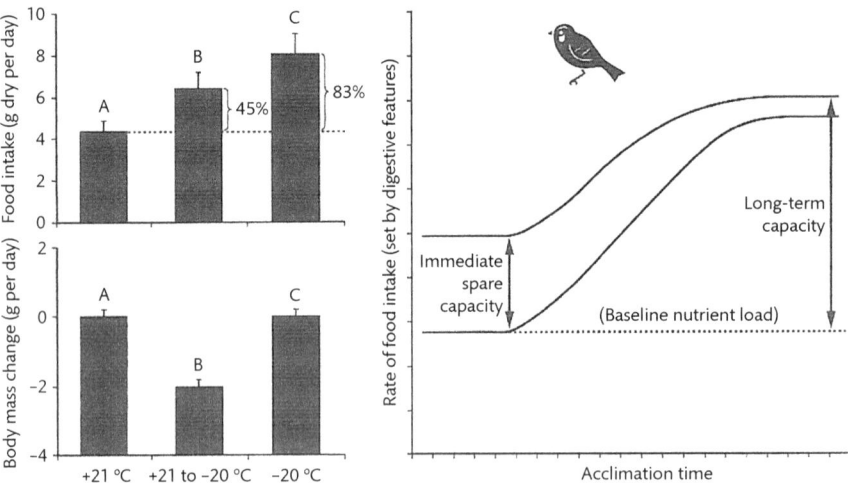

Fig. 4.2 White-throated sparrows (*Zonotrichia albicolis*) suddenly switched to cold temperature (+21 to −20°C) – group B- increased their feeding rate only 45% compared to pre-treatment – group A – and were unable to maintain their body weight. Sparrows acclimated more slowly to −20°C – group C -nearly doubled food intake and maintained body weight. The sparrow can increase its food intake only within limits set by the level of immediate reserve capacity, which decreases as the sparrow approaches its long-term capacity. (From Karasov and Martínez del Rio 2007; with permission from Princeton University Press).

and Hammond 1992). This first limit is flexible as sparrows that were switched more gradually were able to increase their feeding and metabolic capacity by 83%. This capacity can increase even more, since captive White-throated sparrows could tolerate temperatures as low as −29°C (Kontogiannis 1968). At this low ambient temperature metabolizable energy intake was 126% higher than at room temperature. Because the sparrows could not tolerate lower ambient temperatures without loosing body mass, and could not increase further their metabolizable energy intake even when challenged by forced activity (Kontogiannis 1968), their limit to energy flow after long-term acclimation was around 126% above "baseline". The difference between long-term capacity and routine demand can be defined as the sparrow's "long-term reserve capacity".

4.9 Handling and Digestion

The relationship between predation rate (i.e., number of prey eaten per predator per unit time) and prey density is termed "functional response" (Solomon 1949). This relationship is central to many questions in ecology.

Handling prey is an active or "foreground" process whereas digestion is a background process. In contrast to handling prey, digestion does not directly prevent the predator from further searching or handling. Instead, digestion influences the predator's hunger level, which in turn influences the probability that the predator searches for new prey (Jeschke *et al.* 2002). Within this context, it is necessary to discriminate digestion from handling in a functional response model.

The Gause-Ivlev equation (Gause 1934; Ivlev 1961) relates the number of prey eaten in a day (y) and the density of prey (x):

$$y(x) = y_{max}(1 - \exp[-ax]) \tag{11}$$

where a is hunting success (dimension in SI units; m^2 for a two-dimensional system, e.g., a terrestrial system, and m^3 for a three-dimensional system, e.g., an aquatic or aerial system), x is prey density (individuals m^{-2} or individuals m^{-3}, respectively), y is predation rate (s^{-1}), and y_{max} is asymptotic maximum predation rate as x approaches infinity (s^{-1}). Therefore, the digestive system determines y_{max}, and the functional response curve gradually rises to this value. Therefore, an animal does not exceed y_{max} because it becomes satiated, probably because it has a full gut. The Gause-Ivlev equation recognizes the physiological limitations imposed by digestion but does not take into account the limitations that handling prey before its ingestion may impose on feeding rate.

There are a number of models (for a complete account, refer to Jeschke *et al.* 2002) that include handling time but not predator satiation effects (digestive limitations). Holling's (1959) model, for example, assumes that a predator can feed for a time equal to T_t, and that it encounters prey at a rate proportional to prey density x. The model also assumes that it takes the predator a time T_h to handle each prey. Therefore

$$Y(x) = ax(T_t - T_h y), \tag{12}$$

where $Y(x)$ is the number of prey eaten, a is a constant called the instantaneous search rate, and $T_t - T_h y$ is the time that a predator has to search for a prey. The Holling's disk equation is a variation of equation 12:

$$y(x) = Y(x)/T_t = (ax) + aT_h x = (1/T_h)x/(1/aT_h) + x. \tag{13}$$

The curve's gradient at the origin is equal to a, and the asymptotic maximum for x as x approaches infinity is $1/T_h$. Interestingly, Holling's disk equation is mathematically equivalent to the Micahelis-Menten model of enzyme kinetics and the Monod formula for bacterial growth.

However, the Gause-Ivlev equation and the Holling's equation are not good descriptors of what happens to animals, i.e., whether they are limited by digestive post-ingestional processes or by preingestional handling time. In order to integrate these two factors, Jeschke *et al.* (2002) proposed what they called the "steady-state satiation" (SSS) equation:

$$y(x) = \frac{1 + ax(b+c) - \sqrt{1 + ax(2(b+c) + ax(b-c)^2)}}{2abcx}$$

for $a, b, c, x > 0$

$y(x) = ax/(1 + abx)$, for $b > 0$ and $c = 0$

$y(x) = ax/(1 + acx)$, for $b = 0$ and $c > 0$

$y(x) = ax$, for $b = c = 0$

$$y(x) = 0, \text{ for } a = 0 \text{ or } x = 0 \tag{14}$$

with success rate $a = \beta\gamma\epsilon$, corrected handling time $b = t_{att}/\epsilon + t_{eat}$, and corrected digestion time $c = s\, t_{dig}$. β is the encounter rate, γ is the probability that the predator detects encountered prey, ϵ is the efficiency of attack (the percent of attacks that succeed), s is the satiation per prey item, t_{att} is the attacking time per prey, t_{dig} is the digestion time per prey item, and t_{eat} is the eating time per prey item. The complete derivation of equation 14 is presented in Jeschke *et al.* (2002).

SSS equation curves are more flexible than disk equation curves. Therefore, according to Jeschke *et al.* (2002), it is impossible to satisfyingly fit the disk equation to a SSS equation curve (with the exceptions $b = 0$ or $c = 0$). This occurs because in the disk equation, one parameter (b) determines the curve's asymptote, and two parameters (a and b) determine how the curve reaches this asymptote, i.e., the curve's slope. In contrast, in the SSS equation, one parameter (the larger one of the parameters b and c) determines the curve's asymptote, and three parameters (a, b and c) determine how the curve reaches this asymptote (Fig. 4.3 a, b and c).

The SSS equation predicts that handling-time limitations should be rare, and likely to be found in parasitoids, protozoans and drilling gastropods. In contrast, Jeschke *et al.* (2002) found that most of the published studies show evidence of digestion limitation, and that natural predators generally spend a major part of their time in resting. For digestion-limited predators, with an increase in digestion time, the SSS equation predicts that foraging time decreases with increasing prey density if we assume the animal is in steady-state (not growing). This is in accordance with empirical data, for example, from birds: Spotted Sandpipers (*Actitis acularia*; Maxson and Oring 1980), Verdins (*Auriparus flaviceps*; Austin 1978), Oystercatchers (*Haematopus ostralegus*; Drinnan 1957), Yellow-eyed Juncos (*Junco phaeonotus*; Caraco 1979), and the Rufous Hummingbird (Hixon *et al.* 1983).

To illustrate the effects of adding digestive limitations to functional response models, Karasov and Martínez del Rio (2007) used a simpler model than the SSS equation. As an example they used Red Knots which forage on mollusks (clams in the genus *Macoma*) on mudflats along the Wadden Sea in northern Europe, and based their predictions on data in Zwarts *et al.* (1992) and Piersma *et al.* (1995); for details, see Karasov and Martínez del Rio (2007). Red Knots may show a high harvest rate for brief periods but they are not able to sustain it for prolonged periods; instead they frequently stop foraging and preen. Zwarts and Blomert (1992) interpreted such stops as digestive pauses. Even though Jeschke *et al.* (2002) concluded that all studies that have measured both digestion and handling time provide evidence for digestive limitations, Karasov and Martínez del Rio (2007) conclude that Red Knots (and possibly other bird species) are whole-animal limited rather than just digestion limited, unless there is a clear demonstration of a digestive limitation *per se*. That is, other factors such as the need to carry out a conflicting activity, or just becoming tired, may take temporary precedence over feeding.

The importance of Jeschke *et al.*'s (2002) model is that it nicely links digestive and pre-ingestion limitations on intake rate, and because it is consistent with chemical reactor theory (CRT). McWhorter and Martínez del Rio (2000) used a

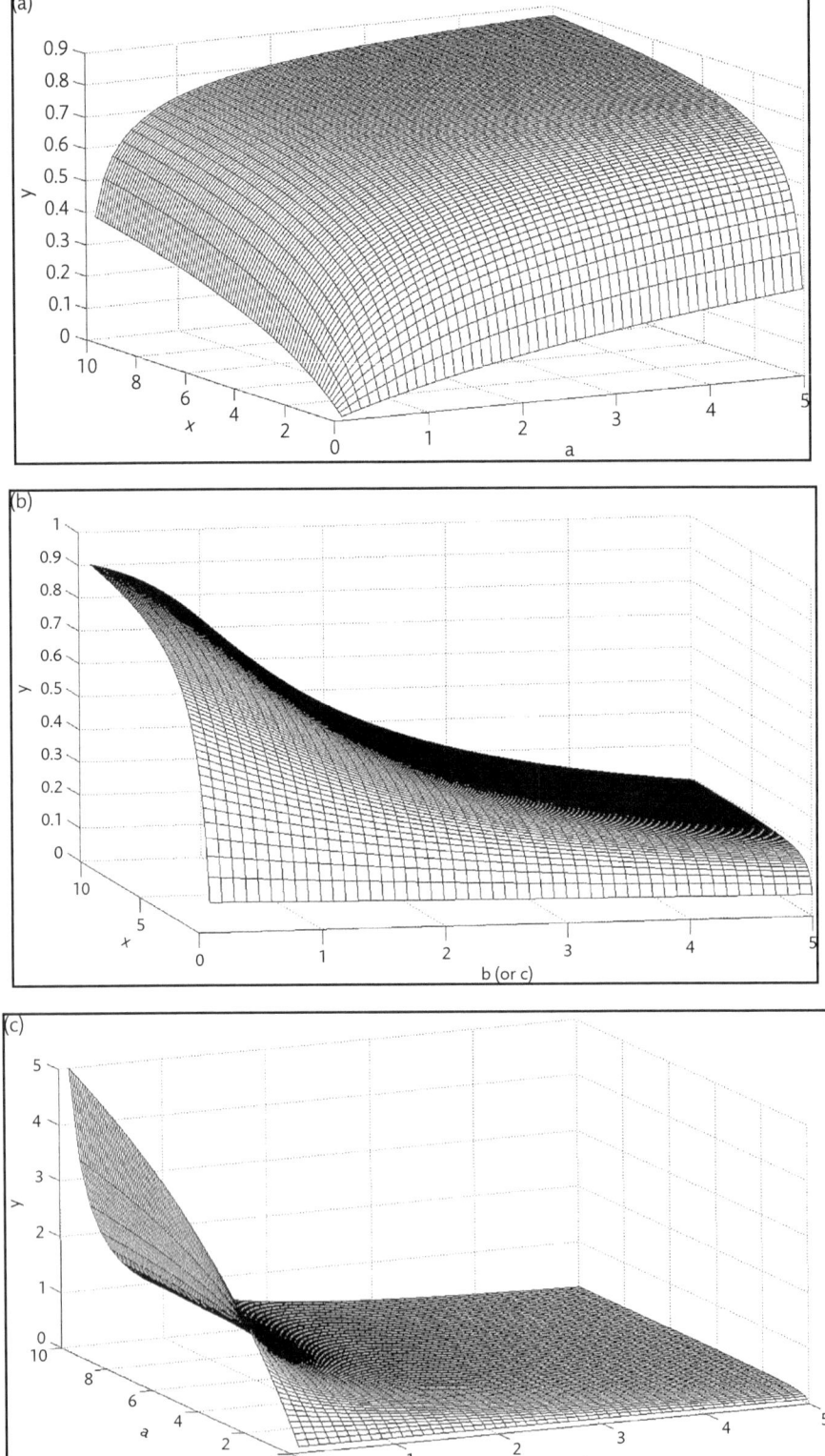

Fig. 4.3 (continued)

Fig. 4.3 a 3-D plot of the function relating prey density (x), number of prey effectively eaten (y) and success rate (a). Note that when the density of prey increases, both the number of prey eaten in a day and the success rate increase.
b Number of prey eaten in a day (y) as a function of density of prey (x). In this case, when handling time (b) and/or digestion time (c) increase, the number of prey eaten in a day is reduced even when the density of prey is high. The maximum number of prey eaten in a day may only be attained when b and/or c are equal do zero.
c Maximal number of prey eaten in a day (y) as function of success rate (a) is only attained when handling time (b) and/or digestion time (c) are equal to zero. Note that in this case y goes to infinity (no explicitly shown in the plot) when a reaches a maximum value, an unreal situation that shows that satiation has been attained. As b and/or c increase, y drops very quickly.
Note: See text for details. All variables are in arbitrary units.

chemical reactor model to predict the maximal rate at which hummingbirds could ingest food. By combining both reactor theory and their simpler model, based on Jeschke *et al.*'s (2002) SSS equation, Karasov and Martínez del Rio (2007) were able to predict hummingbirds' maximal intake rate from a combination of measurements that included the volume of the digestive sections of the gut and biochemical measurements of its hydrolytic capacity. They concluded that hydrolysis rates limited maximal ingestion rate. This illustrates how a combination of guts-as-reactors estimates of maximal digestion rates and the SSS equation can lead to the integration of digestive physiology and functional responses.

4.10 Morphological Adjustments of the Gastrointestinal Tract

In insectivorous House Wrens (*Troglodytes aedon*) acclimated to cold, daily food intake doubled, gut mass and volume increased 25–35%, mouth-to-anus retention time and amino acid uptake rates per unit small intestine did not change significantly, and there was no decrease in digestive efficiency compared to warm-acclimated wrens (Dykstra and Karasov 1992). Similar results were obtained by McWilliams *et al.* (1999) for granivorous White-throated Sparrows and frugivorous Cedar Waxwings. It thus appears that when energy requirements increase greatly, birds eat more while maintaining fairly constant retention time and efficiency, and the primary adjustment is an increase in GI tract mass and volume. Many other studies in birds acclimated to low ambient temperatures (e.g., McWilliams and Karasov 2001) confirm that the most important adjustment to higher feeding rates is mass of the GI tract (and liver), which has the dual effect of keeping retention time relatively constant in the face of higher digesta flow and increasing the intestine's biochemical capacity while holding efficiency constant (Karasov and Martínez del Rio 2007).

It is clear that the avian GI tract exhibits great flexibility in the face of altered food composition and level of intake. Digestive flexibility in birds is ecologically important for what it permits them to do and for how it constrains their feeding ecology. When acclimation/acclimatization occurs, its time course may determine how quickly diet switching can take place (Karasov 1996).

4.11 Adjustments during Reproduction

Up- and down-regulation of GI tract mass may support increased feeding during egg laying in some birds (Karasov 1996). Egg laying appears to increase the daily energy requirements of passerines by an amount equal to about 35% of basal metabolism (Robbins 1993). This would be, however, only about 10% of FMR (field metabolic rate). Parents feeding *their* nestlings tend to have high rates of daily energy expenditure and hence feeding rates. Therefore, gut enlargement might be expected at this time (Karasov 1996). While adult feeding and digestion rates might increase when feeding larger broods, there is evidence in House Wrens (Dykstra and Karasov 1992; Dykstra and Karasov 1993) and Yellow-eyed Juncos (Weathers and Sullivan 1989) that maximal rate of digestion does not limit brood size proximally, because parental energy expenditure while feeding nestlings was below the maximal rate of sustained energy assimilation (A_{max}).

4.12 Adjustments in Migrant Birds

Migrating birds typically carry large quantities of fat to fuel their long flights. A major finding of nutritional ecology studies of migrating birds is that birds obtain fat mainly by hyperphagia. Feeding and assimilation rates during pre-migratory fattening may increase two-fold (Berthold 1975 and 1993; Blem 1980), and may approach A_{max}, as suggested by results from a number of laboratory studies (Karasov 1996). For example, in captive White-crowned Sparrows (*Zonotrichia leucophrys*, 25–30 g body mass) fattening at a rate of 1.7 g day^{-1}, King (1961) measured an assimilation rate of nearly 125 kJ day^{-1}, which is 87% of predicted A_{max}. Captive Garden Warblers (*Sylvia borin*; 17–23 g) gaining mass at 1.2 g day^{-1} had maximum meabolizable energy intake of 111 kJ day^{-1} (Klaassen and Biebach 1994), equal the predicted A_{max}. Data on rates of mass increase at migratory stopover sites indicate that individuals of some species reach A_{max} (Lindstrom 1991). The rate of fat deposition in wild birds is probably only half that of captive birds, but this does not necessarily mean that food availability is limiting wild birds (Meier and Fivizzani 1980). They might have similar energy intake and assimilation rates as captive birds do, but expend much more energy than captives for activity or thermoregulation, and thus allocate less to production (Karasov 1996).

The expected relatively larger GI tracts of migrating birds, because of hyperphagia, are not confirmed by recent evidence. In a small Old World warbler (the Blackcap, *Sylvia atricapilla*; Karasov and Pinshow 1998) and Yellow-rumped Warblers (Afik and Karasov 1995), the intestinal masses of migrants were 25–40% lighter than those of captive individuals feeding *ad libitum*. Smaller GI tracts were also found in migrant shorebirds when compared to non-migrants (Piersma 1998; Piersma *et al.*, 1999b).

The idea of a lighter body prior to migratory departure, thus having the benefits of reduced wing loadings and resting metabolic rates (Battley *et al.* 2000, 2001), which in turn would lead to increased flight distances and a reduction in the number of stops en-route to refuel, is appealing. However, there is an alternative explanation for a decreased GI tract size in migrants. Birds during migration alternate between periods of high feeding rate at migratory stopover sites and periods without feeding as they travel between stopover sites. These intervals without food may be brief (<8 hours) for birds migrating in short stages like Yellow-rumped Warblers, or may last for days for birds migrating over oceans or other large geophysical barriers (e.g., deserts, mountains) like Blackcaps (McWilliams and Karasov 2001). At stopover sites food may not be readily available or may be difficult to find. Laboratory studies with passerines have shown that fasting or food restriction causes proportionally large declines in GI tract mass in comparison with other components such as flight (pectoral) muscles and heart (Klaassen and Biebach 1994; Hume and Biebach 1996; Karasov and Pinshow 1998; Lee *et al.* 2002). Field studies of passerines have revealed that recovery of body condition after arrival at stopover sites is nil or slow for the first 1 or 2 days and then much more rapid (two-step recovery), despite apparently abundant food resources (reviewed in McWilliams and Karasov 2001; Karasov and McWilliams 2005).

Although ecological conditions like food availability influence rates of recovery, there may also be a physiological explanation, because birds exhibit the two-step recovery even when provided food *ad libitum* in the laboratory. In laboratory fasting studies, atrophy of organs important in food processing (e.g., intestine, liver) is associated with lower biochemical digestive capacity (Lee *et al.* 2002) and lower rates of feeding and digestion (Hume and Biebach 1996; Karasov and Pinshow 2000) which, during a migratory stopover, might slow the rate of mass gain. The need for protein to rebuild lean mass and especially atrophied organs important for food digestion and assimilation during stopovers could constrain diet selection and require increased foraging time by migrant birds. The ultimate consequence of a slow mass gain, besides a decrease of an individual's overall migration speed, is a decrease in migration success (Karasov and Pinshow 1998). From theoretical models of optimal migration (Alerstam and Lindstrom 1990), it is apparent that the rate at which a species can add stores or restore a physiological deficit before the next flight is critical (Karasov and Pinshow 1998).

Birds may often experience short-term changes in food quality. Migrants in particular may one day find preferred fruits that are ubiquitous, whereas the next day they may encounter few fruits, but an abundance of insects. In such situations, the diet of a bird may differ dramatically from day to day. These frequent and unpredictable changes in food quality are another challenge faced by migrating birds, and may occur faster than the time scale required for biochemical or morphological adjustments (McWilliams and Karasov 2001).

Physiological mechanisms to explain initial mass loss after arrival at a stopover site are largely unexplored (Berthold 1993; Biebach 1996), but two hypotheses have some support. The nutrient-limitation hypothesis suggests that the initially slow rate of mass gain occurs because birds utilize protein reserves during migration, and recovery of these must occur first and is slow. Only subsequently can lipid reserves be restored but, once initiated, recovery of lipid reserves is rapid (McWilliams and Karasov 2001). This hypothesis is supported by studies of migrating hummingbirds (Carpenter *et al.* 1983), shorebirds (van der Meer and Piersma 1994) a few passerine birds (Kendall *et al.* 1973; Jones and Ward 1976; Marsh 1983, 1984; Selman and Houston 1996b) and fasted domestic geese (LeMaho *et al.* 1981; McLandress and Raveling 1981; Cherel *et al.* 1988).

An organ-specific variant of this idea is the gut-limitation hypothesis, which suggests that the initially slow rate of mass gain at stopover sites occurs because birds loose digestive tract tissue and function during fasting, and rebuilding of The GI tract takes time and resources and itself restricts the supply of energy and nutrients from food (McWilliams and Karasov 2001). This hypothesis is supported by studies that show reduced gut size after fasting in mammals and chickens (reviewed by Karasov 1988) and by a few studies on migratory passerines showing that food intake and gut function are reduced after fasting (Ketterson and King 1977; Klaassen and Biebach 1994; Hume and Biebach 1996; Klaassen *et al.* 1997; Karasov and Pinshow 2000).

These hypotheses are not mutually exclusive provided protein in the GI tract is a primary protein reserve for catabolic energy production (Piersma *et al.* 1993; Piersma and Lindstrom 1997; Piersma 1998). Studies on Blackcaps give the only quantitative support for this idea in which a disproportionate amount (44%) of protein lost during fasts came from small intestine, stomach, and liver tissue (Karasov and Pinshow 1998). However, it is unclear whether this protein was used as a nutrient source or simply lost through disuse and excretion.

The gut-limitation hypothesis has been supported by studies on Blackcaps in which GI tract mass was restored to normal in individuals that had previously been fasted or food restricted after just two days of *ad libitum* feeding (Karasov *et al.* 2004). An important mechanism was cell proliferation (increase in number of enterocytes). It can occur because intestinal mucosal cells have a short cell cycle time during which they are produced in crypts at the base of the villi, migrate up the villi, function for a few days, and eventually die (apoptosis) and are sloughed into the lumen. The size of the mucosal tissue can be manipulated by increases in the rate of cell birth and/or declines in the rate of apoptosis. In small birds the intestinal turnover time is 2–3 days; it may be longer in larger birds (Stark 1996).

Impressively large changes in GI mass under natural conditions have been observed in shorebird migrants like Red Knots (Piersma *et al.* 1999b) and Bar-tailed Godwits (Limosa lapponica; Piersma and Gill 1998; Landys-Ciannelli *et al.* 2003). Bar-tailed Godwits were examined (Landys-Ciannelli *et al.*, 2003) on a stopover site in spring to evaluate changes in lean dry mass of organ and muscle tissue over the period of refueling. It was found that lean dry mass of flight muscles increased during refueling and reached peak size in individuals ready to depart on their next migratory flight, possibly to maximize power output. In contrast, the lean dry mass of stomach, liver, and intestines increased only during the early phases of refueling, indicating a rapid initial growth of organs involved in the support of refueling activities. Landys-Ciannelli *et al.* (2003) also found that in individuals ready for departure, the intestines did not differ in size from those of recent arrivals, indicating a reduction in the size of the GI tract prior to flight. This probably makes flight less costly by reducing the total mass the birds have to lift and carry during long migratory stages.

Phenotypic plasticity in the digestive system of migratory birds is critically important in allowing them to overcome the conflicting physiological challenges of migration. However, phenotypic plasticity has limits that can influence the tempo of migration (i.e., time spent at stopover sites versus time spent flying; McWilliams and Karasov 2001). The examples in this chapter illustrate such limits. Clearly, there is a need for more comprehensive studies of GI tract structure and function in freshly captured migrant birds, experimentally fasted and food-restricted individuals, and normally feeding individuals (Karasov and Martínez del Rio 2007).

4.13 Plant Secondary Metabolites

Many non-nutritive chemicals, or xenobiotics, are present in animal foods; if absorbed, they may not yield any energy and cannot be used as building blocks. Ingestion of xenobiotics may drain energy and/or essential nutrients as the animal expels them from the body, and some are toxic to varying degrees. Through a variety of mechanisms, including the induction of specific enzymes which biotransform xenobiotics, animals attempt to control these effects. However, the energy and material lost in the course of xenobiotic elimination must be replaced, which increases resource demands. If elimination is less than absorption the xenobiotic will accumulate in the body, which can lead to toxicity (Karasov and Martínez del Rio 2007).

Plant secondary metabolites may alter retention time (Murray *et al.* 1994), interactions between dietary components and digestive enzymes or microbes (Robbins 1993; Robbins *et al.* 1991), and nutrient absorption (Karasov *et al.* 1992). Although these effects might reduce digestive efficiency, caution is appropriate when inferring these mechanisms from the observation that a reduction in utilization efficiency is correlated to the presence of a plant secondary metabolite in food (Koenig 1991). An equally plausible mechanism is that energy-rich detoxification products can appear in excreta, thereby inflating energy excretion and lowering the MEC*.

Various secondary metabolites increase or decrease digesta retention time (Cippolini 2000). Such a change theoretically could affect nutrient extraction efficiency. However, examples of this are not known. Some studies have shown how certain phenolics inhibit sugar absorption (Karasov *et al.* 1992; Skopec *et al.* 2004). Examples are the flavonoids phlorezin and phloretin found in apple (*Malus*) bark and foliage, which specifically inhibit the apical membrane glucose transporter (SGLT1) and the basolateral membrane glucose transporter (GLUT2), respectively. Many other flavonoids are known to depress glucose absorption efficiency. Many are found in fruits, and they could have similar digestion altering effects in frugivorous birds and, based on behavioral sensitivity to alterations in digestive efficiency, influence fruit choice (Karasov and Martínez del Rio 2007).

Many absorbed xenobiotics are ultimately excreted conjugated to sugar derivatives, amino acids, peptides, and sulfate. Conjugation with glucoronic acid and maintenance of acid-base status cause catabolism of amino acid and is shown to result in loss of body protein and depletion of glucose. Detoxification of plant allelochemicals by herbivores disrupts acid-base balance and this is an important consequence of allelochemical absorption (Foley *et al.*, 1995). An interaction between allelochemical dose and nutrient has been found, and the ratio of allelochemical to nutrient absorption rate defines the tolerance of the animal to absorbable allelochemicals in foods. This interaction is nonlinear and has ecological implications for foraging behavior and diet. Based on this, theoretical calculations (Illius and Jessop 1995) indicate that the associated energy losses would reduce the metabolizability of food by about 5%. This number varies with the type of compound and its concentration, and empirical studies indicate that the loss can be higher in some animals. The highest losses have been observed in Ruffed Grouse (*Bonasa umbellus*), a herbivorous bird that in winter consumes large amounts of aspen flower buds (Jakubas *et al.* 1993; Guglielmo *et al.* 1996), which contain the phenyl-propanoid ester coniferyl benzoate (CB). Incorporation of The purified compound into an artificial diet reduced the MEC* from 0.54 to 0.48 (11% decrease). All of the reduction could be accounted for by a dilution effect of the compound and its detoxification products in food and excreta. However, there did not appear to be any effect on digestive processes *per se*. Among quaking aspen flower buds (*Populus tremuloides* Michx.) that vary in CB content, MEC* was inversely related to CB content. Absorbed CB and other plant secondary metabolites from aspen buds are detoxified and excreted with conjugates such as glucuronic acid (a glucose conjugate) and ornithine (an amino acid conjugate). The energy lost in glucoronides and ornithine excretion represented 5–15% of metabolizable energy in Ruffed Grouse, compared with 3% or less found in six mammalian species (Guglielmo *et al.* 1996). The generality of these differences, and underlying mechanisms, await further research. Perhaps more importantly, the ornithine excretion of the Ruffed Grouse increased the minimum daily nitrogen requirement by 67-89%, depending on the rate of xenobiotic absorption. This study is a nice example of how detailed measurements are important and necessary to understand the mechanism of action of plant secondary metabolites and their ecological significance in birds.

It is interesting to note that animals can sometimes modulate their food intake to regulate xenobiotic ingestion. Another study on the Ruffed Grouse shows that this bird appears to regulate its food intake according to CB and not energy (Jakubas and Mason 1991; Jakubas *et al.* 1993). Ruffed Grouse eat normal amounts of food as long as it contains no more than 4.5% CB. If food contains more than this the birds reduce daily food intake in response to progressive increases in CB concentration. It seems that if the Ruffed Grouse ate more CB it would suffer intoxication, and thus eating less food and losing weight is the lesser of two poor options.

5

Adaptations: Living in Specific Environments

Birds are found in essentially all terrestrial environments, from lush tropical forests to barren deserts and polar regions, from the oceans to the highest mountains. In this chapter, we describe how birds adapt to this huge array of strikingly different habitats and the evolutionary consequences associated with these habitats and adaptations. Where appropriate we include information on those phylogenetic groups which have evolved unique specializations for particular environments.

5.1 Desert and Arid Environments

Flight enables birds to move long distances quickly and therefore helps migrants reduce their exposure to, or altogether avoid, the harshness of arid or desert environments. However, many non-specialist birds are able to thrive in moderately arid regions by tolerating high temperatures or avoiding activity at the hottest times of the day, or by frequent use of water resources (even though these may be limited, the ability to fly lets birds find and exploit even widely dispersed water sources). However, the very xeric deserts (<150 mm rainfall per year) found on most continents present greater challenges to birds. Nevertheless, some bird species reside in very extreme deserts, and a striking example is the so-called Empty Quarter, or Rub 'al Khali, the largest desert sand sea in the world encompassing the southern third of the Arabian Peninsula. This forbidding place is a primary habitat for Hoopoe Larks (*Alaemon alaudipes*; Williams and Tieleman 2005).

Birds, to a greater extent than mammals, are diurnal and non-fossorial, and hence are routinely exposed to the extremes of daytime heat and dryness. In addition, most are also relatively small, which "couples" them tightly to environmental conditions. A notable exception is the Ostrich (*Struthio camelus*) which may reach almost 100 kg. It so happens that this, the largest of living birds, is a typical arid or desert form. However, despite the advantages of large size in "buffering" extreme temperatures, the Ostrich can survive in deserts only when it can get to succulent vegetation or water, and consequently the species has a difficult time surviving in the harshest deserts.

Reproduction in hot climates raises some problems peculiar to birds. At times the danger to eggs from overheating may be more important that the usual problem of keeping them warm. In particular, if nests are placed on the ground, and many desert birds have this habit, the eggs must be shaded otherwise they become heated to temperatures fatal to the embryo. Likewise, nestlings must be kept cool and provided with the water they need, often a major problem where food with a high water content is not easily available (Schmidt-Nielsen 1964; Dawson 1984). Sandgrouse (*Pterocles*), characteristic of the steppes and deserts of the Old World, never nest out of flying range of water. The distance can, however, exceed 50 km. Sandgrouse are even better fliers than pigeons and since at certain times of the day they accumulate in large flocks at watering-places, they must be coming from considerably long distances (Schmidt-Nielsen 1964). Sandgrouse normally stand in shallow water when drinking, and as their legs are short their breast feathers become saturated with water. They carry this water to their flightless chicks in the hot, bare desert, and the young birds drink directly from the saturated breast plumage (Buxton 1923).

Intercellular lipids of the stratum corneum (SC), the outer layer of the epidermis, form a barrier to water vapor diffusion through the skin. Muñoz-Garica and Williams (2008) showed that desert nestling House Sparrows (*Passer domesticus indicus*) have a greater degree of plasticity in cutaneous water loss (CWL) and lipid composition of the SC than do mesic nestlings House Sparrows (*Passer domesticus domesticus*). Desert and mesic fledglings regulate rates of CWL in response to environmental conditions through modifications of lipid composition of the SC. Desert House Sparrows seem to control more closely the concentration of sphingolipids (ceramides and cerebrosides), whereas mesic House Sparrows prioritize the regulation of the concentration of free fatty acids (FFA). Both desert and mesic fledglings change concentrations of FFA in their SC, but only desert fledglings alter the concentration of ceramides. Ceramides of the SC form the structural backbone of intercellular lipid lamellae and play a major role in preventing water vapor diffusion through the skin (Bouwstra *et al.* 2003).

Although the ability of desert mammals to minimize energy expenditure and water loss is well known (Schmidt-Nielsen and Schmidt-Nielsen 1950; Schmidt-Nielsen 1964; Walsberg 2000), earlier attempts to elucidate similar physiological attributes among desert-dwelling birds were less fruitful. Bartholomew and Cade (1963), after nearly a decade of work on bird species from the semi-arid and desert regions of the Southwestern United States of America, concluded that many avian species living in these deserts have not evolved specific physiological specializations that distinguish them from mesic counterparts. Bartholomew (1972) suggested that "most desert passerines appear to have a basal metabolic rate (BMR) appropriate to their size". A widely accepted axiom is that birds inhabiting deserts are capable of doing so because of characters possessed by all birds - flight, excretion of uric acid, efficient evaporative cooling, and behavioral avoidance of climatic extremes - rather than as a consequence of evolved physiological features specific to a desert environment (Maclean 1996). In other words, birds in general are pre-adapted to the challenges of desert life.

It is unclear if the results summarized by Bartholomew (1972), which are based mainly on North American species, can be generalized to the planet's other desert regions. Climatologists have long recognized that deserts differ in their meteorological parameters, and have emphasized that the environment of a given desert region depends on the interaction of a myriad of factors, including ambient temperature (T_a), amount and timing of rainfall, relative humidity and wind. These differences likely influence the selective pressures imposed by each respective desert (reviewed by Williams and Tieleman 2002).

Williams and Tieleman (2002) emphasize that previous works relied primarily on information gathered from species inhabiting semi-arid habitats of recent geologic origin. They have therefore included data on species from Old World deserts, which are geologically older than those in the New World, and place physiological responses along an aridity axis that includes mesic, semi-arid, arid, and hyper-arid environments. Their work indicates that some birds have, in fact, evolved specific physiological adaptations to desert environments.

5.1.1 Evaporative Water Loss

Because deserts are characterized by low food availability, high ambient temperature extremes, and the absence of drinking water, one might expect that birds that live under such conditions exhibit a lower BMR (to reduce heat loads at rest), reduced total evaporative water loss (TEWL, the sum of respiratory (REWL) and cutaneous (CEWL) evaporative water losses), and greater ability to cope with high air temperatures than their mesic counterparts (Tieleman *et al.* 2002; 2003). To test this hypothesis, and in order to minimize confounding effects of phylogeny, Tieleman and colleagues compared the physiological performance of four closely related species of larks (of the same family) at ambient temperatures from 0°C to 50°C. Hoopoe Larks and Dunn's Larks (*Eremalauda dunni*) live in the hot and dry Arabian desert, whereas Skylarks (*Alauda arvensis*) and Woodlarks (*Lullula arborea*) occur in temperate mesic grassland areas. When their results were adjusted for body mass, both BMR and TEWL were indistinguishable between Hoopoe Lark and Dunn's Lark, and between Skylark and Woodlark. When they grouped the data of the two desert larks in one set and the data of the two mesic larks in another, desert larks showed a 43% lower BMR and 27% lower TEWL values than the mesic species. Although the mechanisms underlying these differences are still

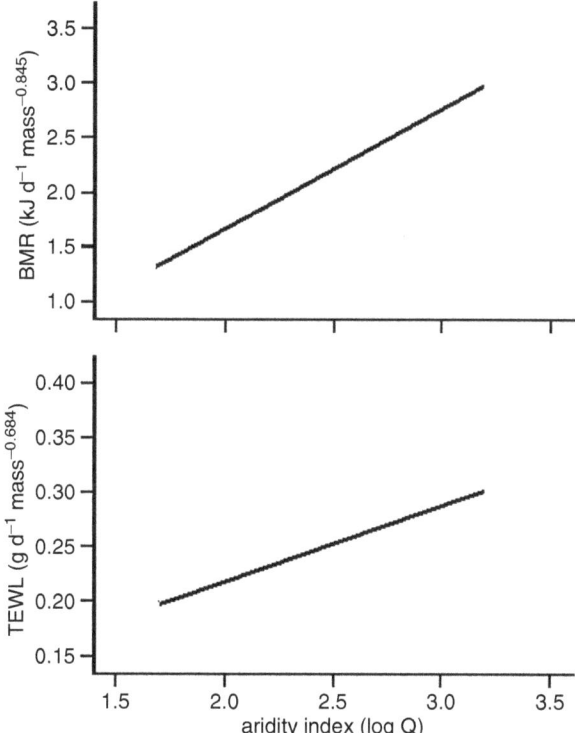

Fig. 5.1 Mass-adjusted BMR (*top*) and TEWL (*bottom*) of 12 species of lark as a function of environmental aridity. (Redrawn from Tieleman *et al.* 2003).

unknown, Williams and Tieleman (2002) attribute these physiological differences between desert and mesic birds to acclimation to different environments and/or to genetic change mediated by selection. (Fig. 5.1)

In another study, Tieleman and Williams (2002) used the same experimental approach to measure the partition between REWL and CEWL in larks from different environments. Their results are consistent with the hypothesis that larks from deserts have reduced CEWL (29% lower than mesic larks at 25°C) at moderate to low ambient temperatures but provided no support for the hypothesis that at high ambient temperatures larks from arid regions rely more on CEWL than larks from mesic environments. Instead, at high ambient temperatures (40°– 45°C) all larks (mesic and desert species) relied on an increase in REWL (more than 6 times higher) for evaporative cooling. Thus, according to these authors, inter-specific differences in CEWL cannot be attributed to acclimation to environmental temperatures, and are possibly the result of phenotypically plastic responses to divergent environments during ontogeny.

McKechnie and Wolf (2004) found a significant variation in CEWL relative to TEWL in response to short-term acclimation in the Western White-winged Dove *Zenaida asiatica mearnsii*. In their study, doves acclimated to high T_a for 2-4 weeks showed high rates of CEWL, and therefore higher CWEL/TEWL ratios, compared to doves kept in a cooler environment. CWEL/TEWL ratios in heat-acclimated Western White-winged Doves found by McKechnie and Wolf are among the highest recorded (Wolf and Walsberg, 1996), and can be compared to the ratios found by Withers and Williams (1990) in Australian Spinnifex pigeons (*Geophaps plumifera*). McKechnie and Wolf found that CEWL increased with increasing ambient temperatures and varied significantly between cool- and heat-acclimated doves. In their study, it has been shown that CEWL in cool-acclimated doves increased threefold between 35°C and 45°C, whereas in heat-acclimated doves CEWL increased fourfold over the same range of air temperatures. McKechnie and Wolf also have shown that in cool acclimated doves, modest increases in cutaneous evaporation were accompanied by a fourfold increase in REWL. In contrast, heat-acclimated doves showed only a twofold increase in REWL, according to McKechnie and Wolf. Despite the fact that CEWL and REWL were partitioned very differently among the treatments followed by McKechnie and Wolf, the rates of total evaporation among the different experimental groups were the same. These differences, which are associated to acclimation in the partitioning of evaporative losses between the skin and the respiratory tract, were accompanied by differences in body temperature (T_b). Body temperature (T_b) in cool-acclimated doves at T_a=45°C was, according to McKechnie and Wolf, higher than in heat-acclimated doves, even though TEWL rates were found to be nearly the same. Despite these increases in T_b, cool-acclimated doves had difficultly compensating for their lower CEWL through increases in REWL (McKechnie and Wolf, 2004) – in other words, evaporation from skin and respiratory surfaces may have different cooling effects per gram of water evaporated. Therefore, the data obtained by McKechnie and Wolf suggest that increases in CEWL in response to short-term thermal acclimation have important thermoregulatory consequences. Based on their observations, "rapid acclimation to changing thermal environments is a critical factor in maintaining thermoregulatory competence at high environmental temperatures, and is also important for water conservation" (Fig. 5.2). The study conducted by McKechnie and Wolf suggests therefore that both the amount of water lost by means of evaporation and the route by which evaporative water is lost have important effects on thermoregulation at high T_a.

Data on TEWL partitioning in Mourning Doves (Hoffman and Walsberg, 1999), in the four species of larks studied by Tieleman and Williams (2002) and in the Western White-winged Doves (McKechnie and Wolf, 2004) support the general view that the patterns of TEWL partitioning observed in

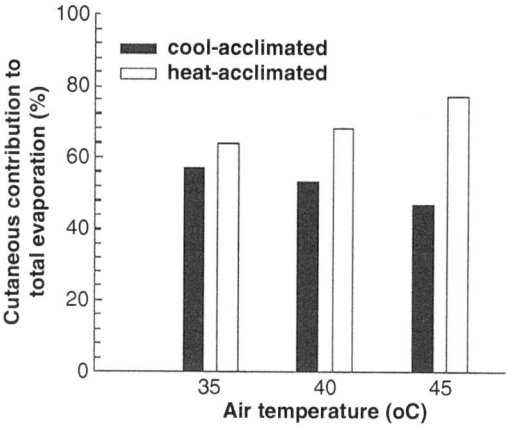

Fig. 5.2 The contribution of cutaneous evaporation to total evaporative water loss at three experimental air temperatures in cool- and heat acclimated Western White-winged Doves *Zenaida asiatica mearnsii*. (Redrawn from McKechnie and Wolf 2004).

birds show considerable taxonomic variation (Wolf and Walsberg, 1996). Columbiformes and Passeriformes, in particular, show pronounced differences in the relative importance of CEWL (Wolf and Walsberg, 1996). In addition, the data obtained by McKechnie and Wolf (2004) suggest that there is considerable variation in the phenotypic plasticity of these patterns. These authors also argue that the ability of Z. a. *mearnsii* to tolerate high T_a and also high heat loads by radiation, while foraging in mid-summer, is at least partially dependent on thermal acclimatization which occurs earlier in the season.

5.1.2 Arid Environment and Phenotypic Plasticity in Birds

Pigliucci (2001) proposed that phenotypic plasticity is an important component of the ecology of organisms and that there are associated energetic costs for the maintenance and production of plastic structures. Intra-individual phenotypic flexibility (Piersma and Drent 2003) is expected to be large in species from environments where the ecological situations change in the course of an individual's life. Hence, the maintenance of inter-individual variation appears dependent upon the frequency of environmental change and on the spatial or temporal nature of environmental heterogeneity. Spatial heterogeneity can maintain genetic variation among populations, while temporal heterogeneity favors phenotypic plasticity, both potentially resulting in phenotypic variation (Hedrick 1986; Schlichting and Pigliucci 1998).

To further explore this important question, Cavieres and Sabat (2008) examined the variability of BMR and TEWL among populations of Rufous-collared

Sparrows (*Zonotrichia capensis*) to assess how much of the observed physiological differences may be explained by phenotypic plasticity. They argue that the variation at the intra-specific level (i.e., physiological flexibility) would be correlated with the magnitude of the seasonal variation in temperature and rainfall, but not with the aridity or predictability of habitats *per se*. They also argue that in the studies carried out by Tieleman and colleagues the species compared belonged to the same family but different genera, and therefore it is possible that the response to acclimation may be influenced by phylogenetic origin. In addition, they argue that the data obtained by Tieleman and colleagues do not support the suggestion that more arid environments are temporally more heterogeneous and unpredictable than mesic localities. However, at least one of those mesic sites (e.g., Northern Cape, South Africa) can be considered periodically heterogeneous, because it exhibits large seasonal climatic fluctuations (Williams *et al.* 1997).

Cavieres and Sabat (2008) tested the hypothesis of a functional correlation between ecological flexibility and phenotypic plasticity at an intra-specific level, and predicted that the magnitude of adjustments in BMR and TEWL would depend on the degree of temporal heterogeneity (e.g., climatic seasonality) of the environment. They studied three populations of Rufous-collared Sparrows in Chile, along an aridity gradient (Copiapo: dry year round; Quebrada de la Plata (Santiago): dry season from November to March; Valdivia: rainy year round). These populations experience substantial habitat differences in climatic seasonality, temperature, and rainfall regime. In this species both BMR and TEWL vary between xeric and mesic environments (Sabat *et al.* 2006a). BMR and TEWL in Rufous-collared Sparrows exhibited a positive association with latitude: birds from the semi-arid locality (Copiapo; lower latitude) had BMR values 9.5% lower than those birds from Santiago and 20% lower than those from Valdivia. TEWL was 29% and 62% lower in sparrows from Copiapo than in sparrows from Santiago and Valdivia, respectively. Even though these data may support the general idea that environments with high ambient temperatures, low water availability and possibly low food availability favor phenotypes with lower energetic requirements and a lower rate of evaporative water loss, they may not be a reflection of field metabolic rate (FMR) or daily water loss in nature. Their study also showed that the three populations of Rufous-collared Sparrows had different responses to thermal acclimation, and the magnitude of the plastic response to thermal acclimation was coupled with environmental temporal heterogeneity. BMR and TEWL of sparrows from Copiapo (the semi-arid locality) were not affected by thermal acclimation, but these traits were affected in birds from mesic (Valdivia) and more seasonal environments (Santiago) (Fig. 5.3). Cavieres and Sabat (2008) argue that the semi-arid condition of Copiapo, characterized by high mean ambient temperatures and low water availability, coupled with a lower physical environment variability, has strongly selected for reduced metabolic and water loss rates and thereby eliminated plasticity from these traits in the local sparrow population.

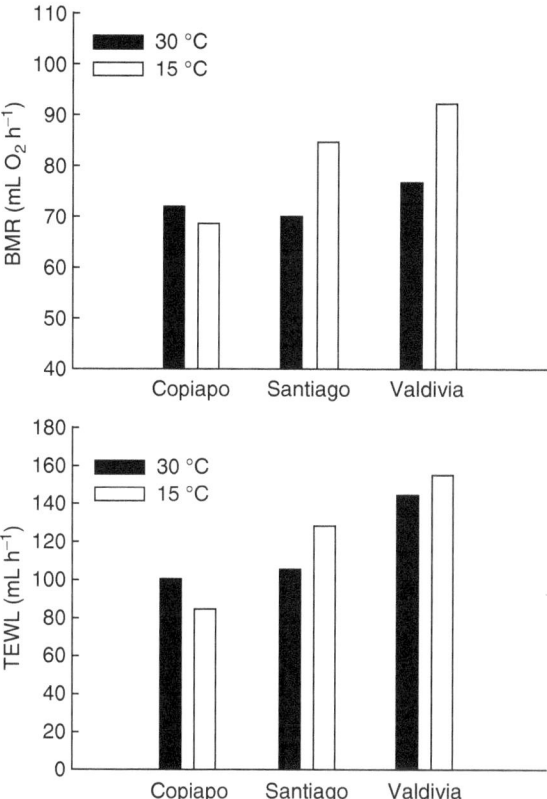

Fig. 5.3 BMR (left) and TEWL (right) of Rufous-collared Sparrows (*Zonotrichia capensis*) from three populations in Chile, acclimated to two contrasting temperatures. (Redrawn from Cavieres and Sabat 2008).

5.1.3 Hyperthermia

A number of authors have suggested that hyperthermia, the elevation of body temperature 2°–4°C above normal, contributes to a reduction in TEWL among birds (Calder and King 1974; Weathers 1981; Dawson and Bartholomew 1968; Withers and Williams 1990). Hypotheses for the potential benefits of hyperthermia have focused on three factors. First, in most ambient temperatures this results in an increased thermal gradient between T_b and T_a, which increases dry heat loss and thereby reduces the need for evaporative cooling (Calder and King 1974). However, the increase in T_b will also elevate CEWL and REWL to some extent. In very hot conditions where $T_a > T_b$, elevated T_b will reduce dry heat gain. Second, heat that is temporarily stored in body tissues during bouts of high T_a could be dissipated by non-evaporative means when T_a becomes more favorable (Schmidt-Nielsen 1964; Dawson and Bartholomew 1968; Calder and King

1974; Dawson and Whittow 2000). And third, Weathers and Schoenbaechler (1976) found that, for some species, T_b increased in the thermoneutral zone (TNZ) while metabolism remained constant. They reasoned that the absence of a Q_{10} effect would reduce evaporative water loss because ventilation rates and metabolic heat production would be lower. Weathers (1981) used the three factors outlined above to estimate that Pyrrhuloxia (*Cardinalis sinautus*) reduce their TEWL by 50% as a result of 2.3°C increase in T_b at a T_a of 28°C.

Tieleman and Williams (1999) did a comprehensive survey of the literature and found that few studies have focused explicitly on the role of hyperthermia in the water economy of birds. According to them, most interpretations of hyperthermia suggest a positive effect on water economy, without considering various subtle and complex features of hyperthermia. Some of these may enhance water loss, instead of reducing it. For example, when birds have an elevated T_b, exhaled air temperature will be higher than at normothermic T_b, and therefore the exhaled air will contain more water vapor, assuming that air in the respiratory tract is fully saturated with water vapor (Schmidt-Nielsen *et al.* 1970; Withers and Williams 1990). When birds become hyperthermic, they start panting. The combination of higher water vapor density and increased ventilation results in increased respiratory water losses, i.e., the opposite of the suggested advantages of hyperthermia. Assessing the benefits and costs of hyperthermia depends on interpretation of whether the "goal" of hyperthermia is to reduce water loss, or to tolerate higher ambient temperatures.

In their review, Tieleman and Williams (1999) partitioned the effects of hyperthermia on water loss into categories of an improved thermal gradient, of heat storage, and of altered respiratory variables. They generated equations describing the effect of each category on water loss. The first two categories tend to reduce TEWL; the latter augments REWL. Their calculations indicate that smaller birds that are hyperthermic for periods up to at least 5 hours save water as a net result of hyperthermia, whereas larger birds only save water as a result of hyperthermia when the hyperthermic bout is short. Large birds increase their TEWL when they are hyperthermic over longer periods. Based on these results, Tieleman and Williams (1999) hypothesize that large birds should maintain their T_b at or near normothermic levels at high T_a's (Fig. 5.4). In support of this hypothesis, Ostriches (100 kg), maintained a normothermic T_b (= 39.3°C) when exposed to a T_a of 51°C for 7.5 hours (Crawford and Schmidt-Nielsen 1967). Also in support of this hypothesis, dehydrated Australian Emus (40 kg) exposed to a T_a of 45°C became slightly hyperthermic (0.4°C increase in T_b to 38.7°C). However, heat storage was maintained for only about 30 minutes until the onset of panting stabilized T_b (Maloney and Dawson 1998).

5.1.4 Cloacal Evaporation

Birds possess three anatomically distinct epithelia from which evaporation can occur: the mouth and pharynx (buccopharyngeal), the skin (cutaneous), and the

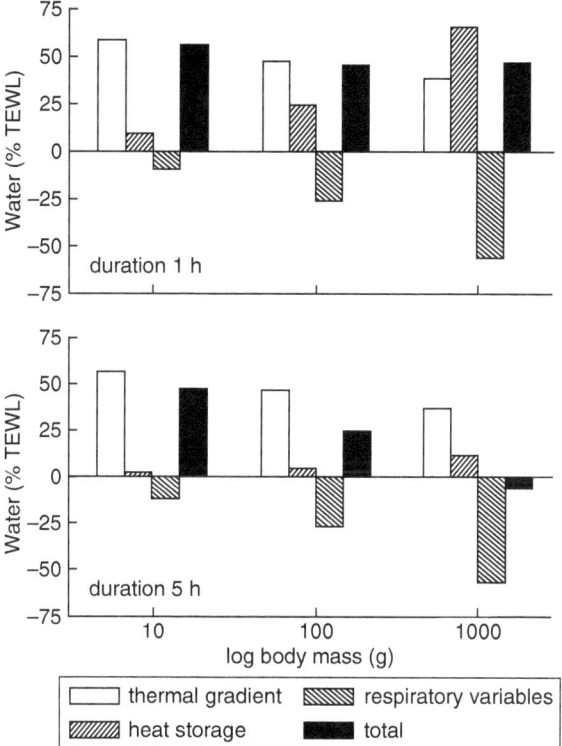

Fig. 5.4 The relative contributions of improved thermal gradient, heat storage, and increased REWL to the total effect of hyperthermia on the water balance of birds of various sizes. Duration of the hyperthermic state is 1 h (*top*). Duration of the hyperthermic state is 5 h (*bottom*). (Redrawn from Tieleman and Williams 1999).

cloaca. Despite the fact that birds do not possess sweat glands, several bird species have been shown to exhibit rates of non-buccopharyngeal evaporation that rival or exceed buccopharyngeal rates of animals with sweat glands (reviewed by Hoffman *et al.* 2007).

Columbiform species withstand high ambient temperatures without panting (Arieli *et al.*, 1988; Marder and Arieli, 1988; Ophir *et al.*, 2002) and show some of the highest non-buccopharyngeal rates of evaporation for any bird (Hoffman and Walsberg, 1999; Marder and Ben-Asher, 1983; McKechnie and Wolf, 2004). Hoffman and Walsberg (1999) demonstrated that Mourning Doves, *Zenaida macroura* Linnaeus, are able to make rapid adjustments to the rate of non-buccopharyngeal evaporation in a response to an experimental suppression of evaporation from the mouth.

Hoffman *et al.* (2007), using more refined techniques, investigated the suppression of buccopharyngeal evaporation in a different Columbiform, the Inca Dove (*Columbina inca*). They were able to quantify the partitioning of the non-buccopharyngeal evaporation into its cutaneous and clocal components. They compared the Inca Dove with a gallinaceous bird, the Eurasian Quail (*Coturnix coturnix*). Both species are widely distributed, occurring in arid and semi-arid habitats, but they represent distinct taxonomic affiliations (different orders). Cloacal evaporation in doves was negligible at T_a's of 30°, 35° and 40°C. However, at 42°C total evaporation in doves was 53.4% cutaneous, 25.4% buccopharyngeal and 21.2% cloacal, with cloacal evaporation shedding 150 mW of heat. In contrast, evaporative partitioning in quail at 32°C (the highest T_a tolerated by this species) was 58.2% cutaneous, 35.4% buccopharyngeal and 6.4% cloacal. The data obtained by Hoffman and colleagues suggest that, for some birds, cloacal evaporation can be controlled and could serve as an important emergency strategy for thermoregulation at high ambient temperatures (Fig. 5.5).

5.1.5 Energy Expenditure and Water Influx in the Field

Laboratory studies provide insights into potentially important physiological mechanisms that enable birds to live in deserts and arid regions, but these findings only achieve ecological and evolutionary meaning if patterns correlate with attributes of organisms in their natural environment (Williams and Tieleman

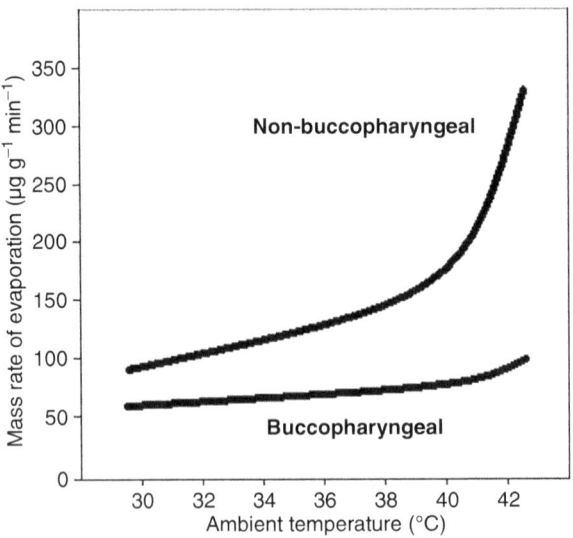

Fig. 5.5 Non-buccopahryngeal and buccoparyngeal rates of evaporation measured in Inca Doves (*Columbina inca*)at different ambient temperatures. (Redrawn from Hoffman *et al.* 2007).

2002). Since the advent of the doubly labeled water method (see Chapter 8) it has been possible to measure FMR and water influx rates (WIR) of free-living birds with reasonable accuracy (Lifson and McClintock 1966; Nagy 1980; Williams and Nagy 1984; Williams 1985; Speakman 1997).

Nagy *et al.* (1999) showed that FMR in desert birds is 48% lower than in non-desert forms. This trend was confirmed by Tieleman and Williams (2000) using both conventional analysis and regressions based on phylogenetic independent contrasts. Factors that might lead to a lower FMR in desert birds include a lower BMR, although the relationship between BMR and FMR is still not completely settled (Ricklefs *et al.* 1996), less energy devoted to thermoregulation or less time spent in energy demanding activities (Fig. 5.6).

Patterns of WIR in desert birds compared to other birds are less clear (Williams and Tieleman 2002). Least-square regression analysis and step-wise multiple regression analysis using phylogenetic independent contrasts led to different conclusions. The first suggested that WIR for desert species are 59% lower than those from mesic habitats (17 desert species; 41 mesic species). However, after phylogenetic correction there were no statistically significant differences. A reduced WIR of desert birds in the field would be consistent with low TEWL rates for desert birds in the laboratory (Williams 1996) and suggests several physiological mechanisms that may have evolved in desert species to reduce both TEWL and WIR.

Extensive FMR data, coupled to field studies of population dynamics, may ultimately help to link individual physiological performance and life-history (Williams and Tieleman 2002). Evolutionary interpretation of life-history variation requires a link between attributes of the individual, such as physiology and behavior, with evolutionary fitness (Ricklefs 2000). As a corollary to the hypothesis that species from arid regions have reduced FMRs, Williams and Tieleman (2002) advocate the idea that increasing aridity results in lower levels of reproductive effort and reduced annual fecundity, important components of life-history. Support for this idea comes from a comparison of clutch size and number of broods per year among larks that live in mesic, semi-arid, and arid habitats. Mesic species have large clutch sizes and raise multiple broods during the breeding season, whereas larks from the desert have small clutch sizes and typically only raise one clutch per year. In addition, during some years when rainfall is very low, larks in the Arabian deserts do not breed (Williams and Tieleman 2002). However, this may or may not be relevant to measured FMRs. To date, most bird FMR data come from breeding individuals because they are easy to recapture, but it is far from clear that this period is necessarily when FMR will be highest in the annual cycle.

5.1.6 Controllable Vascular Thermal Radiator

A recent finding by Tattersal and colleagues (2009) shows that the Toco Toucan (*Ramphastos toco*), the largest member of the toucan family, which possess the

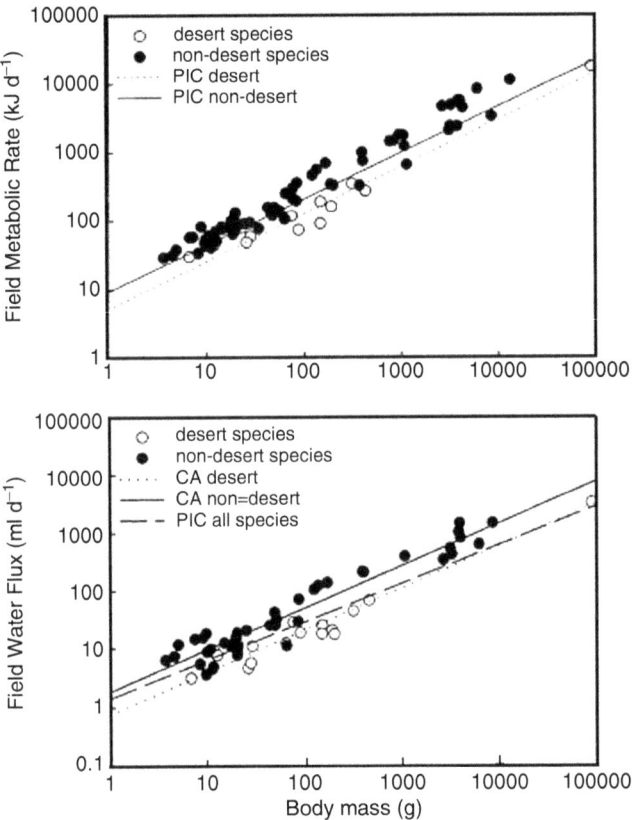

Fig. 5.6 *Top*: Logarithmic plot of field metabolic rate in desert birds (unfilled circles) and nondesert birds (filled circles) versus body mass. The allometric equations obtained by the method of phylogenetically independent contrasts are plotted for desert birds (dotted line) and nondesert species (solid line). *Bottom*: Logarithmic plot of field water flux in birds from deserts (unfilled circles) and nondesert birds (filled circles) versus body mass. The allometric equations generated with conventional least squares regression analysis are plotted for desert (dotted line) and nondesert species (solid line). The equation for all birds obtained with the method of phylogenetically independent contrasts is represented by the dashed line. (From Williams and Tieleman 2002; with permission).

largest beak relative to body size of all birds, and therefore with a significant surface area for heat exchange, uses its beak as a controllable thermal radiator. Using an infrared camera, Tattersal and colleagues were able to demonstrate that the Toco Toucan regulates heat distribution by modifying blood flow, using the bill as a transient thermal radiator. During exercise (flying) and under high ambient temperatures, blood flows to the highly vascularized bill, and excess heat is dissipated to the environment. At low ambient temperatures, vasoconstriction is observed, and body heat is retained. Only adult Toco Toucans are able to modulate heat dissipation through the bill since toucan chicks do not posses such a disproportionate bill in relation to body size. This interesting finding may help us understand the distribution, ecology, and behavior of toucans in general. Toco Toucans typically live in semi-arid areas of Brazil, called "Cerrados," where ambient temperatures vary quite substantially, being very high during the day, and dropping steeply during the night. Tattersal and colleagues suggest that the bill of other birds, despite being much smaller relative to body size when compared to the toucan bill, might also be used for heat dissipation.

5.2 Low Temperatures

Low ambient temperatures represent a potential energetic problem for all homeothermic endotherms, because of a substantial increase in heat loss as the temperature gradient between the environment and the interior of the body increases. For birds living at high latitudes, especially in the northern hemisphere where the major land masses reach far into the Arctic, the winter season will form an especially challenging period. Birds basically have two options to cope with this challenge: either they can leave the area, i.e., migrate to a warmer climate during the winter, or they can stay put and cope with the harsh conditions. The latter option is used by a variety of species. These have adaptively evolved a suite of physiological and behavioral mechanisms enabling them to cope with prolonged low ambient temperatures.

Nearly 60 years ago Per Scholander and co-workers produced a series of landmark papers that have formed the background for much of the thermoregulatory research since. Most noteworthy is their introduction of the equation: $MR = C \cdot (T_b - T_a)$ (Scholander *et al.* 1950). In all its simplicity, this equation points to the fact that in order to maintain a stable internal body temperature, the total heat production of the organism (M) must equal the heat loss, which is the product of the thermal conductance of the body (C) and the temperature gradient between the body (T_b) and the environment (T_a). Given a minimum metabolic rate (basal metabolic rate, or BMR), the equation predicts that below a certain Ta entitled "the lower critical temperature", homeotherms will need to increase their MR (metabolic rate) in a linear fashion in order to maintain a stable T_b. The authors also introduced the concept of the "thermoneutral zone". This model of homeothermic response to cold exposure has dominated the research on homeothermic thermoregulation ever since.

Because of a high surface to volume ratio in small bird species compared to larger species, the potential heat loss from the body is very high in small species. Also, small birds can carry less feather insulation than larger birds. Hence, low ambient temperature is especially challenging for small birds, forcing them to have high intrinsic levels of metabolic heat production, which in turn challenges their ability to maintain a high food intake rate. Alternatively, birds may evolve physiological solutions to decrease the energy expenditure. Because of the high rate of heat loss from small endotherms, there has been a tendency to develop higher body masses in species living at high latitudes. This is known as "Bergman's Rule", and has recently been supported for birds by Olson *et al.* (2009), who found that "species living at high latitudes and in cooler climates tend to be larger-bodied than their relatives living at lower latitudes or in warmer climates".

5.2.1 Metabolic Heat Production

For all homeothermic endotherms, the prerequisite for maintaining a high internal body temperature in cold environments is a metabolic machinery producing the heat needed to compensate for the inevitable heat loss to the environment. Birds potentially have four options for generating enough heat to keep warm in the cold. These are: (i) shivering; (ii) non-shivering thermogenesis (NST); (iii) digestion; and (iv) activity.

5.2.2 Adjustments to Basal Metabolic Rate

All biochemical transformations of the catabolic and anabolic processes in the body release heat, which in a resting bird within the thermoneutral zone amounts to the basal metabolic rate. Shivering thermogenesis will commence when ambient temperature drops below the lower critical temperature, at which point BMR is not enough to keep T_b stable with changes in thermal conductance alone. Shivering in birds consists, as for mammals, of the heat production accompanying involuntary isotonic trembling of skeletal muscles. Shivering will mainly be confined to the larger aerobic, fatigue-resistant skeletal muscles and only a smaller proportion of the muscle fibers may actually be recruited during shivering (Hohtola 2004). This latter may probably be the reason why shivering thermogenesis is able to increase total heat production by roughly five times the level of BMR. In contrast, muscular activity during locomotor movements such as running or flight generates a metabolic power output of up to 15–20 times BMR (Hinds *et al.* 1993). However, bursts of activity may last for only a few minutes (e.g., if escaping from a predator) to several hours (e.g., during migration), but continuous shivering may be necessary for very long periods. Many small northern passerines such as titmice (*Paridae*) and the Goldcrest (*Regulus regulus*), which are resident in the northern temperate zones, will spend most of the winter season at temperatures well below their thermoneutral zone, and consequently must engage in shivering for long durations, perhaps many days. The partial recruitment of subsets of the combined mass of aerobic muscles fibers at any one time may conceivably be a clue to this ability.

It has generally been assumed that BMR is higher in bird species living in cold areas at high latitudes. This also the case for intraspecific comparisons; e.g., Broggi *et al.* (2005) found the winter BMR of the Great Tit (*Parus major*) to be higher for birds living in the north of Scandinavia (Oulu, Finland, 65°N) compared to birds from southern Scandinavia (Lund, Sweden, 55°N). The authors also found, by transporting eggs from Lund to Oulu, raising the chicks in the north and later measuring their BMR during the autumn, that these birds will have a much higher BMR than birds originating from eggs laid in the north (Fig. 5.7). These results demonstrate that in addition to northern Great Tits having a higher BMR than their southern counterparts, birds from the two populations display different metabolic adjustments to winter conditions. The low winter BMR of northern birds, compared to that of southern birds raised in the north, indicate that high latitude metabolic adaptation in Great Tits involves mechanisms that decreases the metabolic needs of the birds (Broggi *et al.* 2005). Such mechanisms could be better thermal insulation of the plumage or lower body temperature.

Results such as these have nonetheless lent support to the general view that a high BMR is beneficial for birds from colder climates. A similar positive relationship between latitude and daily energy expenditure (DEE) has also been described for birds (Anderson and Jetz 2005). Since residual BMR and residual DEE are expected to be positively correlated because a high DEE requires high-capacity and therefore "expensive" basic metabolic "machinery" (Hammond and Diamond 1997), it is difficult to determine whether there has been selection on BMR or on DEE in high latitude species. Moreover, a positive relationship between BMR

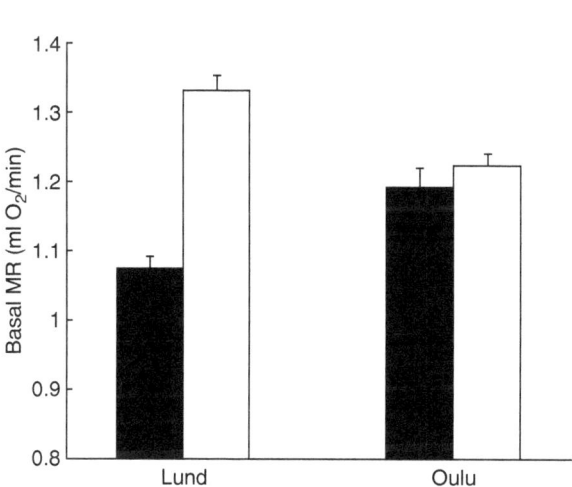

Fig. 5.7 Basal metabolic rate of Great Tits (*Parus major*) from Lund, Sweden, raised and measured in Oulu, Finland (white bars) or Great Tits caught and measured at the respective localities (dark bars). (From Broggi *et al.* 2005).

and latitude might also be caused by an adaptive reduction of BMR in tropical birds (e.g., Wiersma *et al.* 2007).

5.2.3 Non-Shivering Thermogenesis

Non-shivering thermogenesis (NST) is heat production without muscle contraction. During NST in mammals the proton-gradient created over the inner mitochondrial membrane is not used for ATP-production. Instead the protons are diverted back into the mitochondrial matrix through a unique membrane protein, the uncoupling protein (UCP). In this way, the driving force for the electron-transport-chain, viz. the use of protons, will still be present and will continuously drive metabolic heat production, but the production of ATP is decoupled. The UCP is normally in a closed state but can be activated during cold-exposure. When activated, the UCP will open and protons will pass down the concentration gradient without driving phosphorylation of ADP to ATP.

Whether birds have the ability to produce heat by NST has been debated for decades (e.g., Carey 1993). The classical NST, as observed in many mammalian species, is confined to brown adipose tissue (BAT), which contains large amounts of mitochondria expressing a special type of UCP called UCP1. UCP1 seems to be closely associated with NST in mammals, and has not been found in birds (Hughes and Criscuolo 2008; Mezentseva *et al.* 2008). Hence birds do not possess non-shivering thermogenesis, at least of the type found in mammals. Also, true BAT has never been demonstrated in birds. Apparently, birds and reptiles have evolutionarily lost their capacity to produce BAT, and with that the capacity to produce UCP1 (Mezentseva *et al.* 2008). However, birds have uncoupling proteins, but they are of a homologue type, called avUCP (Hughes and Criscuolo 2008). The exact thermoregulatory use of avUCP is still not fully known, although some evidence points to a role in a type of non-shivering thermogenesis different from that found in mammals (Barré *et al.* 1986; Duchamp *et al.* 1999; Vianna *et al.* 2001; Talbot *et al.* 2004). Another mechanism for NST in birds has been suggested to involve a release and re-uptake of Ca^{2+} from the sarcoplasmic reticulum, driving a "futile" pumping of Ca^{2+} which consumes ATP and stimulates mitochondrial respiration (and hence heat production) without producing contractile cycles (Dumonteil *et al.* 1995; Duchamp *et al.* 1999; Bicudo *et al.* 2002).

5.2.4 Heat from Digestion

The heat produced as a by-product of the food-intake process and the subsequent digestive and absorptive processing of foods, collectively known as the heat increment of feeding (HIF; or specific dynamic action, SDA) is also a potential source of heat which could be utilized by birds during cold exposure. HIF is an unavoidable effect of digestion and stems mainly from the postabsorptive biochemical transformation of food. It has long been debated whether HIF may substitute for shivering at low ambient temperatures (Secor 2009). More than 100 years ago, Max Rubner observed that the digestion-related increase in MR at low ambient temperatures was less than at higher ambient temperatures (Rubner 1902), indicating that organisms could make thermoregulatory use of HIF. There have only been a few studies testing the "substitution hypothesis" of Rubner for avian species, and the results have been mixed (as for mammals), with the majority showing either total or partial substitution (Secor 2009). There is so far no consensus as to why these interspecific differences exist.

The Tawny Owl (*Strix aluco*), which is a nonmigratory nocturnal species inhabiting the northern temperate zone of Eurasia, has two feeding peaks per day, one at dusk an one at dawn. The HIF, which lasts for 8–9 hours after a typical meal, completely substitutes for shivering when the owls feed at low ambient temperatures. By utilizing their twice daily feeding activity, they may be able to substitute HIF for considerable shivering activity during most of the 24-hour daily cycle. The energy saved by this substitution has been calculated to be up to 10% of DEE, corresponding to almost 60% of daily thermoregulatory cost (Bech and Præsteng 2004). If, as in the Tawny Owl, HIF can substitute for shivering thermogenesis during cold exposure, birds could get a maximum energetic effect from HIF by "selecting" an appropriate daily food intake routine. In birds having a crop (such as pigeons) food may be stored during daytime and digested at night, thereby maximizing the thermoregulatory effect of the HIF (Rashotte *et al.* 1999).

5.2.5 Heat Derived from Activity

During activity (flight, running, hopping etc.) muscle contractions unavoidably produce heat. This heat is not generated specifically for the purpose of thermoregulation. However, as in the case of the HIF, the "heat increment of activity" (HIA) may also substitute for shivering during cold exposure, and may be a significant contributor to winter energetics in small passerine birds. Experimental evidence shows that the heat produced as a by-product of activity may partly offset shivering requirements below the thermoneutral zone in some small passerines (Paladino and King 1984; Webster and Weathers 1990).

However, excess activity during severely cold conditions is generally regarded as unfavorable in terms of thermoregulation, as it increases the need for food during a period when food is often scarce. Consequently, many small passerine birds in the northern hemisphere apparently reduce their flying activity during severe cold spells in the winter (Haftorn 1988). However, some activity for foraging is necessary. Because heat production during activity (15–20 times BMR) is much greater than during shivering (4–5 times BMR; Hinds *et al.* 1993), it is tempting to speculate that during severe cold stress, it could be advantageous for small birds to be active rather than just perching and shivering.

5.2.6 Thermal Conductance

To conform to the Scholander equation, birds may not only alter heat production, but may also change thermal conductance. The insulatory property of the
plumage is of vital importance in this respect, as it acts as a thermal buffer between
the bird and its cold surroundings. Ptiloerection, which is erection of body feathers to increase the layer of still air in the plumage, is universally used by birds
during exposure to cold ambient temperatures. During sleep, birds may also bury
their head into their plumage and cover their bare feet with feathers, further
decreasing total body thermal conductance.

Of the anatomical adaptations serving to reduce heat loss in the cold, the use
of counter-current heat exchangers is perhaps the best known. The exposure of
extremities to very cold and highly conductive surroundings, such as the feet of a
duck standing on ice, constitutes a potentially immense heat sink from the body
core. To meet this challenge, ducks and probably most other birds, have evolved
an arrangement of the blood vessels in the upper parts of their legs in which
warm arterial blood transfers its heat to the cold venous blood returning to the
body. This arrangement "traps" heat in the central body regions and enables birds
to keep the skin temperatures of their extremities much lower than core body
temperatures without substantial heat loss. The vascular heat exchange in cold
adapted bird species are either in the form of a true *rete mirabile*, (a network of
many smaller arteries and veins, each arising from a larger vessel), or it may be in
the form of the less efficient *venae comitantes* system (with only a few veins which
run along a single artery) (Fig. 5.8). In both type of exchangers, the flux of heat
is from the arteries to the veins.

Birds may at times, e.g., during short flights, produce an excessive amount of
heat, which needs to be dissipated to the surroundings to prevent overheating of
the body. When this occurs at cold ambient temperatures, the counter-current
heat exchangers may be maladaptive. Consequently, in addition to the counter-
current heat exchangers, and arterial and venous vessels which bypass the
counter-current heat exchanger have evolved. These enable arterial blood to pass
the counter-current area without being cooled before reaching the periphery. In
this way, thermal conductance is effectively regulated by active adjustments of
blood flow through the shunt vessels (Fig. 5.8).

5.2.7 Hibernation and Torpor

The capacity to enter hibernation (i.e., to spend a considerable time inactive and
with a depressed metabolic rate) is a well-known strategy used by many mammalian species. However, only a single avian species is known to enter a state
resembling hibernation in mammals. This is the Common Poorwill (*Phalaenoptilus
nuttallii*), a ca. 50 gm North-American nocturnal species. The ability of Poorwills

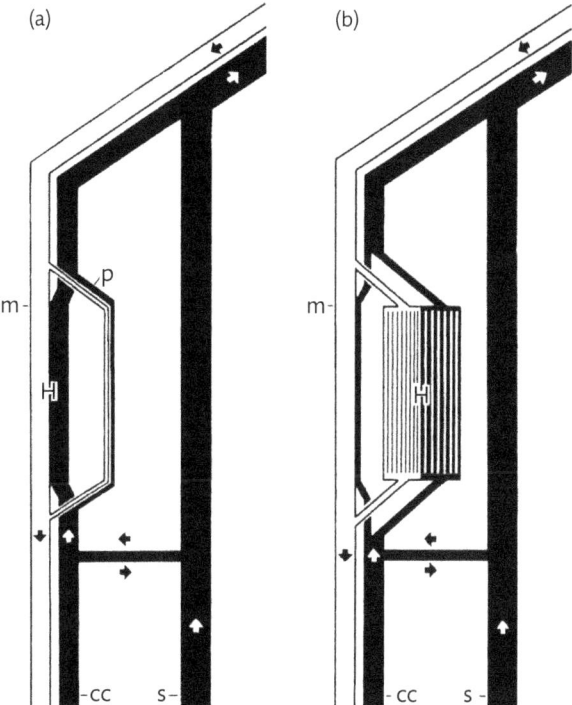

Fig. 5.8 Blood vessels from the legs of birds showing a *venea comitantes* system (A) or a *rete mirabile* system (B). H: Heat exchange system; cc: deep vein return; m: main artery; p: peroneal artery and vein; s: superficial vein. (From Midtgård 1981).

to enter hibernation for a prolonged time was first described anecdotally more than half a century ago (Jaeger 1948, 1949). Subsequent studies have confirmed their hibernation ability (French 1993; Brigham *et al.* 2006). The Poorwill may be inactive for up to several weeks at a time, but will regularly increase body temperature to northermic levels by using the sun as an external heat source. If the birds do not experience sun exposure, they will become normothermic at approximately five day intervals using endogenous heat production to warm up. Hence, this suggests that "periodic arousals may be common to both birds and mammals, but that birds rely more on an exogenous heat source to facilitate re-warming and reduce arousal costs" (Brigham *et al.* 2006).

Torpor, as well as nightly hypothermia, is widely used as an adaptive response to survive unfavorable conditions during cold winter conditions (Reinertsen 1996). In small northern passerine birds, winter body temperature typically drops to between 31°C and 36°C during the night, depending on the prevailing ambient temperatures. In hummingbirds, nighttime body temperature may drop to much lower values, with some declining to as low as 4.3°C (McKechnie and Lovegrove 2002). Many species from different taxa display nightly hypothermia, varying greatly in minimal attained body temperature, which ranges from 4.3 to 38°C. There seems to be no sharp functional distinction between those birds entering only mild hypothermia and those entering deep hypothermia or torpor. The metabolic rate during hypothermia will consequently also vary, ranging between 4.4% and 96% of BMR. Hence the energy saved by entering nightly hypothermia or torpor will vary tremendously between species (McKechnie and Lovegrove 2002).

It has generally been assumed that nightly torpor is used by birds (and mammals) only in situations of energy stress, i.e., when the energy reserves at dusk are not sufficient to keep a high body temperature throughout the night. The physiological trigger responsible for initiating torpor is hence believed to be low energy reserves (Reinertsen 1996). This may not always be the case. Rufous Hummingbirds (*Selasphorus rufus*) may spent the night in torpor even when they are apparently not energy stressed and begin the night with high body mass and apparently sufficient lipid reserves (Carpenter and Hixon 1988). Rufous Hummingbirds migrate through the western part of North America between their breeding grounds in the north (as far north as Alaska) and their wintering areas in Mexico. They will stop for several days on route, presumably to refuel, before continuing their migration. During these stop-overs, lasting for approximately eight days, the birds will initially have a slow mass gain per 24 hours, but near the end of the stop-over the overnight mass loss will decrease, "suggesting that nocturnal torpor facilitated lipid deposition" (Carpenter *et al.* 1993). Although body temperatures were not measured in this study, the changes in overnight mass loss strongly suggest torpor use prior to migration (Fig. 5.9). Torpor in individual Rufous Hummingbirds with high body mass and lipid reserves was observed by Carpenter and Hixon (1988). These results clearly demonstrate that the birds regularly enter nightly torpor even with substantial body fat, probably with the purpose of spending less energy during the night in order to conserve energy reserves in anticipation of continued migration.

It is interesting that mild hypothermia is used by the much larger Barnacle Goose (*Branta leucopsis*) during migration from its Arctic breeding grounds to warmer overwintering sites (Butler and Woakes 2001). These large birds (approximately 2 kg) were equipped with implanted data loggers after the breeding season on Spitsbergen, and the data retrieved later showed that the geese decreased their body temperature before they embarked on their autumn migration south, and that the decrease continued during their migration along the Norwegian coast. Body temperature dropped 4.4°C on average, which will decrease the DEE by 34–39% (Butler and Woakes 2001). The geese benefit from extra fat deposited during the pre-migratory period and by a longer potential migratory distance without feeding. These data, together with that from the Rufous Hummingbird, suggest that torpor and hypothermia in birds is not solely used as

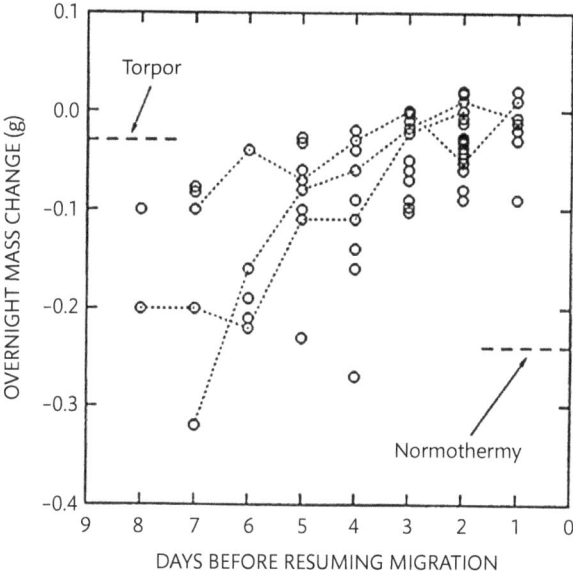

Fig. 5.9 Overnight losses of body mass in individual Rufous Hummingbirds (*Selasphorus rufus*) at a stop-over place in Sierra Nevada during the southward summer migration. The horizontal dashed lines represent the theoretical mass loss during torpor (*left*) and normothermy (*right*). (From Carpenter *et al.* 1993, with permission from the Ecological Society of America).

a response to a compromised energy balance, but may be used by well-fed birds in anticipation of such situations. The physiological cues triggering hypothermia and torpor in these conditions are so far unknown.

5.2.8 Behavioral Adaptations

One well-known behavioral response to seasonal cold conditions is to migrate to a warmer place. However, resident birds that stay behind and survive a cold winter will do so in part by means of behavioral adaptations. The most obvious behavior is to select a microclimate creating the least thermal stress. Hence, small northern titmice carefully choose their nightly roosting place, which most importantly are sheltered from wind. Wind increases convective cooling from the surface of the feathers in addition to penetrating into the feathers and disrupting their insulative property (Walsberg 1986). Somewhat counter intuitively, Walsberg (1986) found that local heating of the roost site by metabolic heat was of minor importance for the heat balance of small birds.

For many species, selection of an appropriate roost-site also involves communal roosting. Communal roosting occurs in some small northern passerine birds, such as the Long-tiled Tit (*Aegithalos caudatus*), Winter Wren (*Troglodytes troglodytes*), and kinglets (*Regulus sp.*) (Haftorn 1988; Heinrich 2003). The advantage is creation of a microclimate which is less thermally stressful. On the other hand, titmice and many other resident northern species do not make use of communal roosting. Haftorn (1988) speculated if this could be related to an increased predation risk.

The reduced foraging activity in some northern titmice during cold periods, mentioned above, is also a behavioral response, although the primary aim may be to economize on the available energy.

5.2.9 Adaptations to Low Food Supply

Birds that live permanently at high latitudes, especially small and medium-sized passerines, will experience a winter-period which often has very low food supply. This is a result of both less food available in the environment and less time available to collect food (due to decreased periods of daylight in high-latitude winters). The solution for many of these species has been to store food during the summer and autumn for later use during the winter. Well-known examples of this behavior are found among the *Paridae* (tits and chickadees) and the Corvidae (nutcrackers, jays, crows). As part of his classic studies on European tits, Svein Haftorn calculated that each individual tit may store as many as 50,000 to 80,000 spruce seeds before the winter (Haftorn 1959).

The North American Gray Jay (*Perisoreus canadensis*) also stores large numbers of small food items during the autumn, when these items (mainly small insects and seeds) are plentiful. Unlike many other food-storing birds, which store food items untreated, the Gray Jay uses saliva to create small balls of food, which are stored on trees. The saliva makes the balls adhesive and helps keep them in place on twigs and small branches (Dow 1965). There is also the possibility that the saliva may help to preserve the food.

5.2.10 Coping with Global Warming?

The earth is warming up at an alarming rate (IPCC 2007). The effect of global warming (natural or anthropogenic) on mean ambient temperature might seem trivial compared to the seasonal variations which many birds will experience in their natural environments. Nevertheless, it is timely to ask whether birds living in cold climates or other cold-adapted birds will be able to genetically adapt to an increase in global temperature. This is especially important to consider because polar regions (especially in the northern hemisphere) will experience the greatest increase in surface ambient temperature, with projected mean annual temperatures as much as 7–8°C higher at the end of the present century compared to the present day (IPCC 2007). Whether organisms can evolutionarily adapt to changing climates depends on: (i) the speed of the climatic changes; and (ii) the genetic ability of organisms to change. A general trend for changes in the biology of organisms, including birds, as an effect of climate change (mainly phenological and population changes), has already been demonstrated (Parmesan and Yohe 2003). For example, a population decline along the southern edge of the Gray Jay's distribution has been linked to warmer autumns. The apparent problem is that stored food supplies will not endure throughout the winter, since this species "relies on natural refrigeration to preserve its hoards" (Waite and Strickland 2006).

Birds (and all other organisms) have experienced changing environmental conditions on the earth throughout their evolutionary histories, and species have either been able to adapt to such changes or have gone extinct. Generally, we know very little about the ability of birds (or any other living organisms) to undergo the evolutionary adaptive changes that may be necessary in response to globally increasing temperatures (Berteaux *et al.* 2004). For birds living at high latitudes experiencing the largest temperature increase, the question is whether the genetic variability of fitness-related physiological traits, combined with appropriate heritabilities, are large enough to permit whatever adaptive evolution of physiological traits becomes necessary in response to changes in environmental conditions.

We know little about the heritability of metabolic traits in birds. As an example, even though BMR is one of the most widely studied metabolic traits in birds, we have very little information about its genetic foundation. As one of the most important physiological measures with respect to cold adaptation, both in its own right and as a proxy of the metabolic scope, information on the genetic basis of BMR could prove important with respect to understanding the adaptability of birds to climate changes. Two recent studies have found significant heritability of BMR in small passerine birds (Rønning *et al.* 2007; Nilsson *et al.* 2009), which indicate that BMR indeed is highly heritable, although the two studies differ with respect to whether BMR can change rapidly independent of body mass.

5.3 High Altitude

The high altitudes experienced by birds that reside or breed in high mountains, or by migrating birds that must cross such ranges present important challenges. Many birds are high-elevation specialists, and a large number of migrants fly at substantial altitude. For example, Bouverot (1985) lists 21 species representing 10 orders of birds that nest between 4,000–6,500 m.

5.3.1 Constraints at High Altitude

The systematic change in O_2 partial pressure is, from a respiratory and metabolic perspective, foremost among the physical changes that occur across elevational gradients. Within the altitude range tolerable for metazoans, air pressure, and hence O_2 partial pressure, decreases by approximately 50% for every 5,500 m of elevation gain above sea level. Hence, some birds spend long periods at elevations with less than half the O_2 partial pressure available to sea level residents. This reduction in O_2 partial pressure is, to some extent, offset by an increase in the gaseous diffusion coefficient, which varies in inverse proportion to total pressure (Reid *et al.* 1987). The high aerobic demands for flight thus lie at odds with reduced O_2 availability, reduced air density, and hence reduced lift, at high altitude (Altshuler and Dudley 2006).

Several other factors that influence avian physiology also change systematically with elevation, such as solar radiation, air temperature, and absolute humidity, the last deriving directly from reduced ambient temperatures. The environmental lapse rate in air temperature through adiabatic cooling and increase in water content is about 0.65°C 100 m^{-1}, and remains linearly so within the troposphere, so that in otherwise similar conditions, air at 4,000 m is approximately 26°C cooler than air at sea level.

Reduced temperatures at high elevations may be important for resident species, particularly for endotherms when they are not active and generating substantial metabolic heat. Two of the most general ecological principles known as Bergmann's and Allen's rules would predict that, at high elevations, body size would be greater and limb lengths would be smaller. However, changes in body and wing size will also influence the power requirements for flight, and several conflicting demands should be considered for volant organisms (Altshuler and Dudley 2006).

For migrating birds, one general suggestion in the avian flight literature is that water loss is reduced at higher elevations because of cooler air temperatures (e.g., Torre-Bueno 1978). However, responses to either alpine or high-elevation thermal regimes may be quite complex (Altshuler and Dudley 2006). Desiccation may result from the reduced water content in cool air at high elevation, especially when coupled with the increased ventilation generally required in hypoxia. It is important to emphasize however that altitude *per se* does not influence water vapor content, but temperature does. Heat loss through convection will decrease in hypobaria, in approximate proportion to air density raised to -1/3 power. Increased metabolic power for flight required by lower air densities yields increased heat production, and the reduced thickness of the atmosphere results in increased solar radiation (particularly potentially harmful UV radiation), although variable cloud cover may alter the situation.

Flight cost analysis is a complex endeavor. Certainly a low air density generates a higher input power for hovering flight. Chai and Dudley (1995) investigated the ability of hummingbirds to hover in thin air. With air densities decreasing from 1.2 kg m^{-1} (normal air at sea level) to 0.54 kg m^{-1} (6,000 m), O_2 consumption of Ruby-throated Hummingbirds (Archilochus colubris) increased from 48.5 to 61.5 mL O_2 g^{-1} h^{-1}.

For forward flight the reduced lift caused by low air densities may be countered by reduced drag, thus allowing birds to maintain higher flight speeds for the same power input. However, how these variables balance in terms of either cost per unit distance or flight metabolic rate is still unknown. Overall, the net outcome of such varied effects is impossible to predict without quantitative knowledge of energy balance during flight (Altshuler and Dudley 2006).

Air density systematically declines with elevation in proportion to the concomitant change in total pressure. Sea-level density of air at 20°C is about 1.21 kg m^{-3}, decreasing to 0.95 kg m^{-3} at 2,000 m elevation and 0.74 kg m^{-3} at 4,000 m, the latter being a 40% reduction relative to sea-level value. By contrast, temperature-dependent variation in air density and viscosity is small, as is that associated with changes in relative humidity (see Denny 1993). Because aerodynamic forces typically vary in linear proportion to air density, morphological and kinematic compensation is necessary to permit flight at different elevations (Altshuler and Dudley 2006).

From an ecological perspective, it is interesting to note that burst performance, instead of sustained performance, mediates competitive ability at high elevation in hummingbirds. Behavioral analysis revealed that short-winged Rufous Hummingbird males are dominant over long-winged Broad-tailed Hummingbird (*Selasphorus platycercus*) males at low elevations, but these roles are reversed at higher elevations (Fig 5.10). A minimum value for burst power may be required for successful competition, but other maneuverability features gain importance when all competitors have sufficient muscle power, as occurs at low elevations (Altshuler 2006).

Behavioral changes in wing and body kinematics occur in birds transiting across high elevations, whereas species-level adaptation to residence at different elevations likely involves concerted changes both in wing morphology and in wing-beat kinematics (Altshuler and Dudley 2006). Changes in air density also alter the mechanical power requirements for flight. The cost of supporting body weight (i.e., the induced power requirements) increase at lower air density, whereas profile drag (on the wings) and parasite drag (on the body), together with their associated power expenditures, will concomitantly decrease (Pennycuick 1975; Norberg 1990; Rayner 1990; Videler 2006).

Another feature of aerodynamic significance is the general trend of increasing wind speed with altitude. For resident avian taxa, high wind speeds may influence numerous behaviors including foraging, sexual displays, nest defense, and roosting (Fisher et al. 2004), as well as increases in thermoregulatory costs under cold conditions. Migrants may be impeded or aided by high wind speeds if the wind is against or in the direction of flight, respectively (Green et al. 2004). Therefore, the overall aerodynamic consequences of flight at high elevation are context-specific and likely depend on both taxon and the particular flight behavior in question (Altshuler and Dudley 2006).

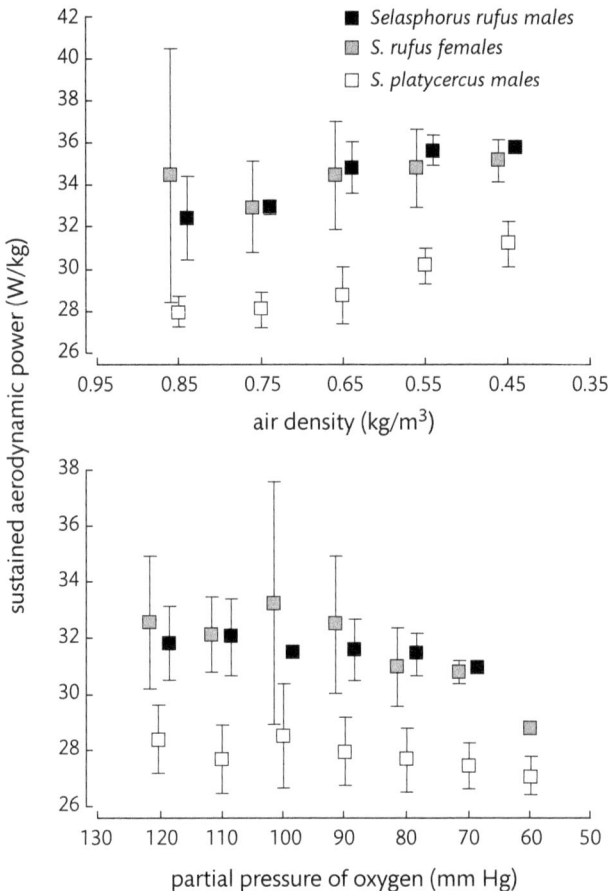

Fig. 5.10 Power requirements of sustained hovering flight did not vary among *Selasphorus* hummingbirds with experimental changes in air density (*top*) and oxygen concentration (*bottom*). Squares represent mean values of aerodynamic power output, with bars representing SE (standard error) about the mean. (From Altshuler 2006, with permission from The University of Chicago Press).

5.3.2 Respiratory Adaptations

Birds exhibit numerous adaptations for enhanced O_2 uptake and delivery from the pulmonary system to the circulating blood and then to the muscle fibers. They are also highly tolerant of hypoxia to levels that are deleterious to most mammals. The anatomical and physiological features of avian respiratory pathways have been reviewed extensively, with several authors focusing on specific adaptations for high-altitude residence and performance (Fedde 1990; Bernstein 1991; Faraci 1991; Maina 2000).

It is well established that gas exchange in the avian pulmonary and circulatory systems is particularly efficient because of a number of adaptations that clearly distinguish birds from their mammalian counterparts (see, e.g., Schmidt-Nielsen 1997). The unidirectional flow of air, with multiple exchanges per inspiration, the convoluted and tubular arrangement of the gas exchange components, as opposed to the spherical and tidally-ventilated alveoli of mammals, and the high O_2 affinity of avian hemoglobin are of particular importance. The result of these adaptations is that the blood leaving the gas exchange interface can have nearly the same O_2 partial pressure as inspired air, indicating that O_2 delivery in birds is not limited by the pulmonary system (Fedde 1990). Nevertheless, Scott and Milsom (2007) found that Bar-headed Geese (*Anser indicus*) had both enhanced ventilation and increased O_2 loading when compared to low-altitude Greylag Geese (*Anser anser*; a close evolutionary relative), under exposure to step decreases in inspired O_2 (Fig. 5.11). Bar-headed Geese are famous for migrating over the highest peaks of the Himalayas, where they are sometimes noted by human mountaineers struggling to ascend on foot. The Himalayan mountains are a formidable barrier to avian migration, and many species that migrate between the northern and southern sides of the range fly around or employ longer routes through riverine valleys (Javed *et al.* 2000; Swan 1961, 1970). Scott and Milsom suggest that these two physiological features of Bar-headed Geese were not caused by neutral evolution, but are related to the species' exceptional hypoxia tolerance. The extreme altitudes achieved by migrating Bar-Headed Geese (well above 8,000 m), strongly suggests that the enhanced effective ventilation and O_2 loading are adaptive.

Other physiological and structural adjustments for O_2 delivery have been observed in birds inhabiting high elevations. Some of these can arise through acclimation or conditioning whereas others are constrained phylogenetically. Changes in muscle ultra-structure have been demonstrated both within and among species and across elevations (Altshuler and Dudley 2006). Muscle capillary-per-fiber number is higher for birds at high elevations in both highly aerobic pectoral muscles and less aerobic leg muscles (Hepple *et al.* 1998; Mathieu-Costello *et al.* 1998). O_2 affinities of avian hemoglobins change in response to experimentally-controlled barometric pressure (Tucker 1968a), but high-elevation taxa, such as the Bar-headed Goose, possess hemoglobin with higher baseline affinity for O_2 (Black and Tenney 1980). These geese also increase O_2 flux to the mitochondria as a result of physical conditioning (Saunders and Fedde 1991). The basic O_2 delivery system of birds functions well across a broad range of O_2 partial pressures and exhibits considerable adaptive plasticity when acutely exposed to deep hypoxia (Shams and Scheid 1993). Altshuler and Dudley (2006) hypothesize that the ultimate explanation for such respiratory flexibility is exposure to varying O_2 partial pressures over geological time (Graham *et al.* 1995). Because flight is one of the most power-demanding types of locomotion, birds need a highly effective O_2 transport system, and that basic requirement may have pre-adapted them for toleration of deep hypoxia (Altshuler and Dudley 2006).

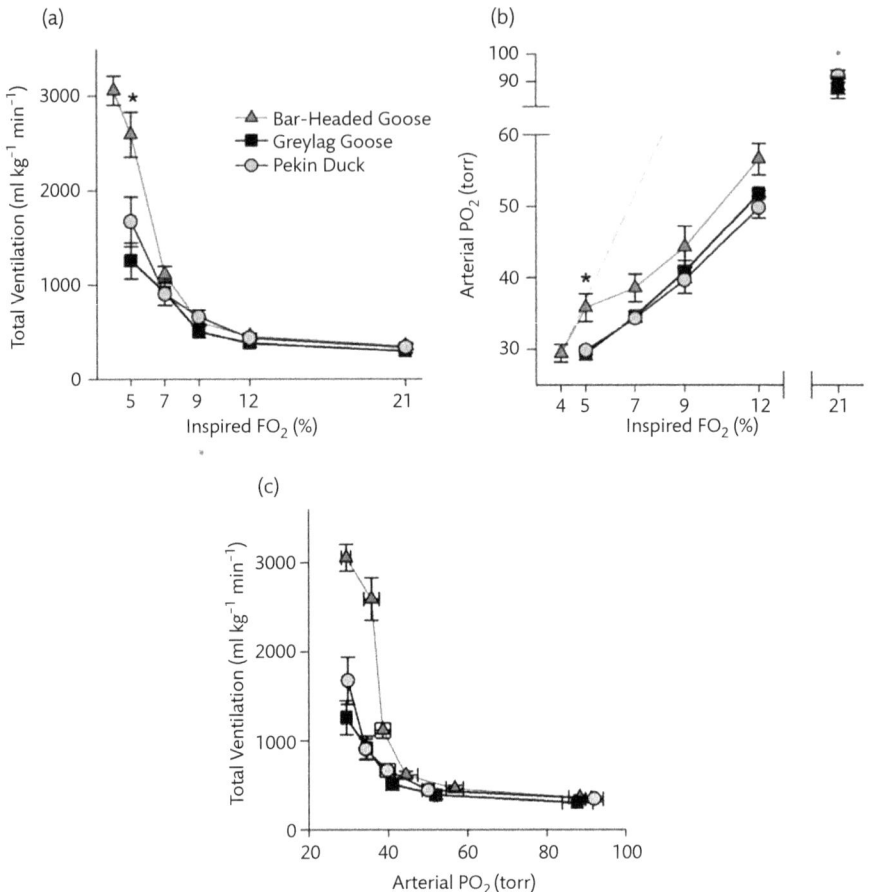

Fig. 5.11 Total ventilation (A) was higher in Bar-headed Geese during severe poikilocapnic (uncontrolled CO_2) hypoxia than in both Greylag Geese and Pekin Ducks. This resulted in higher arterial O_2 tensions (PO_2; B) during reduced inspired O_2 fraction (FO_2). The dashed line in B represents the PO_2 of inspired air. The response of total ventilation to reductions in arterial PO_2 (C) was greater in Bar-headed Geese. Data are means $\pm$ SE. For each species, total ventilation increased significantly and arterial PO_2 decreased significantly after each step reduction of inspired FO_2 ($P \pm 0.05$). *$P \pm 0.05$, significant difference between Bar-headed Geese and both low-altitude species. (From Scott and Milsom 2007; with permission).

5.3.3 Water Loss and Thermoregulation

Migrants also face other altitude-dependent physiological challenges associated with body temperature regulation and the need to minimize evaporative water loss. These processes have been studied with theoretical models, in wind tunnels, and in a limited number of field studies (Altshuler and Dudely 2006). During migratory flights, respiration and active evaporative cooling are the main avenues for water loss (excretory water loss is approximately 10% of total loss, and is largely independent of ambient temperature and humidity; Giladi and Pinshow 1999).

Water loss, via respiration, increases with altitude due to a systematic decrease in absolute humidity, and possibly a requirement for enhanced ventilation at low PO_2. Carmi *et al.* (1992), using a computer-simulation, concluded that respiratory dehydration would ultimately limit flight duration and distance, particularly in birds with sufficient fat stores to fuel their flight. They predicted that birds should fly at low elevations in more humid air to increase flight distance. Although many birds migrate at low altitudes, there have not yet been convincing field studies that demonstrate selection of low altitudes to minimize respiratory water loss (Klaassen 2004).

Birds are most likely to suffer from overheating in direct sunlight and at low elevations (where air temperature is high). Accordingly, migrants that rely on flapping flight have a greater tendency to fly at night than passive gliders (Kerlinger and Moore 1989), and desert migrants will seek shade in the middle of the day to keep their body temperature within tolerable limits. Many migratory flights take place during cloudless nights, thus under conditions where the radiant sky temperature can commonly be 20°C below local air temperature. The sky then acts as a radiative sink, leading objects exposed to it to have a lower surface temperature because less infrared energy is received from the sky than when the radiation environment is isothermic to air.

Léger and Larochelle (2006) estimated radiative heat exchange in the Domestic Pigeon (*Columba livia*) while birds flew in a wind-tunnel in which air and wall temperatures were controlled independently. The main finding of this study was that exposure to a radiative sink comparable to a clear night sky can have significant effects on the thermal budget of a flying bird, particularly at air temperatures warm enough to be considered as limiting for flapping flight. Their results indicate that radiative heat loss, which is much smaller than convective loss during a flight made in a thermally uniform environment (Ward *et al.* 1999), may be the dominant heat dissipation mode for plumage surfaces exposed to a clear night sky. Their data also indicate that exposure to a radiative temperature deficit may also facilitate heat dissipation by reducing the temperature of exposed surfaces (Fig. 5.12). Higher heat loss from these surfaces then offers a bird the possibility to reduce its dependence on evaporative heat loss. Finally, Léger and Larochelle point out that standard isothermal wind tunnels are inappropriate thermal environments to measure the capacity of animals to dissipate heat during nocturnal flight, i.e., under the conditions where many migratory flights take place.

Despite such apparent constraints on thermoregulation and water balance, and the presumptive effect of altitude on these processes, there are few data supporting hypotheses that thermal and water budget processes are regulated through altitude selection (Altshuler and Dudley 2006). Energetic and flight time benefits from wind assistance are likely to motivate migrants to a greater extent (Liechti *et al.* 2000), although obviously this depends on the degree of thermal or water stress (Klaassen and Biebach 2000). Currently, the question of whether birds dehydrate during migratory flights is not well settled. Most laboratory studies have described considerable water loss during flight, and the behavior of many incoming migrants suggest they are dehydrated (e.g., Odum *et al.* 1964). However, other records from incoming migrants at stopover sites suggest these birds have surprisingly high body water content. Landys *et al.* (2000) captured incoming godwits after a three-day nonstop migration flight. When they compared incoming migrants to birds that had already refueled they found no difference in water content as a percentage of body weight.

The strongest evidence for altitude selection related to water balance comes not from migrant birds, but from the nocturnal flights of otherwise diurnally active swifts, sometime called "roosting flights". During these flights, swifts can reach altitudes as high as 3,000m, even though the birds are not flying to move from place to place, and even orient into headwinds to prevent displacement. Instead, flight altitudes are selected according to temperature, with swifts flying at higher altitudes during warmer nights (Backman and Alerstam 2001). Several lines of evidence suggest that environmental changes with altitude can influence temperature regulation, but logistical constraints have prevented coherent tests of these effects using other bird species.

Low ambient temperatures at high elevations are also likely to affect the energy costs of parental care. To test this hypothesis, Weathers *et al.* (2002) used

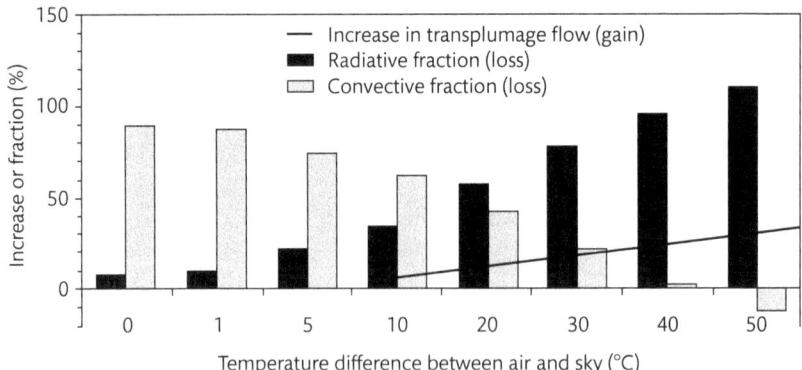

Fig. 5.12 Predicted effects of increasing the temperature difference between air and artificial sky on the thermal budget at the external surface of the insulation plumage in pigeons exposed to an air temperature of 25°C and a wind speed of 10 ms⁻¹. The solid line shows the increase (%) in transplumage heat gain by this surface. Black bars, fractions of the heat loss attributed to radiation; gray bars, fractions attributed to convection, expressed relative to the transplumage gain (solid line; taken as 100%). (Modified from Léger and Larochelle 2006).

the doubly labeled water technique to measure daily energy expenditure during the incubation and nestling feeding stages in two populations of White-crowned Sparrows (*Zonotrichia leucophrys*) – one montane and migratory (*Z. l. oriantha*), the other costal and sedentary (*Z. l. nuttalli*) – that differ in thermal environment and clutch size. They assessed the birds' thermal environment by continuously monitoring operative environmental temperature (Walsberg and Weathers 1986) and wind speed both in the open and within bushes and willow thickets occupied by sparrows. They found that the high elevation populations' daily energy expenditure averaged 24% higher than that of the sea-level population, therefore reflecting both its larger brood size and the colder montane environment.

5.3.4 Wind speeds

As a general trend, wind speed increases with altitude up through the highest elevations where birds have been recorded. This relationship is composed of several interrelated components (Altshuler and Dudley 2006). The planetary boundary layer extends up for approximately 1–2 km above the earth's surface and within this region, wind speeds increase up to free stream velocities with increasing altitude. Global wind speeds are correlated with differences in air temperature across altitudes, and are thus influenced by latitude as well as the time of the year (Stull 2000).

Migrating birds commonly encounter wind speeds ranging from 50–100% of their normal airspeed, and winds are very variable in time and altitude (Altshuler and Dudley 2006). Theoretical analysis also predicts that by taking advantage of wind conditions, birds could double their ground speed and accordingly save as much as half of the energy required for migratory flights (Liechti and Bruderer 1998).

Several studies strongly suggest that birds can minimize the energy costs of migratory flights through wind assistance. One indirect example comes from Western Sandpipers (*Calidris mauri*) during their spring migration, along the Pacific coast of North America, to their breeding grounds in Siberia and Alaska (Iverson *et al.* 1996). Butler *et al.* (1997) compiled several sources of data and calculated that the mass of accumulated fuel measured at stopover sites is insufficient for flight distances without wind assistance. Departures from stopover sites have also been studied in several taxa. In some cases, departure probabilities are strongly correlated with the presence of favorable tailwinds (e.g., Åkesson and Hedenström 2000; Klaassen *et al.* 2004), whereas other taxa only depart when winds are absent or weak regardless of wind direction (e.g., Schaub *et al.* 2004).

Most studies of migrating birds report flights at altitudes within the planetary boundary layer (e.g., Cooper and Ritchie 1995; Klaassen and Biebach 2000;

Klaassen *et al.* 2004), and estimates of mechanical power requirements suggest that ascending to high altitudes can be prohibitively expensive without wind assistance (Pennycuick *et al.* 1996). However, high-altitude flights have occasionally been documented through chance observations by mountaineers and pilots (Stewart 1978), as well as from airplane collisions (Manville 1963; Laybourne 1974). More recently, flight elevations have been recorded using altimeters linked to satellite transmitters mounted on freeflying birds (Weimerskirch *et al.* 2003) and using radar (Bruderer *et al.* 1995; Klaassen and Biebach 2000).

Bruderer *et al.* (1995) have studied migratory flights over the Negev desert in southern Israel. They simultaneously recorded the altitude and wingbeat frequency of individual birds as well as the altitudinal profile of wind speed using radar measurements. They also obtained altitudinal profiles of barometric pressure, temperature, and relative humidity. Of all meteorological variables, Bruderer and colleagues found that only tailwind velocity was significantly correlated with the altitude of migratory flights. Other studies show that some migrants like Bar-headed Geese and the Rüppell's Griffon Vulture (*Gyps rueppellii*), for example, ascend up to 9,000 m to encounter "jet streams" in which they could fly with groundspeeds greater than $45\,\mathrm{m}^{s-1}$ (Liechti and Schaller 1999).

5.4 Marine Environment

About 70% of the earth's surface is covered by the oceans. With an average depth of roughly 3,800 m, the oceans thus comprise most of the space available for life on earth. Because birds are totally dependent on air breathing and the use of the land masses (or, for Emperor Penguins; *Aptenodytes forsteri*, ice shelves) for reproduction, they have never managed to evolve a life-history exclusively confined to the marine environment. Yet, many species of birds make extensive use of the oceans for obtaining food. Some rely on feeding in the oceans to such a degree that they are on land only during breeding periods. These species include some truly oceanic birds such as the tube-nosed birds (Procellariiformes), such as albatrosses, shearwaters, and petrels. These birds, along with some pelecaniform species such as frigate-birds and tropicbirds, fly constantly over the ocean coming ashore only for brief periods during the breeding season. They may be continuously at sea for months or perhaps even years at a time.

The transition zones between the landmasses and the oceans are very productive, and many bird species take advantage of this. Whether birds are truly oceanic or only visit the ocean for short foraging trips, they are exposed to an environment which potentially requires specialized physiological adaptations. In this section we describe some of the physiological mechanisms which adapt seabirds to an oceanic life.

5.4.1 Energetics of Reproduction

Most pelagic seabirds are characterized by a distinctive set of life-history characteristics. These include long lifetime (some albatrosses may live as long as 60 years), small clutch-size (often with only single eggs being laid), long incubation time and slow growth rate of the chicks (Schreiber and Burger 2002). Since seabirds forage over vast areas for prey items that are often very patchily distributed, the feeding rate of the chick may be very irregular. This set of life history characteristics prompted Lack (1968) to advocate that most seabirds are energy constrained and that their unique life-style is an effect of the limited ability of adult seabirds to provide food for the chicks. True oceanic seabirds will simply not be able to raise more than a single chick, because the adult birds are not able to provide food at a rate sufficient to sustain a normal growth rate for more than a one offspring at a time. Only seabirds which are more inshore feeders will be able to raise more than one chick. This "energy-limitation hypothesis", as described by Lack in 1968, has gained much attention and has been a cornerstone for most of the seabird research since then, although direct support for the hypothesis has not been plentiful (Schreiber and Burger 2002).

The often very patchy and widely dispersed distribution of prey in the oceans often requires seabirds to undertake very long feeding trips, and many procellariform birds are renowned for their long-distance flights. During the incubation period large albatrosses may fly nearly 7,500 km during single trips, while more than 4,000 km trips during the chick-guarding period have been recorded (Phalan *et al.* 2007). During the non-breeding season, many seabirds wander over the oceans at even greater distances. A well-known example is the Sooty Shearwater (*Puffinus griseus*), which after the breeding season embarks on a long "figure-eight" flight across the Pacific Ocean, during which they cross the equator twice and may travel an amazing 65 000km during a period of more than 260 days (Shaffer *et al.* 2006).

Because of the long feeding trips, and hence the irregular feeding schedule, many seabird chicks spend considerable time alone in the nest, during which they may be exposed to harsh weather conditions. Since seabirds have established themselves in virtually all suitable habitats on the earth, their chicks may be exposed to both hot as well as cold conditions, depending on the species. Chicks of tropical-breeding seabirds rely on a suite of physiological mechanisms to keep a stable body temperature during the hottest part of the day. This may be especially problematic for chicks of the larger species, such as albatrosses, and to a lesser degree for smaller seabird species (e.g., terns), which find it easier to seek shelter and shade.

Seabird chicks may also enter a state of torpor with decreased body temperature and metabolic rate. This has been described in Fork-tailed Storm Petrels (*Oceanodroma furcata*). Boersma (1986) found that chicks in Newfoundland could reduce body temperature to as low as 25°C.

5.4.2 Thermal Consequences of a Marine Lifestyle

Whether they are diving or swimming on the water surface, seabirds face the possibility of excessive loss of heat because of the much higher thermal capacity of water compared to that of air. Physiological adaptations to a marine lifestyle could include a higher metabolic rate or a lower thermal conductance (higher insulation) in order to counteract the high heat loss. In theory, reduced body temperature could be an evolutionary adaptation in marine birds, but this seems not to be the case, as the resting body temperatures of seabirds do not differ from that of other birds (Prinzinger *et al*. 1991).

Do marine birds have an unusually high metabolic rate? Bennett and Harvey (1987) compiled all the published values of BMR in birds, and tested for correlations with several ecological parameters. One of the few significant correlations they found was a significantly higher BMR in birds with a marine lifestyle. This finding was recently confirmed by McNab (2009). Ellis and Gabrielsen (2002) tested seabird BMR against taxonomy, latitude, ocean regime, season, activity mode, and body mass, and found that latitude was the only parameter which increased the ability of body mass to predict BMR.

Some seabird species have traditionally been considered especially susceptible to heat loss because of a highly wettable plumage with consequently limited insulatory properties. This certainly pertains to the cormorants (Phalacrocoracidae) in which the behavior of wing-drying by basking has been taken as evidence of wettable feathers that must be dried between episodes of diving (Rijka 1968; White *et al*. 2008). An alternative thermoregulatory function for basking has been proposed by Grémillet (1995), who suggested that solar heat gained from wing-spreading would help warm-up the stomach contents after the ingestion of cold fish. However, this explanation has not been supported by more recent studies (e.g., White *et al*. 2008) and does not preclude a drying function for the behavior.

The body plumage of cormorants is very special, in that it consists of an outer part which is "loose" and wettable, and an inner layer which is highly waterproof (Fig. 5.13). Therefore, in spite of their apparently quite wettable feathers, cormorants retain a thermally insulating layer of air deep in their plumage. The unusual structure of cormorant plumage generates low buoyancy – useful to reduce the energy costs of shallow dives – while retaining insulating capacity by means of the

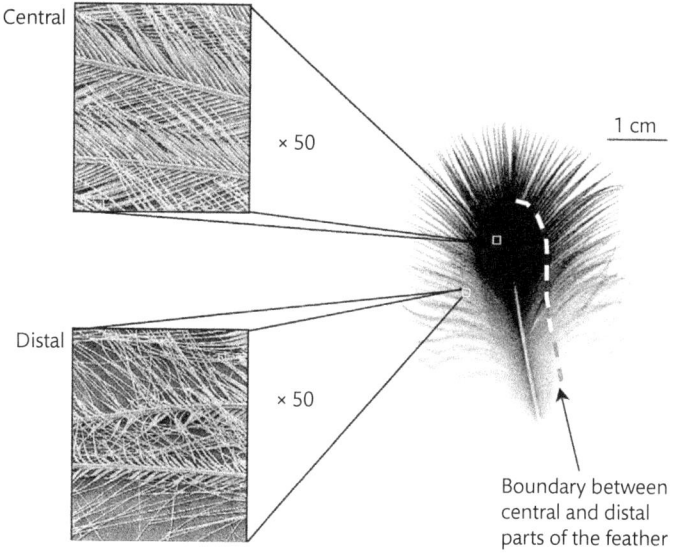

Fig. 5.13 Ventral feather from Great Cormorant (*Phalacrocorax carbo*) showing a regular and waterproof central part and an irregular and wettable distal part. (From Grémillet *et al.* 2005a).

thin inner air layer. In this way, the cormorants are able to have longer and deeper dives (Quintana *et al.* 2007). In the Great Cormorant (*Phalacrocorax carbo*) the partially wettable feathers gradually add water to the plumage during diving, decreasing buoyancy and extending the possible dive time. Retention of water in the plumage at the end of a long (18 minutes) series of dives amount to 6% of the body mass, thereby reducing buoyancy by 18% (Ribak *et al.* 2005b).

Cormorants are so well protected against the cold that some are able to live even in the polar winter in the far north. Populations of the Great Cormorant spend the entire winter in high arctic Greenland (Grémillet *et al.* 2005b). These birds will not experience daylight here during the greater part of the winter (precluding solar basking), and ambient air temperatures will plunge below −30°C while the water temperature is a degree or two below 0°C. For cormorants, the key to surviving such a thermally hostile environment is to increase food intake in order to keep metabolic heat production high (Grémillet *et al.* 2005b). They are helped in this task by readily available prey: cormorants in the high arctic during winter have by far the most efficient prey capture rate recorded in diving seabirds (Fig. 5.14). The need for efficient prey capture is emphasized by the very high energy requirements during diving, with metabolic rates as high as 21 times BMR. A high prey capture rate is a prerequisite for supplying enough food for high and sustained energy requirements, and is made possible in part by a great

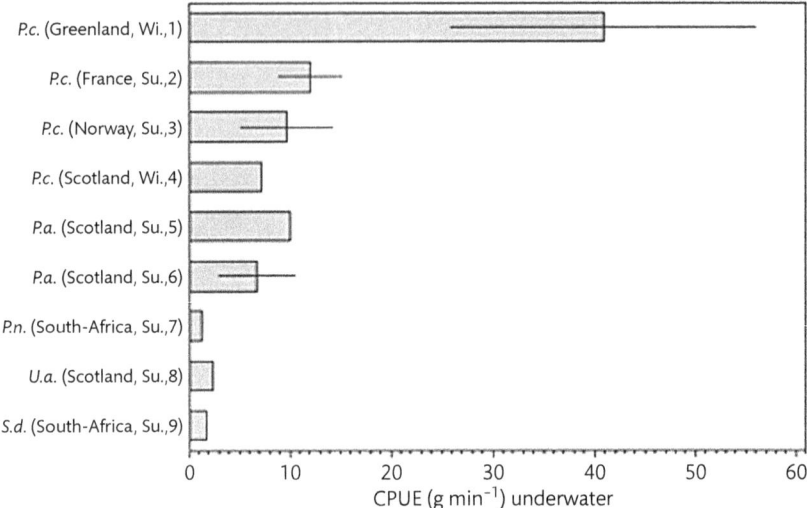

Fig. 5.14 Prey capture rate ("Catch per unit effort") for various seabirds. P.c.: *Phalacrocorax carbo*; P.a.: *P. aristotelis*; P.n.: *P. neglectus*; U.a.: *Uria aalge*; S.d.: *Speniscus demersus*. Wi. and Su. refer to winter and summer respectively. (From Enstipp *et al.* 2007, with permission from Springer Science + Business Media).

plasticity in food choice, enabling the cormorants to switch prey when some species become less abundant (Enstipp *et al.* 2007).

Some apparently harsh weather conditions may not be disadvantageous to seabirds. A species which takes advantage of windy conditions is the Northern Fulmar (*Fulmarus glacialis*), a medium sized procellariiform living in the northern Atlantic. Furness and Bryant (1996) used the doubly labeled water (DLW) method to demonstrate that fulmars spent less energy on a daily basis during windy conditions. DEE varied between 1.4 and 7.8 times BMR with a strong negative correlation between DEE and average wind speed. Flapping flight is presumably much more expensive than gliding flight for any bird species, and the results of Furness and Bryant (1996) strongly suggest that the fulmars make extensive use of windy conditions by changing from a primarily flapping to gliding flight as wind speed increases. In this way they save energy by riding on currents of air.

An even more dramatic dependence on air currents has been described in the Magnificent Frigatebird (*Fregata magnificens*) foraging over tropical waters. Henri Weimerskirch and co-workers equipped frigatebirds in French Guiana with altimeters and satellite transmitters to record the three-dimensional movements of these long-winged seabirds (Weimerskirch *et al.* 2003). They found that the birds could be constantly soaring on up-winds, even though the thermals produced over tropical waters are generally weak compared to thermals over land areas. However, frigatebirds can soar on tropical oceanic upwinds due to their extremely low wing loading (frigatebirds have a larger wing area per unit of body mass than any other bird species (Norberg 2002)). Consequently, frigatebirds are able to soar at very low metabolic costs. Their ability to soar on even the weak thermals that occur during tropical nights enable them to stay in the air during nighttime, a property they share with only one other bird species, the much smaller Common Swift (*Apus apus*; Weimerskirch *et al.* 2003). Low wing loading in combination with high aspect ratios are characteristic of many long-winged seabirds (albatrosses, tropicbirds, frigatebirds etc.), creating a slow and inexpensive flight (Norberg 2002), enabling them to move great distances over open sea at low metabolic costs.

5.4.3 Coping with High Salinity

Seawater is characterized by a high ionic concentration with a total osmolarity in the range of 1,000 mOsm, far above the approximately 300 mOsm in the body fluids of most vertebrates including birds. Therefore, birds drinking seawater incur high solute loads, which in effect act to drive water out of the body. Although it is unclear if seabirds drink seawater, they will unavoidably realize a salt load because of food obtained from the sea. Most fishes defend a hypotonic body osmolarity compared to seawater, so eating fish will not be as problematic as eating marine invertebrates that are usually in osmotic equilibrium with seawater. Nevertheless, the solutes and nitrogenous wastes generated by seabird diets must be disposed of without excessive water loss. To deal with this problem, marine birds excrete the surplus ions as a highly concentrated fluid produced by special "salt glands" located in grooves in the skull dorsal to the orbital openings.

Ornithologists have known about the salt glands of marine birds, or the nasal glands as they usually were called before their true nature was revealed, for a long time. At the beginning of the last century their function was believed to be protecting the nasal mucosa from salt water. However, the true function of the salt gland in seabirds was revealed by Knut Schmidt-Nielsen and co-workers in 1958, when they discovered that Double-Crested Cormorants (*Phalacrocorax auritus*), when experimentally provided with a salt load, would "excrete a highly hypertonic liquid that drips out from the internal nares and collects at the tip of the beak, from which the birds shake the drops with a sudden jerk of the head" (Schmidt-Nielsen *et al.* 1958). Seabirds are especially dependent on salt glands, since the avian kidney generally is less effective in concentrating urine than the mammalian kidney: instead of excreting ion loads in a concentrated urine, seabirds excrete ions in the concentrated salt gland fluid.

Because passerine birds do not possess a salt gland, they are generally absent from the marine habitat. However, a few passerine bird species exploit the marine environment in order to obtain food. Some members of the genus *Cinclodes* (Furnariidae) forage on rocky sea coasts in South-America. These *Cinclodes* species feed on marine invertebrates, which are osmoconformers (e.g., they are in osmotic equilibrium with seawater and hence have a high salt content). For a passerine bird with limited ionic concentration capacity of the kidney, this would pose a substantial risk of osmotic stress. Accordingly, it is reasonable to assume that *Cinclodes* which depend on marine food would have developed effective kidney functions to deal with the osmotic burden. This is indeed the case. Sabat *et al.* (2006b) assessed five *Cinclodes* species with respect to their dependency on marine food using muscle tissue concentration of the carbon isotope $\delta^{13}C$. The concentration of this isotope will be higher in organisms feeding in the marine food web compared to those feeding terrestrially. A comparison of the five species revealed that $\delta^{13}C$ correlated significantly and positively with relative kidney size, with the percentage of the kidney comprised by the medulla, and with the number of medullary cones (Sabat *et al.* 2006b) – all of which are indicators of the kidney's ability to concentrate urine. Hence, those species that rely most heavily on marine food, also have evolved kidneys that are both larger and more effective that those of terrestrial feeding *Cinclodes*. The *Cinclodes* species studied by Sabat and colleagues are an excellent example of physiological phenotypic diversification within a single phylogenetic lineage.

5.5 Diving and Swimming

Birds are widely distributed in aquatic environments. Locomotion and thermoregulation in this medium are constrained primarily by its higher density (3 orders of magnitude), viscosity (3 orders of magnitude), and thermal conductivity (23 times) than air.

Many diving and swimming birds use variants of the same biomechanical systems used for flight. For example, the flight-like swimming of penguins is based on wing motions similar to the wingbeats of flying birds and is sometimes called "aqua-flying". The general pattern consists of generating thrust by muscle-driven cycling of airfoils (or hydrofoils) to generate lift, and adjusting those foils so the lift (or a good portion of it) appears as thrust (see e.g., Vogel 2003).

Aquatic birds maintain a temperature difference with the surrounding water of as much as 40°C, despite the enormously higher thermal conductivity of water compared to air. The feathers of most aquatic birds are well oiled, allowing them to trap and maintain a layer of air around the animal. The low thermal conductivity of this air layer allows the feathers to act as an effective insulator.

Diving activity poses additional constraints because of the need to overcome buoyancy from air spaces (in the feathers and ventilatory system). Deep dives reduce buoyancy problems due to compression of these gas spaces, but because birds power most muscular effort with aerobic metabolism they may face O_2 depletion during long dives.

5.5.1 Surface Swimming versus Sub-surface Swimming

In Tufted Ducks (*Aythya fuligula*) sub-surface swimming is energetically more expensive than surface swimming (Woakes and Butler 1983). Underwater swimming is also energetically costly for cormorants (Schmid *et al.* 1995). Subsurface swimming has been hypothesized to be more expensive in these birds for two reasons. First, although ducks appear to swim effortlessly at the water surface, when diving they must continuously paddle with their feet to overcome buoyancy and stay submerged (Casler 1973: Stephenson *et al.* 1989). Second, in the case of cormorants, although their water-absorbing body plumage (Rijke 1968; Stephenson *et al.* 1989) helps to reduce buoyancy and facilitate underwater prey capture, it also decreases thermal insulation during submersion.

The energetic costs of underwater swimming in cormorants have been studied in detail (Schmid *et al.* 1995). To compare the energetics of surface and underwater activity, and to confirm the previous findings in Tufted Ducks, Ancel *et al.* (2000) determined the metabolic cost of surface swimming in Brandt's Cormorants (*Phalacrocorax penicillatus*). They measured energy requirements while at rest on the water or while swimming on the surface at different speeds. In still water, the birds' mean mass-specific rate of O_2 consumption (VO_2) while floating at the surface was 20.2 mL O_2 min^{-1} kg^{-1}, i.e., 2.1 times the predicted resting metabolic rate. During steady state voluntary swimming against a flow, their VO_2 increased with water speed, reaching 74 mL O_2 min^{-1} kg^{-1} at 1.3 m s^{-1}, which corresponded to an increase in metabolic rate from 11 to 25 W kg^{-1}. The cost of transport decreased with swimming velocity, approaching a minimum of 19 J kg^{-1} m^{-1} for a swimming speed of 1.3 m s^{-1}. Therefore, according to this study, cormorants consume 18% less energy swimming at the water's surface than swimming below the surface, when the cost of transport amounts to 23.2 J kg^{-1} m^{-1} at the same swimming speed, i.e., 1.3 m s^{-1} (Schmid *et al.* 1995). Ancel and colleagues (2000) claim that the difference between the cost of surface and subsurface swimming is due to buoyancy. It is calculated that the average mechanical power output during diving ranges from 1.003 to 1.695 W. Stephenson *et al.* (1989) have reported that that at least 95% of the work performed (forces (N) associated to overcome buoyancy and body drag, during descent, multiplied by traveled distance; i.e., 1.55 m) during a dive of the Lesser Scaup (*Aythya affinis*) is required to overcome buoyancy, and only 3–5% is required to oppose body drag during descent, assuming a constant velocity. Similarly, McPhail and Jones (1998) reported that a 41% reduction in buoyancy of the Lesser Scaup resulted in a 61% reduction in the calculated mechanical power output during the feeding phase of a dive. Within this context, it is important to mention that the cost of overcoming buoyancy is inversely related to depth, as each 10 m of depth adds 1 atmosphere of pressure: gas volume (and hence buoyancy) at 10 m is 50% of that at surface pressure, at 20 m it is 33% that of the surface, at 30 meters it is 25%, and so forth. Thus the cost of overcoming buoyancy decreases rapidly as dive depth increases.

Another factor that may account for differences between surface and sub-surface metabolic rates is wetting of the plumage. According to Ancel and colleagues (2000), only the ventral plumage gets wet while cormorants are swimming at the surface, whereas in diving cormorants the body is completely in contact with water, and the cost of defending body temperature should be greater. Because heat conductivity in water is 23 times greater than that of the air, water contact may increase the thermal conductance of aquatic birds by a factor of 2.2 during swimming and by a factor of 4.8 during diving (De Vries and Van Eerden 1995).

In the swimming behavior termed "burst and glide" (described by Weihs 1974), instead of swimming continuously at a constant speed, the swimmer accelerates to a speed above the desired average speed and then glides (decelerates) until it slows to the initial speed, whereupon it accelerates again to repeat the cycle. The energetic saving from such behavior is calculated as the ratio between the energy required to burst-and-glide and the energy required to swim at the same average speed but at constant speed. Interestingly, burst-and-glide swimming behavior occurs in fish, marine mammals, penguins, water birds, lobsters, plankton and numerous aerial flyers (Haury and Weihs 1976; Hove *et al.* 2001; Lovvorn 2001; Sato *et al.* 2003; Spanier *et al.* 1991; Van Dam *et al.* 2002; Videler and Weihs 1982; Webb and Fairchild 2001; Weihs 2002).

Ribak *et al.* (2005a) explored this behavior in the Great Cormorant (*P.c. sinensis*) a species that feeds by submerged swimming in search and pursuit of fish. Underwater they swim by paddling with both feet simultaneously in a gait that includes long glides between strokes. As mentioned before, at shallow depths cormorants are highly buoyant as a consequence of their aerial lifestyle, and to counter this buoyancy they swim underwater with their body pointed downwards relative to the swimming direction. Ribak and colleagues hypothesized that that this mechanical solution for foraging at shallow depth should increase the cost of swimming by increasing the drag of the birds. Their results indicate that the energy savings from using the burst-and-glide in the paddling cycle is limited to relatively fast swimming speeds (>1.5 m s^{-1}). However, as the birds dive deeper (>1 m, where buoyancy is reduced), the burst-and-glide gait may become beneficial even at lower speeds (Fig. 5.15).

Foot-propelled diving birds can be separated into different groups on the basis of their underwater behavior (Cramp and Simmons 1983; Harper *et al.* 1985). Birds that feed on benthic organisms (plants, mussels, etc.) merely swim under water to a fixed depth, whereas most piscivores must actively pursue their swimming fish prey. An interesting question is whether there are unique features

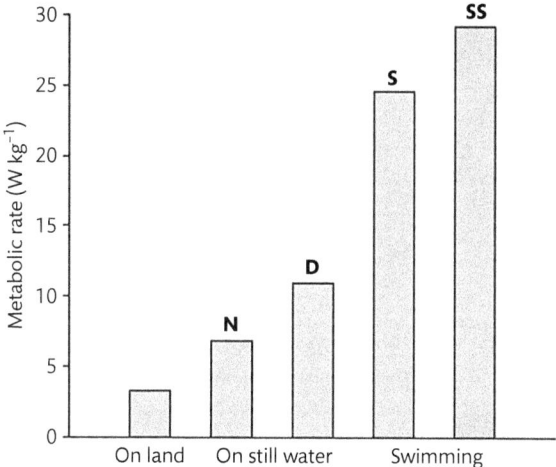

Fig. 5.15 The metabolic rate of Brandt's Cormorants (*Phalacrocorax penicillatus*) under different situations. From left to right: resting on land, resting on still water at 20°C during the night (N), resting on still water at 20°C during the day (D), for a bird swimming at the surface (S) with the minimal cost of transport and for a bird swimming under water (SS). (Modified from Ancel *et al.* 2000).

of the locomotor mechanism of foot-propelled divers that vigorously chase their prey compared with those that take immobile food. Another question is what selection pressures may have acted on fish-eating birds that distinguish them from benthivores. One obvious factor is the need for speed, since chasing fish requires high velocity. Another arguable factor is the need for an energy-efficient propulsion mechanism, since pursuing fish probably requires a higher swimming speed, and thus more energy, than does diving to graze (Johansson and Norberg 2001). Both increased speed and increased efficiency at high speed favor a lift-based oscillating hydrofoil propulsion mechanism, whereas a drag-based paddling mechanism is considered of lower performance (low thrust and low efficiency) and a more primitive form of propulsion than the lift-based mode (see Fish 1996). However, it is important to keep in mind that diving is costly for any bird due to buoyancy, and efficiency ought to be under strong selection in all divers to maximize the time spent at feeding depth and minimize rates of O_2 use.

It has been assumed that the power stroke of foot-propelled divers is drag-based rather than lift-based (Dabelow 1925; Blake 1981; Braun and Reif 1985). The basis for drag-based propulsion is that the propulsive appendages move in a direction opposite to the swimming direction during the power stroke with a speed (relative to the body) greater than the swimming speed (Vogel 1994). In

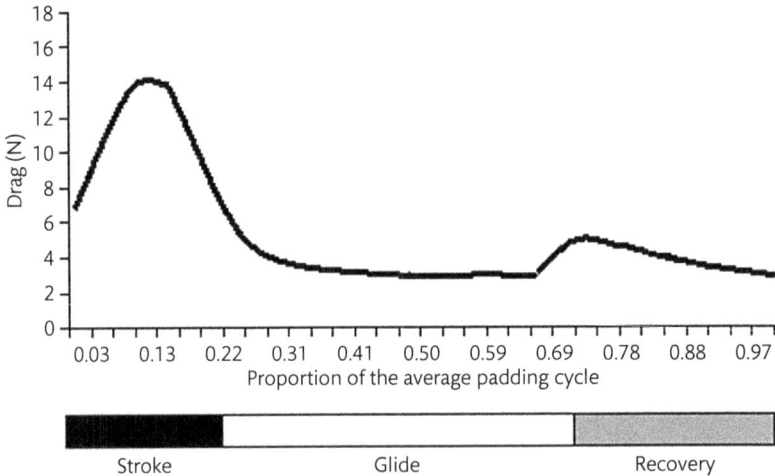

Fig. 5.16 The distribution of the momentary drag along an average paddling cycle of the Great Cormorant (*Phalacrocorax carbo sinensis*). The *x*-axis is the proportion of the cycle duration that is normalized by dividing the period of each stage (stroke, glide and recovery) by the mean duration of that stage. (Modified from Ribak *et al.* 2005).

the Tufted Duck, an example of a drag-based swimmer, the feet move caudally during the power stroke at an average speed of 1.56 times the speed of the bird while swimming at the surface (Woakes and Butler 1983).

Johansson and Norberg (2001) examined the hydrodynamic propulsion mechanism of the diving Great Crested Grebe (*Podiceps cristatus*) a pursuit swimmer, to determine whether the observed propulsion is lift-based. Morphological support for their investigation came from an earlier finding (Johansson and Norberg 2000) that showed that the grebe's toes function as multiple slot hydrofoils during swimming, like the splayed wingtip feathers of hawks and vultures, which increases the efficiency of the power stroke. Their results suggest that the lift-based paddling of grebes considerably increases both maximum swimming speed and energetic efficiency over drag-based propulsion. In addition, the power stroke of the grebe is directed perpendicular to the swimming direction and its toes do not move in a direction opposite to the swimming direction (relative to the water) during the power stroke, hence strongly supporting the hypothesis of lift-based propulsion (Fig. 5.17).

Grebes spend most of their time in or on the water and have almost lost their ability to walk on land (Harrisson and Hollom 1932). The consequence should be that conflicting selection pressures for an ability to both walk and swim would be reduced in favor of selection for the best performance in swimming and diving. The Podicipediformes (grebes) is an ancient lineage of birds with no close living relatives (Sibley and Ahlquist 1990). Their morphological features have

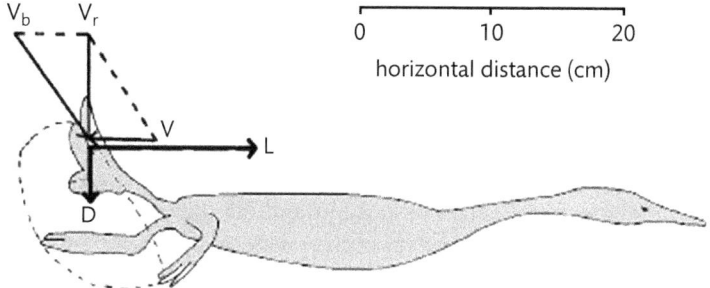

Fig. 5.17 Lateral projection of the foot movements relative to the body. Lift (L) and drag (D) forces and velocities acting on the foot of the Great Crested Grebe (*Podiceps cristatus*) during different phases of the foot stroke. During the power stroke, the toes are spread and the feet move upwards, backwards and inwards, resulting in a forward-directed lift force (L) and an outward and downward directed drag force (D). During the recovery stroke, the toes are folded and moved forwards. At the bottom of the turning phase, the toes begin to move laterally, with the edge of the fourth toe meeting the oncoming water. (Modified from Johansson and Norberg 2001).

remained more or less unchanged since at least the Oligocene (Storer 1960) and maybe even since the end of the Cretaceous (Cooper and Penny 1997), suggesting that grebes have been under stabilizing selection. Lift-based paddling might therefore represent a local evolutionary maximum in foot-propelled swimming (Johansson and Norberg 2001).

Penguins are excellent swimmers that use their flipper-like wings for propulsion (the feet are used for steering, if at all, during swimming). One notable aspect of the swimming behavior of penguins is their ability to leap from the water onto the sea ice. The dependence of penguins on foraging in the marine environment necessitates frequent exits from the ocean onto the sea ice or rocky shores, on which they breed and rest. Because of the presence of predators at the water's edge (leopard seals, orcas, etc.), there is considerable pressure for penguins to exit the sea safely and efficiently. Their ability to attain a given above-water height depends on initial vertical speed when leaving the water, assuming that kinetic energy is converted to gravitational potential energy. In a study of Adélie Penguins (*Pygoscelis adeliae*) exiting the water, Yoda and Ropert-Coudert (2004) concluded that they adjust their take-off angle according to the reflected image of the height of the ice above the water.

Sato *et al.* (2005) hypothesized that Emperor Penguins adjust their swimming speed according to the above-water height of holes. To test their idea, they changed the above-water heights of two ice holes (fast ice off McMurdo Sound, Antarctica) through which Emperor Penguins exited, and by using animal-born recorders, they were able to monitor swim speed, stroke frequency, body angle, and depth. Time of exit and hole choice of penguins were simultaneously monitored by observers on the ice. They also described the behavior of penguins after failed exits onto the sea ice, and determined whether they changed holes after failed exits and/or whether they increased their swimming speed in subsequent attempts.

The data obtained by Sato and colleagues indicate that penguins are able to recognize the above-water height of ice holes and that they do adjust their speeds according to the height required to clear the ice around the hole. Apparently, the penguins chose vertical take-off angles and minimum speeds, particularly for the higher holes (>40 cm), to avoid using excess energy for swimming before leaping (Fig. 5.18). Their study also indicates that leaps by Emperor Penguins were aided by buoyancy and that they can sometimes exit through the ice hole without any stroking effort before the leap. Penguins might know that they do not need to stroke if they reach enough speed. They also increase swim speed for subsequent attempts after failed exits, instead of selecting lower holes.

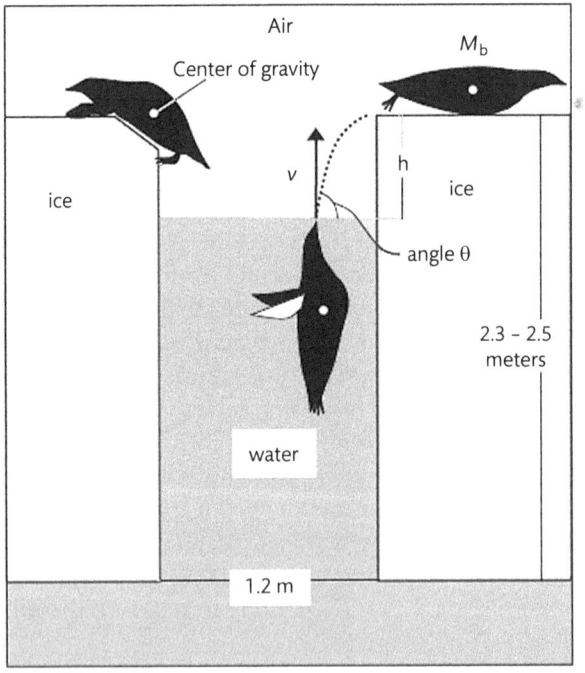

Fig. 5.18 Diagram of exit options for Emperor Penguins (*Aptenodytes forsteri*): (1) the bird is able to climb onto the ice if the center of gravity reaches the height of the ice around the hole (left); or (2) the exit speed is sufficient to project the bird onto the ice (right). (Modified from Sato et al. 2005).

5.5.2 Thermal Exchanges while Swimming

One of the most extreme thermal challenges for endothermic animals such as seabirds is the movement between water and air, because water has a thermal conductivity 23 times that of air. This poses problems for temperature regulation since potential differences in heat loss between water and air differ by orders of magnitude. This situation is particularly extreme for those species that inhabit low-latitude upwelling regions which generally have a warm climate but are characterized by cold water rising to the sea surface owing to offshore winds (Eckman 1905).

Wilson and Grémillet (1996) used remote-sensing logging units (Wilson *et al*. 1992a) to investigate changes in body temperature as a function of activity in breeding Bank Cormorants (*Phalacrocorax neglectus*) and African Penguins (*Spheniscus demersus*) in an upwelling system. Both species are medium sized (Bank Cormorant 1950 g; African penguin, 3,000 g; Cooper 1972 1985) pursuit divers which breed throughout the year on South Africa's west coast (Cooper 1981; Rand 1960). The Benguela upwelling current maintains water temperatures at approximately 13.4°C although air temperatures may reach 35°C, with substantial radiant heat loads from sunlight on clear days. In air, both the cormorants and penguins are likely to be in their thermoneutral zone (or above it). In water the situation is more complex since cormorants and penguins have different thermoregulatory capacities owing to substantial differences in insulation. Cormorants have little plumage air (Wilson *et al*. 1992b) because of a family-specific feather structure which leads to feather wetting (Rijke 1968). The lack of large quantities of trapped air minimizes energy expenditure during swimming by reducing buoyancy, although heat loss is believed to be high (Wilson *et al*. 1992b). Penguins on the other hand have quite robust air layers under their feathers, and often have very little subcutaneous fat. Even Antarctic-breeding Adélie Penguins, (~4 kg, water temperature −1°C) have little or no subcutaneous fat for most of the breeding season. And, while the skin is thick relative to other birds, it will provide little insulation (Personal observation made by of one of us; Mark Chappell).

Wilson and Grémillet (1996) tested the hypothesis that the efficiency (energy gain from prey ingestion per unit time/energy costs per unit time) of foraging birds (see Nagy *et al*. 1984) which experience such dramatic changes in ambient temperatures is dependent on their activity and their capacity to regulate body temperature within limits. They found that body temperature in these endothermic seabirds was environmentally and activity-dependent and varied in the case of the cormorant by over 5°C. Heat flux considerations showed that such flexibility confers considerable energetic advantages: by allowing body temperature to drop when the heat loss to the environment is high, such as in water, birds may save the energy that would normally be necessary to compensate metabolically for this drop. During normal foraging bouts of approximately 50 minutes during which cormorants were observed to allow their body temperature to vary up to 5.2°C they were, according to Wilson and Grémillet (1996), 150% more efficient then they would be if their body temperature were not allowed to vary. However, this number should be taken with caution because it is hard to predict such efficiency given that a low body temperature may also reduce muscle power. Thus, if cormorants did not reduce body temperature, heat transfer to the environment would be higher, but on the other hand they might have caught prey at a higher rate.

In cormorants, a lower body temperature resulting from extended time in the water can subsequently be elevated to higher levels using solar energy when the birds return to land (basking). In penguins, body temperature varied between 38.7°C and 40.0°C, and muscle-generated heat during swimming was supposedly used to re-elevate low body temperature. Continued swimming eventually caused body temperature to rise above normal resting levels. Therefore, metabolic rate could be reduced immediately post-exercise, allowing body temperature to drop to the resting level before any increase in metabolic rate is necessary to maintain it. However, the problem is more complex than it appears, and, up to the present, there is no solid data on the cost of transport during foraging dives for making such predictions. There are some data for shallow swimming in flumes, but that is likely different from deep dives if only from buoyancy considerations. Furthermore, it is not really known how much substitution of exercise heat for thermogenic heat there is in penguins. At present, understanding of how heat generation is regulated, and how heat is transported among tissues during exercise, digestion, thermal challenge, and breath holding, is inadequate for predicting substitution and aerobic efficiencies without direct measurements for conditions of interest (Lovvorn 2007).

Heat transfer as a by-product of digestion (referred to as SDA, or HIF) might be important for maintaining thermal balance in diving birds. SDA is the increase in resting metabolic rate observed after a meal, associated with heat production during the process of digestion, assimilation and nutrient interconversion (Brody 1945). The magnitude of SDA depends on meal size (Janes and Chappell 1995; Kaseloo and Lovvorn 2003; Green *et al.* 2006). Because of differences in intermediary metabolism, high-protein foods tend to elicit a greater SDA than food containing mainly lipid or carbohydrate (Blaxter 1989). Studies, which closely replicated ecologically relevant conditions, carried out by Kaseloo and Lovvorn (2003 2005 2006), using Mallard (*Anas platyrhynchos*) and Lesser Scaup ducks confirmed earlier findings of the presence of SDA in Brünnichs Guillemots (*Uria lomvia*; Hawkins *et al.* 1997) and Little Penguins (*Eudyptula minor*; Green *et al.* 2006) when they ingested food voluntarily while resting in air.

Some cormorant species inhabit thermally challenging environments. A small population of Great Cormorants (*P. c. carbo*) winters in West Greenland near the

Arctic Circle, encountering water temperatures below 0°C and air temperatures as low as −30°C (Grémillet *et al.* 2005a). These birds continue to dive throughout the winter for up to several hours per day (Grémillet *et al.* 2005b). Thermoregulatory costs might therefore account for a substantial part of their overall daily energy budget. Recording O_2 consumption (VO_2), respiratory exchange ratio (RER) and stomach temperature in the laboratory, Enstipp *et al.* (2008) studied Double-crested Cormorants, which are closely related to Great Cormorants, to estimate the magnitude and time course of SDA following a standard meal (100g herring, *Clupea palasi*) at themoneutral conditions (~21°C ambient temperature). They conducted a second set of trials at sub-thermoneutral temperatures (~5.5°C ambient temperature) to test the hypothesis that Double-crested Cormorants use transferred heat by SDA to substitute for shivering thermogenesis. The effect of meal size on SDA was also investigated. Enstipp and colleagues found that following the ingestion of a 100 g herring meal at themoneutral conditions, VO_2 was elevated for an average of 328 min, during which time birds consumed 2,697 mL of O_2 in excess of their resting rates. At sub-thermoneutral conditions the duration and magnitude of VO_2 elevation (228 min and 1,391 mL of O_2 respectively) were both significantly reduced. When meal size was altered, during sub-thermoneutral trials, SDA was significantly greater after larger food intake. Based on these experimental results, Enstipp and colleagues calculated that substitution from SDA might reduce the daily thermoregulatory costs of wintering Double-crested Cormorants by approximately 38%, and they remark that the magnitude of SDA and its potential for thermal substitution should be integrated into bioenergetic models to avoid overestimating energy expenditure in these top predators (Fig. 5.19).

5.5.3 Diving and Foraging Behavior

Penguins are important components of many marine ecosystems, and their large size, dense populations, tolerance of human presence and handling, and nest fidelity permit detailed analysis of diving and foraging, as well as other aspects of reproductive behavior. Feeding takes place underwater and often far from land, but the birds are sufficiently big and robust to carry sophisticated recording devices that provide detailed records of foraging trips. Consequently, we know more about the details of foraging behavior and diving physiology in penguins than for most other birds. Nevertheless, some examples show how, despite these technical advances, many unanswered questions remain about the remarkable ability of these birds to dive repeatedly for long periods.

Adélie Penguins are one of the most abundant penguins, with a population of 2–3 million breeding pairs distributed around the Antarctic continent and neighboring islands (Croxall and Linshman 1987). Foraging energetics is important from an ecological perspective, since food requirements are central to

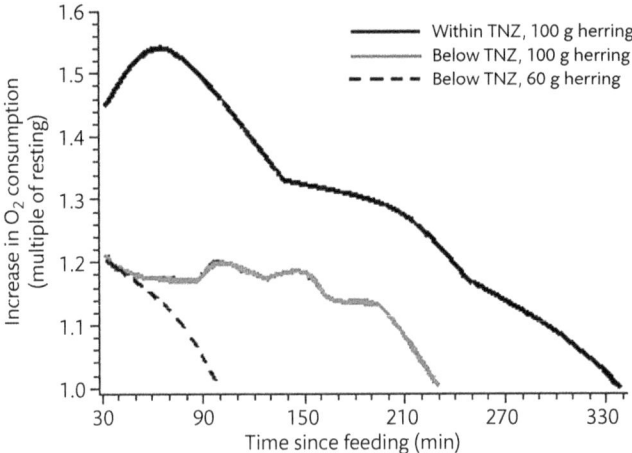

Fig. 5.19 Changes in oxygen consumption rate of Double-crested Cormorants (*Phalacrocorax auritus*) after voluntary ingestion of a single herring (mass 60 or 100g) when resting in air at temperatures within or below their thermoneutral zone (TNZ) (lower critical temperature ~9°C). (Modified from Enstipp *et al.* 2008).

trophic relationships. Chappell *et al.* (1993a) studied the temporal patterns of foraging, the depth distribution of hunting effort (cumulative bottom time), and physiological aspects of diving in Adélie Penguins. They were particularly interested in the relationship between hunting dive depths, duration, surface recovery periods, and time use efficiency (bottom time/[dive duration + surface interval]) during foraging bouts. They also examined the potential time and energy costs of anaerobiosis during repetitive diving. In the course of their investigation, Chappell and colleagues used electronic recorders attached to the center lower back of Adélie Penguins to estimate diving times and depths. Plasma volumes of freshly captured penguins were measured with the Evans Blue dilution method (Linden and Mary 1983) for estimating O_2 stores in the circulatory system. Hematocrit, hemoglobin and myoglobin concentrations, skeletal muscle mass and lung-air sac volumes were also estimated to calculate the total O_2 stores available during dives.

Chappell and colleagues observed that most hunting dives consisted of a rapid descent to depth, a period of bottom time at near-constant depth, and a rapid ascent to the surface. Most hunting activity occurred at depths between 3 and 98 m, with a mean of 26 m. Adélies' overall hunting effort was concentrated between 0500 and 2,100 hrs at depths between 10 and 40 m. Although bottom time decreased slightly with increasing depth, they found that the correlation was weak. On the other hand, dive duration was positively correlated with depth. Maximum dive duration was 160 s, and most dives lasted between 60 and 90 s. Post-dive surface intervals averaged approximately 50% of time duration. Time use efficiency during dive bouts decreased with increasing dive depth. Based on the O_2 stores estimated by Chappell and colleagues, together with measurements of diving metabolic rates (Chappell *et al.* 1993b), the aerobic dive limit of Adélies is 46–68 s, and most hunting dives apparently require some anaerobic metabolism. They conclude that the use of anaerobiosis engenders an energy penalty and probably affects both the behavior and energetics of foraging in these diving birds. Reliance on anaerobiosis will increase energy costs during foraging, and according to Chappell and colleagues it may also influence the birds' time budgeting and selection of foraging depths, since the need for anaerobiosis increases with increasing dive duration. Near Palmer Station, in Antarctica, where their investigation was conducted, most foraging occurs at depths between 10 and 40 m, considerably shallower than the maximum depths attained by Adélies. The restricted depth range probably reflects both depth-related physiological constraints to foraging efficiency and the depth distribution of the Adélies' euphausid prey.

Kooyman *et al.* (1992a) performed a thorough study of diving behavior and energetics during foraging cycles in King Penguins (*Aptendyctes patagonicus*). King Penguins are the second largest of all diving birds and share with their congener, Emperor Penguins, breeding habits strikingly different from other penguins. In their study, penguins were selected from pairs engaged in exchanging incubation and brooding duties at Saint Andrews Bay, South Georgia, and in Crozet Archipelago, at Marin Bay, Possession Island. A depth recorder and a radio were attached to the birds, and once they were released, their presence or absence from the colony was monitored by radio. Kooyman and colleagues also estimated swim speed and metabolic rate (doubly labeled water method; Nagy 1980) while penguins were at sea, as well as their stomach contents upon their return. The average foraging cycle of King Penguins is around 6 days, during which the mass gain is about 2 kg. They pursue their prey day and night, a feature that apparently distinguishes them from other penguin species. However, it seems likely that high-latitude penguins feed at night during the winter, when day length is very short, but studies carried out during that season have not been reported so far. The daytime feeding was confined to the depth range of 100–300 m, and the night feeding confined to the range of 4–20 m. The frequency distribution of dive depth is bimodal, with few dives between 40–100 m. Apparently, the prey species depth distribution rises towards the surface at night and descends during the day. Maximum dive duration time was 7.7 min. Swim speeds when a bird was at sea averaged 2.1 m.s^{-1}; descent and ascent rates of change in depth averaged 0.6 m.s^{-1} for dives less than 60 m deep and 1.4 m.s^{-1} for dives greater than 150 m deep. The distance to the hunting area is 28 km assuming that travel velocity is 2.0 m.s^{-1}. While resting on shore, metabolic rate was 3.3 W kg^{-1}, 1.6 times the predicted standard metabolic rate of birds with the same body mass (SMR; see Lasiewski and Dawson 1967). The average metabolic rate while away from the colony was $10 \, W \, kg^{-1}$, or 4.6 times SMR.

The main prey of King Penguins, by number, are myctophid fish (*Krefftichthys anderssoni* and *Electrona carlsbergi*), and the average energy, based on their stomach contents, was $24.6 \, kJ \, g^{-1}$ dry mass, determined by bomb calorimetry. Time use efficiency (see above) ranged from 0.44 during shallow dives to 0.15 during deep dives. According to Kooyman and colleagues, King Penguins mix deep dives with shallow dives, the later functioning most likely as recovery. Their analysis also shows that the deep dives incur a high risk factor (probability of not catching anything) and investment in energy. Probably, the type and depth distribution of prey or catch rates must have a strong influence on the preference for shallow or deep dives in this species. Like Adélies, King Penguins also face the problem of exceeding their aerobic dive limit (ADL), apparently relying extensively on anaerobic metabolism during deep dives. Muscle morphometry and biochemistry studies performed on Emperor and King Penguins have shown that muscle fiber diameter, pH buffering capacity, and distribution of lactate dehydrogenase isoenzymes indicate that these penguins are poised for prolonged anaerobic work (Baldwin 1988).

These two studies, on Adélie and King Penguins, because of their thoroughness, are illustrative of how the question of anaerobic diving still remains unclear, and significantly corroborate many other studies devoted to this subject in which, by standard estimates, a substantial proportion of dives are too long for the calculated ADL. This problem is highly dependent on the energy costs during underwater versus on the surface swimming, and as it has been pointed out earlier, underwater swimming cost is not yet fully understood. To date, we still have difficulties in understanding how these birds can do such sustained repetitive diving bouts with standard anaerobic mechanisms working on a large proportion of their dives.

5.5.4 Dive Capacity

Emperor Penguins are premier avian divers with routine dive durations of 5–12 min and a reported maximum dive duration of 22 min (Kooyman and Kooyman 1995; Robertson 1995). Such breath-hold capacity is dependent, in large part, on utilization and depletion of the blood O_2 store (Scholander 1940). Near McMurdo Sound, Antarctica, Emperor Penguins primarily feed on the sub-ice fish *Pagothenia borchgrevinki* (Ponganis *et al.* 2000). Dive depths are usually less than 100 m. During these dives, Emperor Penguins undergo variable bradycardias (Kooyman *et al.* 1992b), maintain aortic and vena cava temperatures near 37–39°C (Ponganis *et al.* 2001; Ponganis *et al.* 2003; Ponganis *et al.* 2004), and have an aerobic dive limit (dive duration associated with post-dive blood lactate accumulation) of 5.6 min (Ponganis *et al.* 1997).

Successful application of an air sac O_2 electrode and a backpack recorder to diving Emperor Penguins by Stockard *et al.* (2005) has allowed the recording of O_2 partial pressure (PO_2) profiles. About 42% of the voluntary dives had end-of-dive air sac PO_2 values less than 20 torr (1 torr = 133.3 Pa [0.133 kPa]; for comparison PO_2 in the atmosphere is about 150–160 torr at sea level depending on temperature and humidity). Such low PO_2 values are remarkable in comparison with other birds for several reasons. First, the lowest of these air sac values in Emperor Penguins is less than inspired air values (23 torr) of birds at altitudes as high as ~9,000 m (Black and Tenney 1980). Second, these values in free-diving emperor penguins are also lower than the air sac PO_2 values (~30 torr) of Pekin Ducks (*Anas platyrhynchos*) forcibly submerged to the point of "imminent cardiovascular collapse" (Hudson and Jones 1986). Third, because air sac PO_2 represents the maximum arterial PO_2 and is usually greater than the integrated arterial value (Powell 2000; Weinstein *et al.* 1985), these low air sac values imply that blood PO_2 is commonly less than 20 torr at the conclusion of dives. This is quite remarkable when we consider that blood O_2 content at a PO_2 of 22 torr is already quite low (less than 5 mL O_2 dL^{-1} blood, or about 1/4 that of fully saturated blood) in a high-altitude bird like the Bar-headed Goose and is even lower than the PO_2 values of Pekin Ducks and

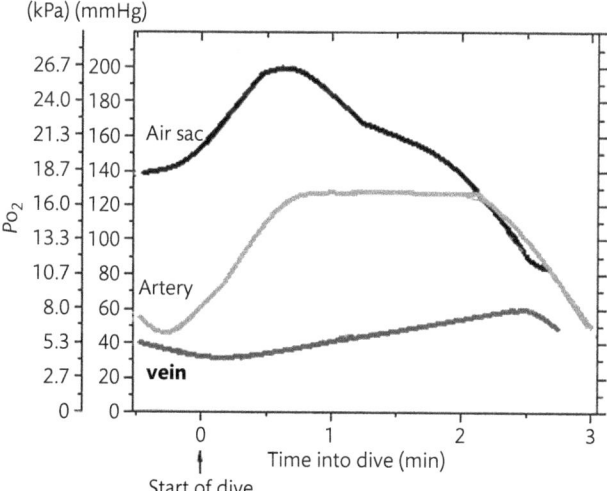

Fig. 5.20 Air sac, arterial and venous PO_2 profiles during shallow (less than 30m), ~3-min dives. PO_2 profiles from the air sac and aorta shows initial compression hyperoxia and then a gradual decline in PO_2 secondary to air sac O_2 depletion and the decrease in ambient pressure during ascent. Vena cava PO_2 slowly increases during the dive. (Modified from Ponganis *et al.* 2007).

pigeons (*Columba livia*) under acute hypoxia (Black and Tenney 1980; Hudson and Jones 1986; Weinstein *et al.* 1985).

In order to further investigate hypoxemic tolerance and blood O_2 depletion, Ponganis *et al.* (2007) equipped Emperor Penguins with intravascular PO_2 and backpack recorders. The intravascular/air sac PO_2 profiles during diving revealed that their remarkable capacity for long, deep dives is at least partially achieved by careful management of the blood/respiratory O_2 stores and an extreme hypoxemic tolerance (Fig. 5.20). A hemoglobin molecule with a notably high O_2 affinity (low P_{50}) in penguins, given that all hemoglobins store four O_2 molecules at most, is essential to enhance blood O_2 content during hypoxemia by increasing the depletion of the respiratory O_2 stores, which in Emperor Penguins constitutes 19% of the total O_2 store (Kooyman and Ponganis 1998). This allows Emperor Penguins to sustain very long dives without jeopardizing proper oxygenation of vital organs like the heart and the brain. An interesting question is how aerobic tissues continue to function at the very low PO_2 needed to extract O_2 from a 20 torr blood PO_2.

6

Adaptations: Neural and Sensory

As we have seen in the previous chapters, birds are able to perform numerous tasks with a high degree of complexity. It is not apparent, in contrast to previous dogma, that the structures associated with the cognitive ability of the bird brain are in many ways comparable to those of their mammalian counterparts.

As with other parts of the body, the brains of distantly related species tend to be derived from the same basic elements found in the common ancestor — they exhibit homology. So although the common ancestor of birds and mammals lived approximately 300 million years ago, studies of extant reptiles have revealed that the reptilian (therapsid and sauropsid) forebrain is pallial (cortical-like) in origin and therefore the common ancestor should also have shared this trait. If so, the forebrain of modern birds and mammals is expected to be pallial as well. This seems to be the case (Emery and Clayton 2005).

Recent discoveries have challenged the classical view of bird brain evolution based on a theory of linear and progressive evolution, where the vertebrate cerebrum was argued to have evolved from lower to higher anatomical stages, with birds in an intermediate stage (i.e., more "primitive" than the mammals). As such, the avian cerebrum was thought to consist mostly of regions associated with basal ganglia, and these were thought to control mostly primitive behaviors. The avian pallial (or cortical-like) domain is organized differently than the mammalian pallium (or cortex) in that the avian is nuclear and the mammalian is layered. First proposed by Karten (1969; 1991), nuclear-to-layered hypotheses propose that similarities in connectivity between the so-called hyperstriatum, neostriatum, and archistriatum of birds and the neocortex of mammals stem from a common origin of these structures, i.e., they are homologous. Karten proposed that the common ancestor of birds, reptiles, and mammals possessed a nuclear pallium that was transformed into a laminar pallium early in the mammalian lineage, maintaining the connectivity of the ancestral nuclear network. Karten argued that the avian pallium is divided into three groups of serially connected neuron types — thalamorecipient neurons (field L2, ectostriatum and basalis), pallio-pallial neurons (neostriatum), and extratelencephalic projection neurons (archistriatum), with cell types and interconnectivity that resemble those of mammalian cortical layers IV, II–III, and V–VI, respectively (Fig. 6.1).

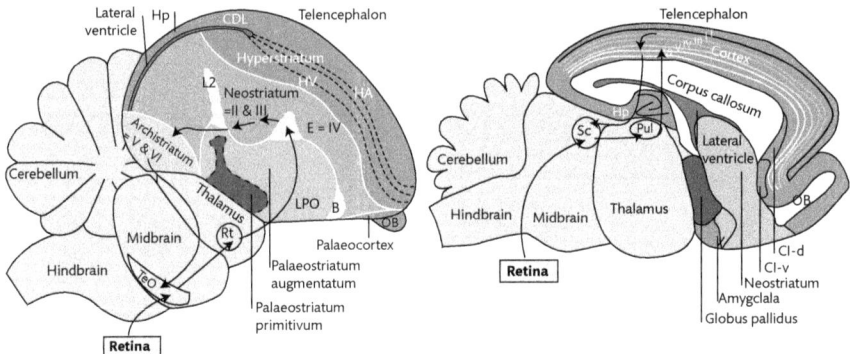

Fig. 6.1 Example of a nuclear-to-layered hypothesis (Karten 1991). Different shades of gray indicate proposed homologies between birds (on the right) and mammals (on the left). I-VI, cortical layers I-VI; B, nucleus basalis; CDL, dorsal lateral corticoid area; Cl-d, claustrum, dorsal part; Cl-v, claustrum, ventral part; E, ectostriatum; HA, hyperstriatum accessorium; Hp, hippocampus; HV, hyperstriatum ventrale; L2, field L2; LPO, lobus parolfactorius; OB; olfactory bulb; Pul, pulvinar nucleus; RT; nucleus rotundus; Sc, superior colliculus; TeO, optic tectum. (From Jarvis *et al.* 2005, with permission from Macmillan Publishers Ltd).

Similar arguments were later made for the avian hyperstriatum, which also has serially connected neuron types that resemble those found in the mammalian neocortex. In this hypothesis, avian L2 neurons are homologous to layer IV neurons of mammalian primary auditory cortex, basalis neurons to layer IV of primary somatosensory cortex, ectostriatal neurons to layer IV of extrastriate visual cortex, and the interstitial hyperstriatum accessorium to layer IV of striate visual cortex. Recent gene expression studies give support to this hypothesis (Wada *et al.* 2004; Mello and Clayton 1995; Dugas-Ford and Ragsdale 2003). Although the avian pallium is not organized into layers, its nuclear subdivisions bear marked similarities in connectivity and molecular profile to different layers of the mammalian neocortex, and the pallia of both groups perform similar functions (Jarvis *et al.* 2005). The avian striatal and pallidal domains are organized more similarly to those of mammals and perform similar functions in both groups (Jarvis 2009).

Based on this current view, the adult avian pallium, as in mammals, comprises approximately 75% of the telencephaic volume. The realization of such a relatively large and well-developed pallium that processes information in a similar manner to mammalian sensory and motor cortices may help explain some of the sophisticated cognitive abilities of birds.

As summarized by the Avian Brain Nomenclature Consortium (Jarvis *et al.* 2005), Pigeons (*Columba livia*) can memorize up to 725 different visual patterns, learn to categorically discriminate objects as "human-made" versus "natural," discriminate cubistic and impressionistic styles of painting, communicate using visual symbols, rank patterns using transitive inferential logic, and occasionally "lie". New Caledonian Crows (*Corvus moneduloides*) make tools out of *Pandana* leaves or novel human-made material, use them appropriately to retrieve food, and are believed to pass this knowledge on to other crows through social learning. Magpies (*Pica pica*) develop an understanding of object constancy at an earlier relative age in their life span than any other organism thus far tested and can use this skill to the same extent as humans. Western Scrub-jays (*Aphelocoma californica*) show episodic memory, recall for events that take place at a specific time or place, once thought to be unique to humans. This same species modifies its food-storing strategy according to the possibility of future stealing by other birds and therefore displays a behavior that would qualify as theory of mind. Barn Owls (*Tyto alba*) have a highly sophisticated capacity for sound localization, used for nocturnal hunting, that rivals that of humans and that is developed through learning. Parrots, hummingbirds, and oscine passerines (songbirds) display vocal learning. This trait is a critical substrate in human spoken language, and with the exceptions of cetaceans and possibly bats and elephants, it is not found in any other mammal. Parrots, in addition, can learn human words and use them to communicate reciprocally with humans. African Gray Parrots (*Psittacus erithacus*) in particular can use human words in numerical and relational concepts — abilities once thought unique to humans. These vocal behaviors are controlled by vocal learning pathways through both pallial and subpallial regions.

Emery and Clayton (2005) argue that one relevant example of convergence in the neural systems of birds and mammals, related to the issue of intelligence, is the suggestion that birds may have a functionally equivalent structure to the mammalian prefrontal cortex. In mammals, the prefrontal cortex contributes to the organization, planning, and flexibility of behavior based on previously acquired information. As there is good evidence now that some birds display these complex cognitive traits, these birds may also have functionally equivalent areas located in the telencephalon. The strongest candidate, according to Emery and Clayton, is the caudolateral nidopallium (CDNL). Recent neurobiological studies focused on the CDNL of Pigeons have revealed similarities in connectivity, neurochemistry, neurophysiology, and function to the mammalian dorsalateral prefrontal cortex (Emery 2006; Emery and Clayton 2004).

6.1 Response of Birds to Novel Environments

Much of our current knowledge and understanding of the evolution of large brains focuses on the brain's function in cognition and information processing (Jerison 1973; Kaas 2000; Allman 2000). Empirical evidence and theoretical models suggest that large brains can process, integrate, and store more information about the physical and social environment, enhancing the propensity of individuals to modify or invent new behaviors in potentially adaptive ways (Jerison 1973; Kaas 2000; Allman 2000; Barton 1996; Lefebvre *et al.* 1998; Lefebvre *et al.* 1997; Madden 2001; Reader and Laland 2002; Timmermans *et al.* 2001; Burish *et al.* 2004; Byrne and Corp 2004; Iwaniuk *et al.* 2001; Marino 1996; Seyfarth and Cheney 2002; Reader and MacDonald 2003; Nicolakakis and Lefebvre 2000; Byrne and Whiten 1988). A quite remarkable recent finding is that relative brain size is positively correlated with the ability for behavioral innovation and learning in both birds and mammals (Lefebvre *et al.* 1998; Lefebvre *et al.* 1997; Reader and Laland 2002; Timmermans *et al.* 2001; Nicolakakis and Lefebvre 2000; Byrne and Whiten 1988).

The most widely accepted hypothesis about the evolution of brain size is that large brains have evolved as an adaptation to cope with novel or altered conditions, i.e., unpredictable environments (Allman 2000; Reader and MacDonald 2003; Reader 2004; Potts 1998; Vrba 1985). When animals are faced with such situations, the ability to produce innovative behavior and store it in the repertoire through individual or social learning may have a critical impact on survival and fitness (Reader and MacDonald 2003; Dukas 1998; Dukas and Bernays 2000; Lee 2003; Lee 1991). Examples include the development of effective responses to novel predators (Berger *et al.* 2001), adoption of new food resources when the traditional ones become scarce (Estes *et al.* 1998), or adjustments of breeding behavior to prevailing ecological conditions (Brooker *et al.* 1998).

Environmental uncertainty and behavioral complexity are central in many of the social and ecological hypotheses proposed to explain the evolution of large brains (Madden 2001; Burish *et al.* 2004). Sol and colleagues (2005) tested the hypothesis that enlarged brains of birds have evolved as an adaptation to cope with novel or altered environmental conditions. They examined whether large-brained species show higher survival than small-brained species when introduced to nonnative locations. They used a previously compiled global database documenting the outcome of more than 600 human-mediated introductions of birds to new locations (Cassey *et al.* 2004). They defined an introduction as the release of individuals of a species to either an island or an area within a continental mainland that is outside the species' native range (Cassey *et al.* 2004). Multiple releases to the same island or location were counted as one introduction. Sol and colleagues considered an introduction to be successful if it resulted in the establishment of a persistent or probably persistent population. Their operational measure of cognitive capacity was the frequency of feeding innovations, defined as any food type or feeding technique described by an observer as novel for the species. Frequency of feeding innovations reported in the literature is a practical and commonly used index of cognitive skills in birds and primates (Lefebvre *et al.* 2004). Assuming that innovation rate is an accurate measure of cognitive ability,

Sol and colleagues confirmed that avian species with larger brains relative to their body mass tend to be more successful at establishing themselves in novel environments (Fig. 6.2). Furthermore, they also provided evidence that larger brains help birds respond to novel conditions by enhancing their innovation propensity rather than indirectly through noncognitive mechanisms (Fig. 6.3).

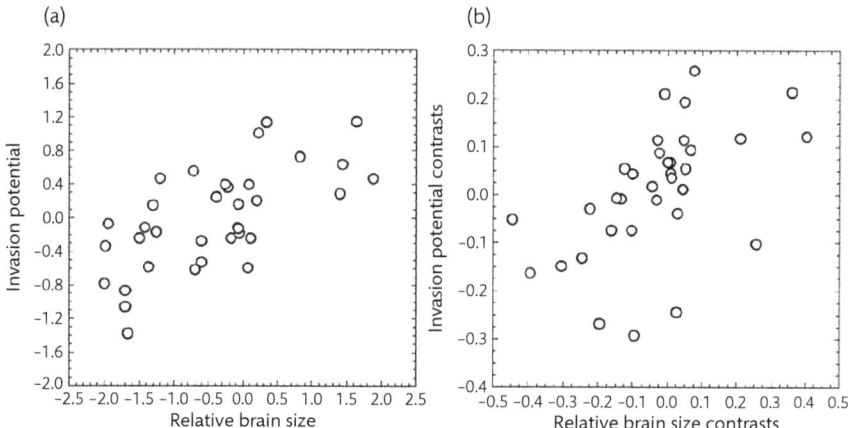

Fig. 6.2 Relationship between mean relative brain size and invasion potential for worldwide avian families. The relationship is shown without (A) and with (B) control for phylogenetic effects by using independent contrast analysis. (From Sol *et al.* 2005; with permission from National Academy of Sciences).

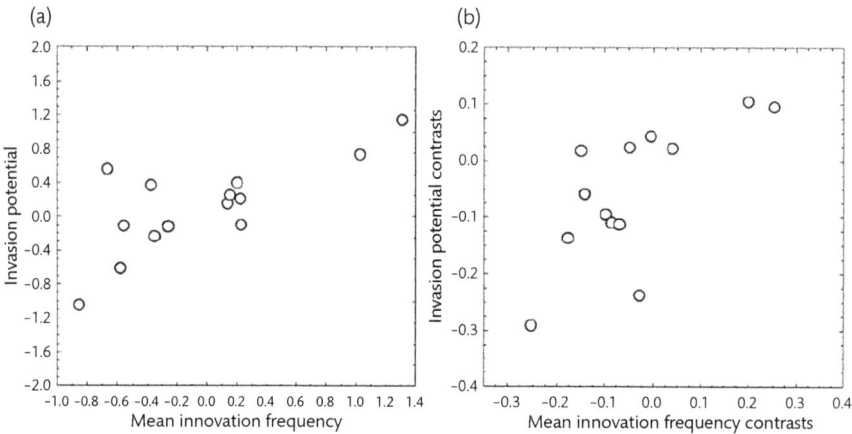

Fig. 6.3 Relationship between mean feeding innovation propensity and invasion potential for Palearctic avian families. The relationship is shown without (A) and with (B) control for phylogenetic effects by using independent contrast analysis. (From Sol *et al.* 2005; with permission from National Academy of Sciences).

6.2 Chemical Senses

Although the ability to detect chemical cues is widespread in many organisms, it is surprising how little is known about the role of chemical communication in avian biology. Growing evidence suggests that birds can use olfaction in several contexts and with different functions (Amo *et al.* 2008). We discuss two topics of interest, "Predator Detection" and "Foraging, Orientation and Navigation", because of their ecological and environmental implications.

6.2.1 Predator Detection

For numerous prey species, the ability to detect predators is an important component of antipredatory behavior. The use of chemical cues to the presence of predators for the assessment of predation risk can be important in many bird species, particularly in those that use habitats in which visual or auditory detection of predators is constrained (Amo *et al.* 2008).

For accurate assessment of predation risk before entering a cavity, chemical cues can be more effective than visual ones, and may allow birds to avoid a risky encounter with a predator inside the cavity. Amo *et al.* (2008), in a pioneering investigation, explored the use of chemical cues by the cavity-nesting Blue Tit (*Cyanistes caeruleus*), which may encounter predators lurking unseen in their nest sites. They performed their experimental study in a nest-box breeding population subjected to predation pressure by mustelids (Díaz and Carrascal 2006). To test whether Blue Tits were able to recognize the chemical cues of mustelids and to use them to assess the risk of predation, Amo and colleagues added predator scent (urine and glandular secretions) to nest-boxes in which birds were feeding 8-day-old nestlings. They simulated a situation where the parents return to the nest with food and find the chemical cues of the predator inside the nest-box. Two control scents, odorous (quail scent) and odourless (water), were added to other nest-boxes. To assess the short-term response of parents to the predator scent, parent provisioning behavior was recorded, after adding the treatments to the nest boxes. Body condition measurements of nestlings were performed at days 8 and 13 of age to examine differences in growth rate of nestlings in relation to nest scent, and also to examine possible consequences of parent behavior in relation to the presence of chemical cues of the predator in the long-term.

Blue Tits were able to detect the chemical cues of the predator and exhibited behaviors to decrease predation risk. Birds delayed their entry to the nest-box, and they perched on the hole of the nest-box and refused to enter more times in the presence of predator scent than in the presence of control scents inside the nest-box. Birds also decreased the time spent inside the predator-scented nest-box. Even though Blue Tits decreased the time spent in the nest-box, they did not decrease the number of feeding events. Therefore, according to Amo and

colleagues, Blue Tits solved the trade-off between avoiding predation and reproduction by maintaining constant feeding rates but decreasing time exposed to
predation when entering the next-box to feed the nestlings.

In their study, Amo and colleagues referred mainly to the capacity of birds for
chemical cues detection. However, trigeminal chemoreceptive receptors also
detect volatile compounds (Mason and Clark 2000; Hagelin 2007) and could be
involved in the detection of chemical cues. Their bias, nevertheless, is supported
by the results obtained by Steiger *et al.* (2008), in a comparative genomic study,
in which the olfactory receptor genes of nine bird species from seven orders were
examined. Their results support the growing body of evidence that the importance of the sense of smell for birds has been greatly underestimated. In particular, the estimated olfactory receptor gene repertoire size, and the proportion of
olfactory receptor genes that is potentially functional, contradict the general view
that avian olfactory ability is poorly developed.

6.2.2 Foraging, Orientation, and Navigation

The tube-nosed seabirds (Procellariiformes) are known for their wide-ranging,
pelagic lifestyle. This order includes storm-petrels, albatrosses, gadfly petrels, fulmars, prions, and shearwaters (Warham 1990; Warham 1996). These birds spend
most of their lives in flight over the ocean, and are tied to land for only a few
months each year or every other year to breed and rear a single offspring.
Procellariiformes are renowned for their strong, musky scent (Bonadonna *et al.*
2007), which perfumes their oily plumage, their nest material and even their
eggs. Procellariiformes have among the largest olfactory bulbs of birds (Bang
1966), and their neuroanatomy suggests a highly developed sense of smell in the
few species that have been examined at the cellular level (Nevitt 2008). The olfactory bulbs of Northern Fulmars (*Fulmarus glacialis*) have twice as many mitral
cells as rats ($120,000$ vs. $60,000$) and six times as many as mice (Wenzel and
Meisami 1987). These cells are fundamental to olfactory processing and also play
a key role in odor contrast enhancement (reviewed by Shepherd *et al.* 2007).

Most Procellariiformes forage over vast areas of the ocean for patchily distributed prey, including various species of fish, squid, and krill. "Their survival
depends on finding the proverbial needle in a haystack on a daily basis" (Nevitt
2008). Procellariiform diets can be highly variable, and can change with respect
to prey availability (Reid *et al.* 1997; Reid *et al.* 1996) or time of the year (Ainley
et al. 1984). During the breeding season, they are temporally constrained to their
nest either to relieve a mate or to feed a hungry chick (Stephens and Krebs
1986).

Data from satellite tracking of larger species have revealed that they use different strategies to accomplish this task, including opportunistic and commuter
foraging strategies (Weimerskirch 1998). Opportunistic foragers may rely on

unpredictably distributed prey, while commuters travel between nest sites and predictably reliable foraging areas. For example, opportunistic foragers such as the Wandering Albatross (*Diomedea exulans*) hunt for prey along continuous, looped paths covering many thousands of kilometers of pelagic and neritic water even on a single foraging trip (Nevitt 2008). These birds forage mainly on fish (myctophids) and various squid species. A large fraction of their diet tends to be squid in the form of carrion (Croxall and Prince 1994), which they are able to track using a combination of visual and olfactory cues (Nevitt *et al*. 2008). In contrast, Black-browed Albatross (*Thalassarche melanophrys*), a commuter species, may travel thousands of kilometers more-or-less directly to a shelf break or seamount where prey are likely to be concentrated. Upon arrival, the bird engages in area-restricted search (ARS) to locate prey (reviewed by Nevitt 2000). Black-browed Albatrosses forage on a combination of squid, krill, and fish in roughly equal proportions during the breeding season (Rodhouse and Prince 1993). Therefore, foraging strategies occur over different spatial scales. At larger scales, the task is to localize productive areas within the vast oceanic environment where prey is likely to be found, whereas at small scales, birds must pinpoint and capture prey using whatever proximate cues are available (Nevitt 2008).

Nevitt (2000) proposed that scent cues in the marine environment present guideposts to help seabirds in foraging and navigation. At large scales, Nevitt *et al*. (1995) suggested that seabirds use changes in the olfactory landscape to recognize potential foraging opportunities as they fly over them. These changes in the olfactory landscape reflect bathymetric features, which tend to accumulate phytoplankton and therefore increase prey availability. Nevitt speculates that birds build up a map of these features over time. Thus, foraging birds might navigate to a historically rich productive area, i.e., a shelf break, a seamount or an upwelling zone, using mechanisms that are still unknown. Birds presumably know that they have arrived by a predictable variation in the way the ocean smells. This change in the background scent triggers the birds to begin ARS at a much smaller scale (tens to hundreds of square kilometers). For ARS, birds might use olfactory, visual or a combination of signals, including the foraging activity of other seabirds, to locate and capture prey (Nevitt 1999a; Nevitt *et al*. 2008).

This conceptual model arises from the discovery that seabirds and other predators can smell trace concentrations of sulfur compounds, which are associated with oceanic features where prey tend to aggregate. Dimethyl sulfide (DMS), and its precursor dimethylsulfoniopropionate (DMSP), can be detected by a variety of marine organisms including procellariiforms (Nevitt and Haberman 2003; Nevitt *et al*. 1995) and also some species of fish (DeBose *et al*. 2008). This work has been extended to include harbor seals (Kowalewsky *et al*. 2006) and whale sharks (Martin 2007).

DMS is largely produced as a byproduct of the metabolic decomposition of DMSP in marine phytoplankton (notably *Phaeocystis* in the Southern ocean) and other marine algae (including zooxanthale in coral reefs) (Nevitt 2000). In the Southern ocean, DMS is frequently associated with oceanic features where phytoplankton are abundant, including upwelling zones, seamounts, and shelfbreaks (Berresheim *et al.* 1989; Daly and DiTullio 1996; McTaggart and Burton 1992). These are areas where seabirds and other marine predators aggregate and forage (Nevitt and Bonadonna 2005; Nevitt *et al.* 1995). DMSP is released when phytoplankton cells are crushed and it is then rapidly converted to DMS by processes within the marine microbial food web (reviewed by Pohnert *et al.* 2007). It is now well established that DMS emissions increase when phytoplankton are grazed by both protozoans (Wolfe and Steinke 1996), and metazoans (Dacey and Wakeham 1986) including krill (Daly and DiTullio 1996), suggesting that local elevations in DMS may alert higher order predators, including birds, to rapidly accumulating aggregations of zooplankton (e.g., krill) and zooplankton predators (squid and fish) (Nevitt 2008).

To test these ideas experimentally, Nevitt and colleagues (Nevitt 1999a; Nevitt 1999b; Nevitt *et al.* 2004; Nevitt *et al.* 1995) conducted a multi-year study of small-scale foraging in the Atlantic sector of the Southern ocean, which confirmed that birds are attracted not to prey scents *per se* but rather to odors such as DMS that are released during feeding interactions. Predators are messy eaters, and procellariiform species are adapted to pay attention to who is eating whom (Hay and Kubanek 2002; Nevitt 1999b; Nevitt *et al.* 2004). When DMS was presented to seabirds in controlled experimental trials conducted at sea, several species of storm-petrels (*Oceanodroma* sp), prions (*Pachyptila* sp), and gadfly petrels (*Procellaria* sp) responded by tracking this odor to its source, using a zigzag (Fig. 6.4), upwind search behavior characteristic of olfactory tracking in organisms as diverse as fish, moths, and crustaceans (DeBose and Nevitt 2008; Montgomery *et al.* 1999; Moore and Crimaldi 2004; Nevitt *et al.* 1995; Willis

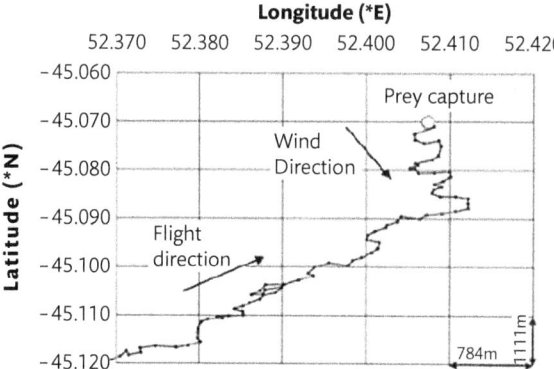

Fig. 6.4 GPS track of a Wandering Albatross (*Diomedea exulans*) attracted to a source of DMS. (From DeBose and Nevitt, 2008, with permission from Springer Science + Business Media).

2005; Zimmer-Faust *et al.* 1995). Interestingly, these species ignored krill odors, even though krill contributes significantly to their diets. By contrast, larger and more conspicuous seabird species recruited to visual cues and to odors associated with crushed krill (pyrazines). Nearly every species recruited to fishy scents, possibly through conditioning to fishing boats (Nevitt 1999b; Nevitt *et al.* 2004; Nevitt *et al.* 1995).

These findings suggest that procellariiforms within this sub-Antarctic assemblage exploit at least two fundamentally different sensory strategies for ARS (Nevitt 2008). DMS responders are adapted to forage opportunistically on small or less concentrated prey patches, whereas more aggressive species (e.g., albatross, Diomedeidade, and giant petrels, *Macronectes*) are better adapted to exploit multi-modal cues, which include scents from crushed prey and visual cues associated with the activity of other birds and marine predators (Nevitt 1999b; Nevitt and Bonadonna 2005; Nevitt *et al.* 2004).

Phylogenetic analysis shows that odor responsiveness is linked to life history strategy (Van Buskirk and Nevitt 2008). The early environment that procellariiform chicks experience may be linked to the evolution of different sensory-based foraging strategies. In comparison with other birds, procellariiforms have a lengthy chick-rearing period that can last from 6 weeks in some species to nearly a year in some larger albatrosses (Warham 1990). Species that are either small or produce more vulnerable chicks (e.g., shearwaters, diving petrels, prions, and storm petrels) construct deep nesting burrows that offer protection from predators. Larger species (e.g., albatross and giant petrels) nest at the surface, since their chicks are less vulnerable to avian predation and thermoregulatory stress because of their larger size. Burrow-reared chicks spend their early life in a dark underground nest, where odors are likely to dominate their early sensory experience. Because predation on chicks happens mostly in breeding colonies, burrow-nesting chicks usually remain deep underground in the dark until just before fledging. Chicks reared above ground or in surface crevices, on the other hand, grow up with early access to light and are exposed to an array of different stimuli, including visual, auditory and olfactory inputs.

Using comparative methods, Van Buskirk and Nevitt (2008) found that, in contrast to surface nesting, burrow nesting was significantly correlated to DMS tracking behavior, but not to a behavioral attraction to odors more directly associated with macerated krill or fish, including pyrazines or trimethylamine. Ancestral trait reconstruction also indicated that procellariiforms arose from a burrow-nesting lineage, with the albatrosses and fulmarine petrels independently adopting a surface-nesting strategy. Burrow nesting appears to be an ancestral trait, suggesting that some degree of DMS sensitivity may also have been an ancestral condition for the Procellariiformes and their sister order, the Sphenisciformes (penguins) (Nevitt 2008). Penguins are not considered to be particularly olfactory-oriented birds. However, they also forage in productive areas which are characterized by high concentrations of DMS (Culik

2001). Interestingly, Little Penguins (*Eudyptula minor*) not only nest underground but also show tube-like structures on their nares during development (Kinsky 1960).

It has been shown that, even as chicks, Blue Petrels (*Halobaena caerulea*) can detect and are attracted to prey-related odor cues well before they leave the nest. Nevitt and colleagues (Bonadonna *et al.* 2006; Nevitt *et al.* 2006) demonstrated that chicks can detect both DMS and ammonia, a urinary byproduct of most marine organisms, at concentrations ranging from 10^{-12} to 10^{-16} mol L^{-1}. Bonadonna *et al.* (2006) tested the response of Blue Petrel fledglings to DMS at very low concentrations (10^{-12} mol L^{-1}) just days before they were to leave their nests to forage. Using a simple Y-maze design, Bonadonna and colleagues found that Blue Petrel fledglings were attracted to DMS over a control odor, suggesting that a preference for DMS is already established before chicks leave their nests.

Foraging opportunities also involve interactions with hetero- or conspecifics (Silverman *et al.* 2004; Ward and Zahavi 1973). As the fledgling gains more foraging experience, it acquires a working knowledge of potential foraging sites by associating foraging success with other cues (e.g., scent cues associated with prey and productivity, visual cues provided by other seabirds, geomagnetic references associated with foraging sites). A bird might therefore develop a map of foraging sites linked to spatially fixed features such as shelf breaks or seamounts, for example. However, in the marine environment, productive foraging locations are dynamic and unpredictable. Foraging opportunities may vary both temporally and spatially over large areas such as upwelling and convergence zones, suggesting that any map cannot be strictly or entirely spatially explicit, but must be used with proximate cues which allow birds to recognize when they have arrived in a profitable location (Nevitt 2000). Whether the foraging site is spatially fixed (e.g., seamount) or transient (e.g., upwelling or convergence zone), scent cues associated with trophic interactions would provide a foraging petrel with immediate feedback as to whether foraging is likely to be successful at that particular time and location (Nevitt 2008).

Because DMS is produced by phytoplankton, which often occurs in spatially predictable locations, Nevitt (2008) suggests that the ability to recognize predictable features such as shelf breaks or seamounts by scent may be also adaptive for navigation as the bird matures. However, navigation systems usually involve multimodal cues, and the spatial scales over which birds operate suggest that geomagnetic or astronomical cues might be useful for cross-referencing positional information of olfactory features that are spatially predictable.

6.3 Magnetoreception

Approximately 50 species, including birds, mammals, reptiles, amphibians, fish, crustaceans, and insects are known to use the Earth's magnetic field for orientation and navigation (reviewed by Johnsen and Lohmann 2005). Birds in particular have been intensively studied because they can use a magnetic compass for orientation during their migratory flights covering thousands of kilometers. However the physiological mechanisms underlying the avian magnetic compass are still poorly understood.

Two types of potential magnetoreception mechanisms have been suggested over the past decades: one mechanism that is based on magnetite particles, and another mechanism that is based on photoreceptors, forming radical pair intermediates (summary by Ritz *et al.* 2000). The latter mechanism is well established as the source of a variety of magnetic effects on free radical reactions *in vitro* (Brocklehurst 2002; Timmel and Heinbest 2004).

The magnetic compass of migratory passerines is an inclination compass, i.e., it detects the axis but not the polarity of the magnetic field lines (Wiltschko and Wiltschko 1972; Wiltschko and Wiltschko 1995a). Furthermore, magnetic orientation in migratory passerines depends on the wavelength of ambient light (Wiltschko *et al.* 1993; Wiltschko and Wiltschko 1995b; Wiltschko and Wiltschko 2002; Muheim *et al.* 2002). Migratory passerines are active and orient magnetically under dim blue and green light, whereas they are active but disoriented under dim red light (Wiltschko *et al.* 1993; Wiltschko and Wiltschko 1995b; Wiltschko and Wiltschko 2002). These results strongly suggest that photoreceptor molecules in the eye are involved in magnetoreception, and that these photoreceptor molecules should absorb in the blue and green range of the spectrum. The involvement of photoreceptors in the eye is further supported by the finding that birds with their right eye covered are unable to perform magnetic orientation (Wiltschko *et al.* 2002).

Based on behavioral evidence and theoretical considerations (Schulten *et al.* 1978; Ritz *et al.* 2000), it has been proposed that magneto-sensitive radical pairs formed by light-induced intramolecular electron transfer reactions in an array of aligned photoreceptor molecules in the retina could enable migratory birds to perceive the magnetic field as visual patterns. The cryptochromes (CRYs) (Sancar 2000; Sancar 2003; Sancar 2004; Fu *et al.* 2002; Haque *et al.* 2002; Cashmore *et al.* 1999) have been suggested as the most likely candidate class of molecules because they are blue-green photoreceptors in plants and because closely related 6,4-photolyases have been shown to form radical pairs when photoexcited (Giovani *et al.* 2003). CRYs are the only currently known class of molecules found in the retina of vertebrates that are likely to fulfill the physical and chemical characteristics that are required for function as the primary magnetic sensor (Ritz *et al.* 2000).

Mouritsen *et al.* (2004) investigated the retina of night-migratory passerines (Garden Warbler, *Sylvia borin*), and showed that at least one CRY1 and one CRY2 exist in the Garden Warbler retina, and that CRY1 (gwCRY1) is cytosolic. They also showed that gwCRY1 is concentrated in specific cells, particularly in ganglion cells and in large displaced ganglion cells, which also showed high levels of neuronal activity at night, when the Garden Warblers performed magnetic orientation in orientation cages. Furthermore, there are striking differences in CRY1 expression between Garden Warblers and Zebra Finches (*Taeniopygia guttata*, a non-migratory passerine), used as controls. The difference in CRY1 expression between migrants and non-migrants is particularly pronounced in the large displaced ganglion cells, which are likely to be a site for magnetoreception because they project exclusively to the nucleus of the basal optic root (Fite *et al.* 1981; Britto *et al.* 2004), where magnetically sensitive neurons have been reported (Semm and Demaine 1986), and visual flow-fields arising from self-motion are processed (Wylie *et al.* 1998). Although these results indicate differences between a migrant species and a non-migrant species, a two-species comparative study cannot robustly demonstrate adaptive differences. Moreover, the very large phylogenetic distance between finches and warblers makes this comparison unconvincing.

Evidence of magnetically sensitive free radical reactions is gaining support, despite the fact that no chemical reaction *in vitro* has been shown to respond to magnetic fields as weak as the Earth's ($\sim 50\,\mu T$) or to be sensitive to the direction of such a field. Based on the fact that a photochemical reaction can act as a magnetic compass, Maeda *et al.* (2008) used spectroscopic observation of a carotenoid-porphyrin-fullerene model system and showed that the lifetime of a photochemically formed radical pair is changed by the application of $\leq 50\,\mu T$ magnetic fields. Therefore, in principle, greater sensitivity to weak magnetic fields would be possible for radical pairs formed in a specialized photoreceptor as CRYs. In order to function as a chemical compass, a radical pair magnetoreceptor must respond anisotropically to an external field. Maeda and colleagues were also able to demonstrate the existence of such a property in their model system. Measurements of the kind described by Maeda and colleagues could be performed on isolated CRYs subject to applied magnetic fields.

6.3.1 Field studies

Billions of passerines migrate several thousand kilometers from breeding to wintering sites each year and are challenged with crossing ecological barriers and dealing with displacement by the wind along the route. The navigational tasks faced by adults and juveniles are fundamentally different because only adults migrate toward wintering grounds known from the previous year. Using radio tracking from small aircraft, Thorup *et al.* (2007) were able to demonstrate that only adult, and not juvenile, long distance migrating White-crowned Sparrows (*Zonotrichia leucophrys gambelii*) rapidly recognize and correct for a continent-wide displacement when they were experimentally relocated 3,700 km, mid-migration, from the west coast of North America to previously unvisited areas on the east coast. Their results show that the learned navigational map used by adult long-distance passerine migrants extends at least on a continental scale and that adults were able to correct for the relocation and arrive precisely at their wintering areas. Juveniles, with less experience, rely on their innate or "distance and direction" orientation mechanism to find their distant wintering sites, and when displaced, they continue to migrate in the innate direction without correcting for displacement.

When tested in captivity, night migratory passerines can use the stars, the sun, the geomagnetic field, and polarized light for orientation. Cochran *et al.* (2004) investigated the interaction of magnetic, stellar, and twilight orientation cues in free-flying passerines. They exposed *Catharus* thrushes to eastward-turned magnetic fields during the twilight period before takeoff and then followed the birds for 1,100 kilometers. Instead of heading north, experimental thrushes flew westward. On subsequent nights, the same individuals migrated northward again. Based on these field-based results, Cochran and colleagues suggest that migrant birds orient themselves with a magnetic compass calibrated daily from twilight cues. According to them, this could explain how birds cross the magnetic equator and deal with declination (drift).

6.4 Visual capacity

6.4.1 Night Vision

Night-migratory passerines use both a geomagnetic and a star compass to orient during migration (Emlen 1967; Wiltschko and Wiltschko 1972; Mouritsen 1998; Wiltschko and Wiltschko 2002; Cochran *et al.* 2004; Mouritsen and Larsen 2001), but the underlying brain circuits are still unknown. Star-compass orientation requires vision in dim light for processing constellations of the starry night sky. Birds are hypothesized to use a time-independent star compass based on learned geometrical star configurations to pinpoint the rotational point of the starry sky (north in the northern hemisphere, south in the southern hemisphere) (Mouritsen and Larsen 2001). As discussed earlier, magnetic-compass orientation also seems to require night vision because magnetic-compass orientation is dependent on the wave-length of the dim ambient light. These requirements lead one to predict that processing of both magnetic- and star-compass information during night-time migration requires specialized night-time visual processing in night-migrants.

To test this hypothesis, Mouritsen *et al.* (2005) examined two distantly related species of wild-caught night-migratory passerines (Garden Warblers and European Robins, *Erithachus rubecula*) and two distantly related species of nonmigratory passerines (Zebra Finches and Canaries, *Serinus canaria*) used as controls. They measured the expression of ZENK (acronym for zif268 Egr-1, NGF-1A, and Krox-24) and c-fos immediate early genes in the brain. Expression of ZENK and c-fos mRNA in the brain is driven by neuronal activity and can be detected in neurons approximately 5 min after onset of neural firing with peak expression after 30–45 min. Mouritsen and colleagues discovered that night-migrants possess a tight cluster of brain regions which are highly active only during night vision. They named this "cluster N", and it is located at the dorsal surface of the brain adjacent to a known visual pathway. This area is fairly large, taking up approximately 40% of the hyperpallium and dorsal mesopallium. The strong night-time activation in cluster N observed in the migratory passerine species was not found in the two nonmigratory species (controls), and it disappeared in migrants when both their eyes were covered. Mouritsen and colleagues therefore suggest that, in night-migratory passerines, cluster N is involved in enhanced night vision, and it could be integrating vision-mediated magnetic and/or star compass information for night-time navigation.

6.4.2 Eye Size and Foraging Methods

Most birds are active during daylight and use vision to detect prey (Martin 1990). Among shorebirds (Suborder Charadrii, excluding sandgrouse Pteroclidae, and gulls and auks Laridae) there is considerable variation in both the daily pattern of activity and the sensory cues used in foraging (reviewed by McNeil 1991; Barbosa 1995; del Hoyo *et al.* 1996; Barbosa and Moren 1999). Many shorebirds detect prey by touch, probing with their bills in soft ground, or by sweeping their bills through shallow water. Others detect prey by sight, and some use a mixture of tactile and visual cues. Some species forage mainly at night while others forage mainly during daylight. Because many shorebirds forage in the intertidal zone, both the daily tidal cycle and the light-dark cycle may affect food availability (Thomas *et al.* 2006).

The visual capabilities of animals are influenced by their eyes' size (Walls 1942; Hughes 1977; Martin 1993; Motani *et al.* 1999; Land and Nilsson 2002). The larger the eye, the larger the maximum pupil aperture that can be attained, and hence the more photons reaching the retina. The number of photoreceptors and the focal length are also both affected by the size of the eye. The animal's ability to detect light of low intensity (visual sensitivity) and its ability to distinguish detail at a given light intensity (visual resolution) is positively correlated with eye size. Evolutionary increases in eye size are predicted to be associated with behaviors requiring high spatial resolution and/or high sensitivity (Land and Nilsson 2002).

Despite the benefits of large eyes, there are potential associated costs. These include the energetic cost of manufacturing and maintaining the additional nerve cells and other ocular structures (Laughlin 1995; Laughlin *et al* al. 1998), and possibly an increased risk of damaging the eye (e.g., Harper 1988). Therefore, the size of eyes is likely to be a reflection of these costs and benefits, which in turn depends on the animal's ecology (Walls 1942; Motani *et al.* 1999; Land and Nilsson 2002).

Birds that forage using visual cues are likely to require high visual acuity, and those which forage at night are likely to require high visual sensitivity and acuity at low light intensities. Based on these assumptions, Thomas and colleagues (2006) predicted that eye size is related to both foraging strategy and the time of the day at which birds are active. They tested the hypothesis that eye size in shorebirds has evolved in relation to foraging technique, and foraging activity patterns.

Their results show that species of shorebirds that forage at night possess larger eyes than diurnal species. Their comparative analyses suggest that shorebird eye morphology and the time of foraging have evolved in relation to each other. However, species using vision to detect their prey did not have larger eyes than species that detect their prey by touch. Thomas and colleagues did not find evidence that nocturnal foragers are more likely than diurnal foragers to use tactile rather than visual cues to detect prey. Their study indicates that evolutionary changes toward nocturnal habits are associated with increases in eye size, and therefore with increases in aspects of visual capacity, but that foraging technique may not be the prime selective force in this taxon. Links between visual ecology and behavior can be quite complex. For example, fast flight requires good visual acuity, so eye size is predicted to increase with flight speed (Hughes 1977; Martin 1985). Species with especially fast flight may possess large eyes regardless of their foraging method. Predation risk may also be a factor. Shorebirds are frequently attacked by fast-flying aerial predators (falcons) and hence would benefit from an excellent visual system to detect these predators at a safe distance. Such "constraints" may thus contribute to mask any effect of foraging behavior on the overall costs and benefits determining eye size.

6.5 Vocalization

Vocal learning is the ability to imitate the vocalizations of others. It is a rare trait found to date in five distantly related groups of mammals (humans, bats, and cetaceans; Janik and Slater 1997), seals (Janik and Slater 1997; Sanvito *et al.* 2007) and elephants (Poole *et al.* 2005)), and in three distantly related groups of birds (Fig. 6.5), i.e., passerines, hummingbirds, and parrots (Vates and Nottebohm 1995).

Among birds, vocalization-driven immediate early gene (IEG) expression experiments revealed that each taxon exhibiting vocal learning possesses seven comparable cerebral (telencephalic) song nuclei, also called vocal nuclei (Jarvis *et al.* 1998; Jarvis

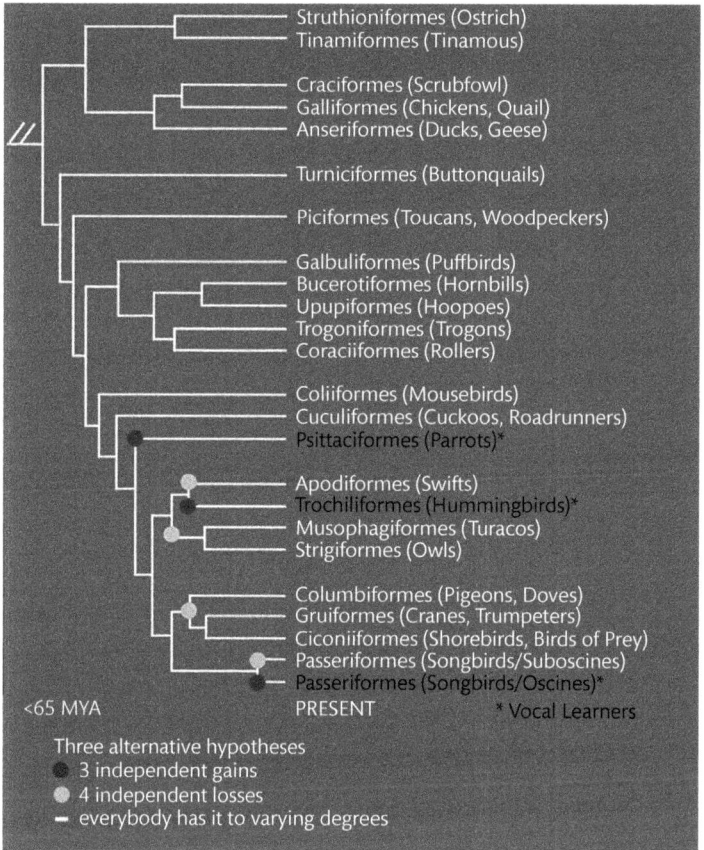

Fig. 6.5 Phylogenetic relationships of avian vocal learners. One view of the phylogenetic relationships of living birds (Sibley and Ahlquist 1990). Vocal learners are in black. Black dots: possible independent gains of vocal learning; light gray dots: alternatively, possible independent losses. (From Feenders *et al.* 2008).

and Mello 2000; Jarvis *et al.* 2000). Three of these nuclei make up part of an anterior vocal pathway which in passerines is necessary for vocal learning (Bottjer *et al.* 1984; Scharff and Nottebohm 1991; Jarvis 2004). The other four nuclei form a posterior vocal pathway of which both the passerines' vocal nucleus and robust nucleus of the arcopallium are necessary for the production of the learned vocalizations (Jarvis 2004; Nottebohm *et al.* 1976; Simpson and Vicario 1990). To date, none of these cerebral vocal nuclei have been found in vocal non-learners, such as in the suboscine passerines closely related to vocally-learning oscine passerines (Kroodsma and Konishi 1991; Brenowitz 1991), the interrelated doves (Wada *et al.* 2004; Haesler *et al.* 2004), and the distantly related galliforms (Nottebohm 1980).

Both vocal learners and non-learners have brainstem vocal nuclei which are responsible for the production of innate vocalizations (Jarvis *et al.* 2000; Wild 1997), and an auditory cerebral pathway responsible for processing species-specific vocalizations and auditory learning (Jarvis *et al.* 1995; Chew *et al.* 1995; Vates and Nottebohm 1995; Vates *et al.* 1996). Thus, according the dominant hypothesis (Nottebohm 1972; Brenowitz 1991), within the past 65 million years 3 out of 23 avian orders independently evolved seven similar cerebral vocal nuclei for complex behavior (Jarvis *et al.* 2000). These similarities suggest that the evolution of brain structures for vocal learning is under strong genetic or epigenetic constraints, or that given ancestral brain genetics, the evolutionary pathway of vocal learning is narrowly canalized such that multiple independent evolutionary events led to essentially identical solutions.

Feenders *et al* (2008) conducted a series of experiments using behavioral molecular mapping, and discovered that in passerines, parrots, and hummingbirds, all cerebral vocal nuclei are adjacent to discrete brain areas active during limb and body movements. Similar to the relationships between vocal nuclei activation and singing, activation in the adjacent areas correlated with the amount of movement performed and was independent of auditory and visual input. These same movement-associated brain areas were also present in female passerines that do not learn vocalizations and have atrophied cerebral vocal nuclei, and in Ringed Turtle-Doves (*Streptopelia risoria*) that are vocal non-learners and do not posses vocal nuclei. These findings may explain why the cerebral vocal systems are similar across distantly related vocal learning birds.

Feenders and colleagues advocate a modified view of the independent evolution hypothesis, this being that the three vocal learning bird groups independently evolved similar vocal systems but were constrained by a previously genetically determined motor system inherited from their common ancestor. Also, according to these authors, this pre-existing motor system may be a basic characteristic of the avian brain that consists of different areas (possibly seven nuclei) distributed in two pathways (posterior and anterior), which in parallel incorporate portions of the cerebral subdivisions (mesopallium, nidopallium, arcopallium, and striatum), each sub-serving a specific function.

If this is true, then such a basic posterior/anterior motor system that controls different non-vocal muscles in parallel pathways via pre-motor neurons in the brainstem could be used as a template for the evolution of a vocal/motor learning system that controls muscles of the syrinx, taking over sites which normally control innate vocalizations (Feenders *et al.* 2008).

Feenders and colleagues argue that what makes vocal learning, and spoken language for that matter, special is the existence of a cerebral motor system that controls the vocal apparatus. In other words, vocal learners and non-learners have similar auditory pathways, but vocal learners have a unique vocal motor system that gives them the ability to translate auditory inputs into vocal signal outputs.

Vocalization and associated behaviors were investigated by Ferreira *et al.* (2006) in two species of hummingbirds (vocal learners): Sombre Hummingbird (*Aphantochroa cirrhochloris*) and Rufous-breasted Hermit (*Glaucis hirsutus*). These are the only hummingbirds in which the brain areas activated by singing have been demonstrated. They are also among the basal species of their respective subfamilies, Trochilinae and Phaethornithinae, and therefore represent early stages in the evolution of hummingbird vocal communication.

Sombre Hummingbirds and Rufous-breasted Hermits differed markedly in vocal behavioral contexts. Sombre Hummingbird song was highly stereotyped in syllable structure and syntax (with an orderly temporal arrangement of acoustic units within the song), and calls were used mostly in defense of a food-centered territory. A direct advantage of having territorial vocalizations would be the assurance of access to food and thus reproductive success; good territories make males more attractive to females (Wolf and Stiles 1970). On the other hand, Rufous-breasted Hermit song was highly variable, produced mostly while birds were perched, and appears not to be used for food-territory defense, because this species is a trapliner (a non-territorial forager; Snow and Snow 1972). Ferreira and colleagues suggest that Rufous-breasted Hermit song may be used to attract mates. In Trinidad, song in a population of Rufous-breasted Hermits was restricted almost exclusively to the breeding season (Snow 1973). According to Ferreira and colleagues, the incorporation of calls in Rufous-breasted Hermits' song suggests that this species may sing to communicate various behavioral states. By using different combinations of call syllables, Rufous-breasted Hermits may be sending information concerning their location, copulation disposition, and aggressive behavior.

It has been hypothesized that the Trochilinae and Phaethornithinae subfamilies originated in the lowlands of the American tropics, possibly through competitive interactions of their ancestors, and then the Trochilinae colonized diverse habitats from tropical to temperate areas, while the Phaethornithinae stayed in the tropics and colonized its forests (Bleiweiss 1988). In this context, Ferreira and colleagues argue that the pressures of diverse habitats on the Trochilinae selected for territorial vocal behaviors (calls and songs) to make defense of food-centered territories more effective. Vocalizations produced during territorial defense have been found in basal tropical species (Ferreira *et al.* 2006) as well as in more derived tropical (Ornelas *et al.* 2002) and temperate trochiline species (Pitelka 1942; Stiles 1982; Goldberg and Ewald 1991; Rusch *et al.* 1996 2001). The more constant habitat with widely-dispersed resources of the Phaethornithinae may have selected for vocal behavior that makes long-distance interactions possible between conspecifics which are constantly moving through forests. These interactions would be male-female and, for those species which engage in lek behavior (Stiles and Wolf 1979), male-male as well. Vocalizations associated with location, bond-maintenance between mates, and female attraction, have been found in basal

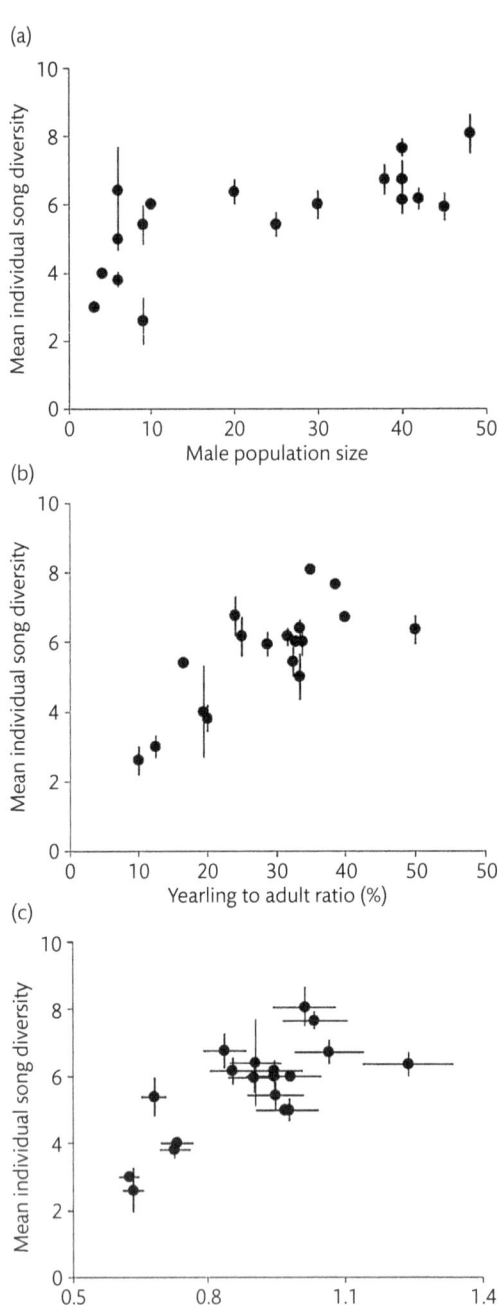

Fig. 6.6 Relationship between mean individual song diversity and (A) male population size, (B) patch productivity (yearling to adult ratio), and (C) population viability (expressed as the mean annual rate of population change (λ) under different scenarios). (From Laiolo *et al.* 2008).

species (Ferreira *et al.* 2006; Snow 1973) as well as in more derived phaethorniti-nae species (Snow 1968; Wiley 1971; Stiles and Wolf 1979).

Geographic differences in naturally occurring learned song syntax within a species are also behaviorally salient to both male and female birds. The possibility of culturally transmitted intraspecific differences in song syntax has implications for the process of conspecific song perception and may be involved in the regulation of genetic exchange between large populations of Swamp Sparrows (*Melospiza georgiana*), as has been demonstrated by Balaban (1988).

Laiolo *et al.* (2008) explored the potential of learned behaviors, specifically acoustic signals, to predict the persistence over time of fragmented bird populations. Recent studies revealed that the communication systems of Dupont's Lark (*Chersophilus duponti*) are disrupted by habitat fragmentation, which reduced population sizes and created barriers to individual movements and cultural transmission (Laiolo and Tella 2007; Laiolo and Tella 2005; Laiolo and Tella 2006). Laiolo and colleagues investigated two parameters of singing performance that appear to be reliable signals of individual quality in many oscine species: song rate and song diversity (repertoire size) (Collins 2004). These traits are thought to be costly because of their correlation with brain size and nutritional condition during learning (song repertoire size). Laiolo and colleagues analyzed the relationship between these characters and parameters related to population social milieu and variability (number of neighbor territories, male population size, productivity, annual rate of population change) to test whether a culturally acquired behavior, i.e., birdsong, can reveal the demographic trends of population, apart from exerting its sexual function among individuals (Gil and Gahr 2002). They found an association between male song diversity and the annual rate of population change, population productivity and population size, resulting in birds singing small repertoires in populations more prone to extinction (Fig. 6.6). This is the first demonstration that population viability can be predicted by a cultural trait acquired through social learning. These findings emphasize that cultural attributes can reflect not only individual-level characteristics, but also emergent population-level properties.

7

Adaptations: Developmental Physiology

Breeding birds must allocate their energy optimally between offspring and self maintenance so that lifetime reproductive success is maximized (Stearns 1976; Case 1978). First and foremost, this requires an investment of resources in eggs by the female. In many taxa the chemical potential energy invested in individual eggs in terms of lipids and proteins is only a small fraction of the total female's reproductive costs, and is generally deposited in the yolk follicles over a period of days prior to egg laying (Astheimer and Grau 1985). However, eggshell formation requires that the female bird must mobilize minerals (calcium and magnesium) from internal stores, while still maintaining mineral balance, or seek out calcium rich foods in the period before and during egg laying (Larison *et al.* 2001). Next, breeding also requires that one or both parents spend time protecting the vulnerable eggs from predators during incubation. However, in many taxa the largest energetic investments in the young are incurred during the so-called nestling periods, when food provision and self maintenance can be expensive.

7.1 Reproduction and Development

Birds and mammals have independently evolved endothermy and homeothermic patterns of body temperature regulation that have enabled them to not only inhabit harsh environments, but to also breed in most of these places. Unlike mammals, birds never evolved egg retention. Avian development is external to the maternal body and therefore the environmental conditions the young experience can vary considerably between species. The many varied patterns of parental care and nest construction among avian taxa modify the ambient conditions in which birds develop and therefore the energy cost of development. There is no simple correlation between the pattern of incubation (single vs. both parents) and nest attentiveness and therefore, the extent of exposure of the eggs and young. Some species with biparental incubation may be frequently absent from the nest and some uniparental incubators can be very attentive. This aside, in general the least amount of exposure of the young to unfavorable environmental conditions that can increase the risk or predation or lethal cooling is associated with taxa that employ biparental incubation.

In birds, rapid early development has evolved in association with higher embryonic metabolic rates than that maintained by similarly sized reptiles, throughout most of the embryonic and neonatal periods (Vleck and Vleck 1987; Vleck and Hoyt 1991). Rapid development and the early acquisition of physiological endothermy in modern birds are now necessarily linked to the maintenance of egg temperatures within a narrow range, above that experienced by most reptilian young. Nonetheless, there is remarkable variation within the Class Aves in the developmental times and developmental modes associated with differences in the regularity of feeding, frequency of exposure to extreme environmental temperatures and conditions, and the timing of nest departure. It is these external influences and parent-offspring behaviors during development that are considered in this chapter.

The interested reader is directed to Deeming (2002b) for a detailed discussion of aspects of incubation biology, including nest construction and thermal exchanges, embryonic development and parental effort during incubation. A comparison of development in the vertebrates will be the subject of another volume in this Oxford University Press Series. This chapter considers other areas of incubation physiology relevant to the adaptations of breeding birds and their young in different environments.

7.1.1 Nest Environments, Eggs, and Incubation

The vast majority of birds make nests of some kind. In some species, in particular seabirds, the nest may comprise a simple scrape or a collection of stones because materials are scarce at communal nesting grounds. Exceptionally, the Emperor (*Aptenodytes forsteri*) and King Penguins (*A. patagonica*) have evolved to hold their single egg above the ground, on top of their feet, under a flap of skin, and no nest is made. Many birds will collect feathers, lichens, mosses or soft grasses to cushion the floor of the nest. In doing so, the parent not only provides a softer surface to minimize the chance of the egg being punctured or rolling out of the nest, but provides a second source of resistance that affects egg water loss and therefore hydration state of the embryo during development. At the same time nest materials can reduce heat loss from eggs to cold substrates beneath and around the nest.

In some species the nest may be an elaborate construction or within an earthen tunnel or tree cavity, which it may or may not construct itself. In these cases the nest can provide energetic benefits to the incubating adult by serving as a thermal buffer to environmental conditions. With the evolution of nest making behavior, birds have modified the gas and heat exchanges between the egg, the parent and the environment. Here, we wish to briefly note how embryonic development is affected by nest design and how this in turn influences parental energy expenditure. In avoiding predators, certain nest designs can become susceptible to problems that create adverse incubation environments. Water logging can be a problem in aquatic nesting species such as grebes, swans, and in the humus mounds built by megapodes. Hypoxia can develop within hollow logs or tunnels, often in association with ammonia accumulation from the nitrogen rich faeces (White *et al.* 1978; Ar *et al.* 2004). This is potentially a problem in species with large clutches of young or in species where helpers from earlier clutches may spend the night within the same hollow (e.g., woodpeckers). Both water logging and hypoxia act to reduce O_2 transport, and therefore retard embryonic development.

Egg temperature and nest humidity are determined by the incubating parent's metabolism, egg water loss, and the nest construction. The nest must allow sufficient exchange of respiratory gases between the egg and the environment. Most of the heat that warms the avian egg comes from direct contact with the incubating adult, and exceptionally from decaying organic matter or even sand or soil heated by geothermal activity in the mounds constructed by megapodes. The incubating adult bird is able to sense surface egg temperature through peripheral skin receptors and increase heat transfer to the egg to maintain an average egg temperature above that of the ambient surroundings. Heat from the incubating bird results in egg water loss, which raises the humidity of the nest environment. Although it is not known how much the adult's cutaneous water loss might also contribute to humidification, it is likely to be small. Nest humidity is also modulated by ambient humidity and the resistance of the nest materials to diffusion (Ar and Rahn 1980). Hence, nest design has an impact on nest humidity.

In contrast to mammals, and even the monotremes, birds must make large investments in their eggs during the brief periods of egg formation. Every egg is provided with at least enough energy, water, and minerals for the embryo to develop to the neonatal stage. The unique anatomy of birds, and in particular the pelvic girdle, places a constraint on the maximum diameter of the rigid eggshell that the adult female can form within the oviduct and pass through the cloaca safely. The evolution of flight has also placed constraints on the net weight of yolk that can be accumulated in a mature ovum and the albumen capsule secreted around the yolk. No longer constrained by flight the Brown Kiwi (*Apteryx australis*) lays an egg that comprises more than 30% of the female body mass. Interestingly, other land bound ratites which are the largest avian taxa do not lay such proportionally large eggs. The large egg of the Ostrich (*Struthio camelus*) weighs an impressive 1800g, but comprises only 2% of the female body mass.

Larger eggs are associated with longer developmental times, and hence require a greater proportion of energy rich lipids in the yolk solids to fuel the total cost of development (Ar *et al.* 1987). Conversely, smaller eggs require a smaller energy investment by the female, while the male parent can contribute to food provisioning if the neonate hatches after a shorter incubation period (Vleck and

Vleck 1987). Not surprisingly, most extant avian taxa lay eggs that are proportionally considerably smaller than that of the kiwi, and frequently invest in multiple eggs and raising more than one offspring in a breeding attempt. Furthermore, there is good evidence that the evolution of smaller, less energy dense eggs has evolved from a likely precocial ancestor independently in multiple phylogenetic lineages (Bucher 1987). From an evolutionary perspective the length of incubation has placed strong selective pressures on egg composition and shell structure, particularly eggshell porosity (Ar *et al.* 1974). More important to the focus of this chapter, environmental pressures have modified these interdependent relationships. As embryonic development is entirely dependent on diffusion of O_2 and CO_2 through the eggshell pores, water is always lost from the egg contents as vapour through the same pores, because the airspaces between the eggshell and shell membranes are saturated with vapor. Although the reasons are still not clear, eggs that lose too much water or too little during incubation have higher mortality in the hatching period.

Egg shell porosity, and to a lesser extent the permeability of the shell membranes, determine gas conductance. Since pore size does not change appreciably during incubation the porosity of the shell and the resistance to gas movement from the egg is fixed during shell deposition *in utero* by the number of pores formed. Water loss from the egg is determined by the product of eggshell conductance and the water vapor pressure gradient between the egg and the nest (see Box 7.1). In dry climates (low absolute humidity) the fraction of the total resistance to egg water loss offered by a well lined nest can be as much as 35%, in the case of the Eider Duck (*Somateria mollissima*; Rahn 1984). In spite of the low ambient humidity in the Arctic nesting grounds of Eider Ducks, eider eggs lose water at rates similar to predictions based on allometry for eggs raised in temperate climates. In fact, most bird eggs have eggshell conductances within the predicted range of allometric relations between water vapor conductance and fresh egg mass in both moist and dry climates. Not surprisingly, eggshell porosity is adapted in some taxa to extreme nesting environments. For example, the water vapor conductance of eggs raised in the very moist environment of floating nests of birds like grebes are notably many times higher than predicted.

Birds long ago evolved a dependence on high and stable incubation temperatures to achieve continued embryonic development. In doing so, all birds have achieved rapid development times, which have allowed certain taxa to breed more than once in a season. Indeed, in some species such as the nomadic Zebra Finch (*Taeniopygia guttata*) the rapid development of the altricial neonate allows the young to achieve flight within two weeks of hatching and a successful breeding attempt can be completed within a month. Thus adult Zebra Finches and their offspring are able to move with feeding opportunities, tracking the ephemeral grasses as they spring up in the arid interior following good rains. Three or four breeding attempts are often possible in years of above average rainfall in

Box 7.1 Factors controlling water vapor loss from eggs

Hermann Rahn in his many publications with other renowned colleagues described the effect of egg size on the development of embryonic metabolism in birds, and the factors controlling water loss from eggs during incubation. His works have provided us with the best techniques to evaluate how climate influences embryonic development and egg water budgets. Birds hatch with similar water content to that initially invested in the freshly laid egg, and water is produced as a metabolic byproduct of lipid oxidation. In order to maintain an appropriate hydration state at the end of incubation, eggs lose water continuously through thousands or tens of thousands of pores in the calcareous shell (Ar and Rahn 1980).

Incubation time is therefore an important determinant of pore number and daily water loss. Daily water loss rate is the quotient of the water vapor pressure gradient and the total resistance between the egg and the environment. If an egg were prevented from losing water during incubation the embryo's metabolism would result in an increase in egg water content, about 5% in a precocial egg. However, assuming that lipids are the primary fuel for embryonic metabolism, a 20% reduction in egg solids would need to be balanced by a 20% reduction in water (since the C:H ratio for most egg lipids is 1:2.0) (Rahn 1984). Therefore, it is no coincidence that in general bird eggs lose 14–15% of the fresh egg mass as water vapor by the time lung ventilation commences shortly before hatching.

The figure below illustrates two examples of typical temperature and water vapor pressures in the egg (P_e), the nest (P_n), ambient environment (P_a), and the

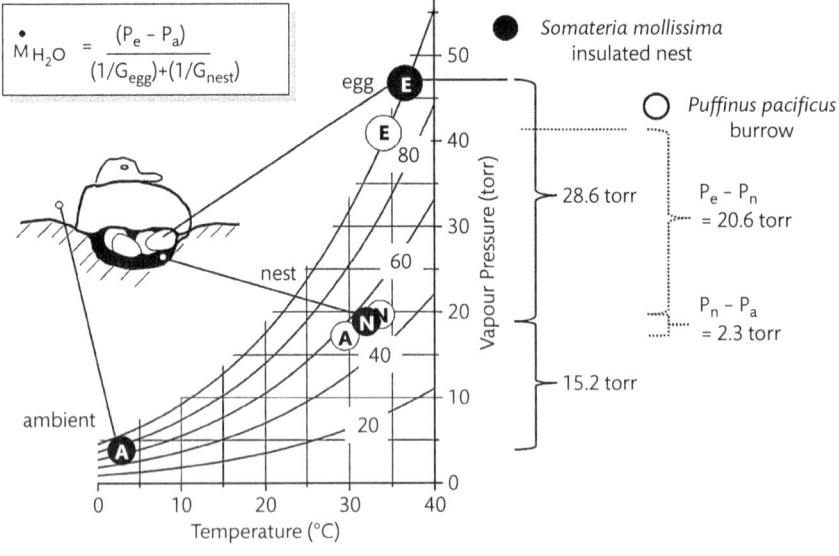

Fig. 7.1 Egg and nest water vapor loss gradients.

gradients of water loss between the egg and the environment (Adapted from Rahn 1984). The uppermost curve indicates 100% relative humidity or saturation water vapor pressure at various temperatures. Curves below this indicate isopleths for the indicated lower relative humidities. Although the major resistance to water loss is the eggshell itself the nest resistance to water loss is increasingly large in cold dry climates (e.g., Arctic and Antarctic). Simple stick nests and burrows, like that of the Wedge-tailed Shearwater (*Puffinus pacificus*), offer very little resistance to water loss. Hence, moderation of the incubation environment by nest insulation is most important in regions of low ambient humidity. It should also be evident that in dry climates the sudden loss of nest resistance (e.g., loss of nest lining) would increase daily water loss of the egg, and potentially result in desiccation of the egg as the net water vapor pressure gradient does not change ($P_e - P_a$). Eggshell water vapor conductance is essentially constant throughout incubation in most birds. Only in large eggs is there appreciable eggshell thinning and a significant increase in gas conductance. Birds do not appear to regulate nest humidity to influence egg water loss. For this reason water loss rates of all avian eggs can be adversely affected by changes in ambient water vapor pressure.

opportunistic species such as the Zebra Finch and Budgerigar (*Melopsittacus undulatus*) that are tuned to seasonal cues.

One consequence of the accelerated development of birds is that most taxa that are continuously incubated are not able to tolerate prolonged egg neglect because allowing eggs to cool to low ambient temperatures generally results in the death of the embryo. Moderate egg cooling is well tolerated during parental absences generally during the first days of incubation, when incubation behaviours are not yet enhanced by the hormonal changes that reinforce nest attentiveness and commitment to the current reproduction effort (see Box 7.2 – Hormonal control of incubation). Conversely, in arid environments that exceed 45°C, direct exposure of the egg to solar radiation when the parent bird is absent from the nest can lead to death of the embryonic bird, as the embryo is incapable of cooling itself to avoid organ failure and proteolysis.

The thermal tolerances of avian embryos in nature are rarely examined quantitatively because of the difficulty in making accurate long term recordings of temperatures in the egg and nest. Webb's (1987) review of incubation temperature regimes suggests that certain taxa, like the penguins, hummingbirds and passerines, have lower mean egg temperatures over the course of incubation and a greater range of temperature exposures than other birds. The procellariiforms as a group on the other hand, do not have low mean egg temperatures as originally perceived, although some of the long-distance foraging species (e.g., Diving Petrel *Pelecanoides georgicus* and Fork-tailed Storm Petrel *Oceanodroma furcata*)

Box 7.2 Hormonal control of incubation

The neural and hormonal changes during breeding have been well studied in domesticated birds and more recently in penguins, waders, and some passerines. These findings are discussed in detail in relation to the different patterns of incubation by C. Vleck (Deeming, 2002b; Chapter 5) and the brood patch by R.W. Lea and H. Klandorf (Deeming, 2002b; Chapter 8). The pituitary hormone prolactin increases in adult breeding birds before eggs are laid and generally peaks during early incubation, but then declines during the post hatching period at rates that appear to differ between altricial and precocial modes of development. Prolactin is not only important for reinforcing incubation behavior and the return of feeding birds to the nest, but is also responsible for inducing the formation of the brood patch, crop milk secretion in columbiforms and possibly regulating sex steroid release.

---

are reported to experience ambient temperatures as low as 10–15°C for prolonged periods. All birds appear to be better able to tolerate hypothermia than hyperthermia, and the estimated tolerance range of species that do not have prolonged interruptions in incubation is 16–41°C. In those species with interrupted incubation patterns, which are able to tolerate egg cooling for several hours a day, there does not appear to be any obvious adverse impact on development, other than delayed hatching. However, as a general rule for those taxa with continuous incubation, avian development, once started cannot be arrested for any significant length of time in terms of days.

Although moderate cooling of the eggs is common among birds the impact it has on growth and development has not been investigated adequately across a range of avian taxa. Many researchers have observed that regular brief interruptions to incubation, resulting in egg cooling, extend the normal incubation period. Therefore it is expected that the cumulative cost of development should increase. Olson *et al.* (2006) performed an experiment to investigate whether there are any additional costs associated with frequent exposure to short term egg cooling. Using artificially incubated Zebra Finch eggs they simulated the periodic cooling that might occur in a species with uniparental incubation or if a Zebra Finch lost its mate (ordinarily a species that utilizes biparental care). In this experiment the eggs were exposed to cooling to 20°C once per hour during daylight hours. They found that periodic cooling delayed development and significantly increased metabolic costs; i.e., mass-specific metabolic rates measured at a constant incubation temperature were elevated. Furthermore, growth efficiency and neonate size were reduced. The authors suggest that elevated maintenance costs might be incurred when egg temperature is returned to a stable high temperature.

Conceivably elevated costs might be associated with the production of multiple isozymes of key enzymes that can operate at wider thermal ranges, increased embryonic metabolic intensity (mitochondrial density) and or a change in lipid metabolic pathways. How cold acclimation is achieved in embryonic birds remains to be elucidated.

Birds in the tropics are long-lived and produce smaller clutch sizes and fewer breeding attempts than in temperate regions (Ricklefs 2000). The young of tropical birds typically grow more slowly and reach maturity later, suggesting that the pace of life is slower in the tropics. Indeed the basal metabolic rates across a large sample of adult birds in the tropics were shown to be lower than similarly sized birds in temperate regions (Wiersma *et al.* 2007). Interestingly, other studies now suggest that parental attentiveness and egg temperatures might be as large a determinant of developmental time as egg mass in tropical passerines (Martin *et al.* 2007; Martin 2008). The relation between relative egg size and incubation duration was compared in conspecific passerines from tropical Venezuela, subtropical Argentina, and temperate Arizona. Longer development times were observed in the tropical species as these species more frequently allowed egg temperature to fall during long nest absences than temperate species of similar egg mass (Martin 2008). Larger egg size in tropical passerines might reflect the increased energetic requirements of frequent egg cooling during development. However, extrapolation of this finding to non-passerine orders might be premature, as there is no supporting evidence from other within-order comparisons. For example, parrots are found from the tropics to temperate regions and all lay relatively small eggs with long incubation periods in comparison to other altricial orders (Bucher 1987; Pearson 1999), and therefore do not appear to fit this hypothesis.

Two distinct modes of heat transfer are often employed by birds to warm their eggs and reduce heat loss: the specialized brood patch and webbed feet. Many terrestrial birds will develop a brood patch on the abdomen of one or both sexes soon after egg laying. Highly insulative feathers have low thermal conductivity and present a barrier to heat transfer between the adult body and the egg or young. Thus, in terrestrial birds the breeding adult will lose feathers from the tracts in the brood patch region. Efficient heat transfer is achieved by two processes. First, the skin develops loose folds in the regions of contact between the ventral surface of the adult and the upper egg surface. Nonetheless, the amount of direct contact between the brood patch and any individual egg may be a small fraction of the egg surface. Heat transfer is further improved by changes in the circulation to the brood patch, including increased vascularization of the brood patch skin (Midtgård *et al.* 1985), increased sensitivity of the brood patch to cold, and selective redistribution of blood flow to this region (Brummermann and Reinertsen 1991, 1992). Several studies have shown that at least some species are capable of increasing blood flow to the brood patch in proportion to the size of the thermal gradient between the adult core temperature and the egg (Gabrielsen and Steen 1979; Tøien *et al.* 1986; Tøien 1993). How is heat transfer from the incubating adult bird affected by metabolic heat from the embryo as its thermogenic capacity develops? Since only a small fraction of the egg's surface may be in contact with the brood patch, heat enters the egg through a small area but leaves the egg through areas not in contact with the adult bird. With development the embryo's circulation increases across the inner surface of the eggshell and so does the area through which heat is lost from the egg. Consequently, parental heat production to warm eggs must increase during incubation (see Box 7.3 Modeling of heat flow during incubation).

It is noteworthy that bird embryos develop in thermal regimes that differ from that of placental mammals. Placental mammalian embryos will develop under temperatures close to the core temperature of the adult mammal. In contrast, due to lapses in incubation the avian embryo may experience fluctuating temperatures throughout most of its development (Whittow and Tazawa 1991). In multiple egg clutches the temperature of individual eggs will vary with the degree of brood patch contact. The presence of nest insulation and plumage around the brood patch traps a layer of air around the egg, reducing the extent of heat loss from the sides and bottom of the eggs (Turner 1987). Nevertheless, if environmental temperature is much lower than the adult core temperature then mean egg temperature will of course be lower than that of the adult bird.

Box 7.3 Modeling of heat flow during incubation

Scott Turner's modeling studies demonstrate why unidirectional heat flow vectors fail to explain heat flow in the contact incubated egg (Deeming, 2002b; Chapter 9). Heat flow is affected by the distribution of egg contents within the egg and the extraembryonic circulation through the vessels of the chorioallantoic membrane. A one-dimensional temperature gradient assumes that heat from the brood patch leaves the egg uniformly over the egg's surface. In fact, a closer approximation is a two dimensional temperature field within the egg. Early in incubation the embryo floats close to the uppermost surface in contact with the brood patch, due to the higher specific gravity of the allantoic fluid. In a contact incubated egg more of the heat imparted to the egg is lost from the exposed surfaces of the egg adjacent the brood patch than from surfaces on the opposite side of the egg. As the chorioallantoic membrane spreads across the egg's inner surface, heat transfer within the egg increases, thus increasing heat loss from the surfaces further from the brood patch. Earlier models of incubation thermodynamics erroneously presumed that as heat production by the embryo increased with development, parental energy expenditure in maintaining a steady state incubation temperature should decrease. In fact, Turner's model predicts that heat requirements from the adult increase during incubation because of the changing temperature field within the egg.

Not all birds develop specialized vascularized brood patches. Some species undergo localized defeathering (e.g., parrots, Cassin's Auklet *Ptychoramphus aleuticus*, Bank Swallow *Riparia riparia*). It is not clear whether the efficiency of heat transfer is reduced in species that do not form vascularized brood patches, as accurate recordings of thermal profiles within eggs, at the brood patch surface, and within the nest are lacking for a diverse range of avian taxa.

Many diving species in the Pelecaniformes, such as the boobies and pelicans, do not develop brood patches. Some have argued that the maintenance of a highly insulative plumage for species feeding in the water for prolonged periods has selected against the seasonal development of brood patches (R.W. Lea and H. Klandorf in Deeming, 2002b). Instead these birds, which breed across all ocean coastlines but the polar regions, have developed the ability to incubate their eggs by close contact with their webbed feet. However, this reasoning cannot explain why other specialist aquatic taxa, such as the alcids and penguins (excluding *Aptenodytes* spp.), do develop highly vascularized brood patches despite frequent immersion in cold waters. Regardless, such incubation behavior in pelecaniforms is closely correlated with fine regulation of blood flow to the highly vascularized skin webbing.

Higher incubation temperatures and higher metabolic rates have resulted in more rapid growth and development in birds in comparison to reptiles, but the maintenance of stable thermal incubation environments for their eggs occurs at the expense of reduced opportunities for parental foraging. Although activity at the nest of the incubating adult is reduced, continuous incubation can be energetically expensive because eggs lose heat to the environment and the embryo is unable to contribute appreciable heat. Furthermore, rewarming of cool eggs can transiently increase parental energy expenditure, as evidenced by the high heart rates, increased heat production and metabolic rates in a range of small and large species with multiple egg clutches (Tøien 1993).

Measurements of daily energy expenditure with doubly labeled water (DLW, $^3H_2^{17}O$) in breeding birds suggest that incubation can be more energetically expensive than the post hatching period in small terrestrial birds that experience low ambient temperatures (J.M. Tinbergen and J.B. Williams, Chapter 20 in Deeming, 2002b). Even in temperate regions, small birds such as sunbirds, Tree Swallows (*Tachycineta bicolor*), tits, and honeyeaters frequently experience very low temperatures at night and in some cases the energetic cost of maintaining high stable egg temperatures is further elevated by large clutch sizes relative to adult body mass. In contrast, studies of larger birds indicate that energy expenditure is higher in the post hatching period than during incubation. For example, metabolic rates interpolated from 24 h field energy expenditures of incubating Adélie Penguins (*Pygoscelis adeliae*) at near zero ambient temperatures are no different from the basal metabolic rates of non-breeding birds (Chappell *et al.* 1990).

7.1.2 Neonate and Development

Hatching relieves the physical constraint of the egg and opens up opportunities for investment of chemical potential energy beyond that initially invested in the egg, from food provided by one or both parents or (in precocial species) from the hatchling's own foraging behavior. Notably, it also places new challenges on the neonate to maintain thermal, energy, and water budgets. The altricial-precocial spectrum of avian development is strongly influenced by environmental conditions and has been the subject of extensive investigation. The influence of physiological and behavioral maturity during early development on postnatal growth was reviewed by Starck and Ricklefs (1998). The reader is referred to this earlier work for much of the details about the neurohumoral basis of thermoregulation, the diversity of responses to cold challenges and the origins of altriciality in birds.

As stated earlier in this section, one long standing hypothesis is that growth rate is inversely related to functional maturity (Ricklefs 1979). During embryonic development relative growth rates of embryo body mass do indeed appear to be inversely correlated with the increase in tissue dry matter. However, recent findings now suggest that the trade-off between growth rate and functional maturity in birds during the post hatching period might be resolved through different strategies in the chicks of small and large species (Krijgsveld *et al.* 2001; Williams *et al.* 2007). Here, we describe how the timing of the development of endothermy in birds affects post hatching growth and energy expenditure in different climates. Finally, we briefly consider the evidence for alternate strategies for growth and tissue maturation in some taxa.

Most chicks hatch with weak thermoregulatory abilities and so even moderate thermal environments can be lethal to the neonate. While all avian neonates attain a high metabolic intensity shortly before or at hatching, resting metabolic rates are much lower than adults of the same body mass. Within the first 24 h after hatching precocial neonates activate temperature regulation mechanisms and become capable of modest thermogenic responses to cooling. The weak thermogenic heat production of many precocial and semi-precocial wader and gull chicks allow them to remain active until body temperature drops down to ~30°C. Distress calls from the chilled chicks elicit parental brooding and prevent lethal hypothermia. The tolerance of chicks to hypothermia varies considerably among taxa and probably reflects adaptations to the irregularity of chick feeding in some taxa. Chicks of the Fork-tailed Storm Petrel regularly experience severe hypothermia (body temperature ~10°C) in the safety of their burrows during prolonged parental absences without any apparent adverse effect (Boersma 1986). Altricial neonates do not increase metabolic rate during cold challenges and are totally reliant on parental brooding. On the basis of a very limited number of studies, altricial neonates do not appear to be any more tolerant of low ambient temperatures than other birds.

For some taxa the early development of thermogenic powers and fine control of body temperature is essential for development because they are either exposed to cold when moving to feeding grounds or the open sea, or they are exposed to low temperatures during prolonged parental absences. In the Ancient Murrelet (*Synthiliboramphus antiquus*) and Xantus' Murrelet (*S. hypoleucus*), chicks leave their nest within one or two days of hatching to join the male parents on the open ocean, where he brings them food on the water's surface. At hatching (about 25 g body mass), murrelet chicks are endothermic and capable of large increases in metabolic rate in response to cold (~4 times resting metabolic rate) and within days are able to reduce the rate of heat loss to the cold ocean waters presumably through more effective control of their peripheral circulation (Eppley 1984) (see Box 7.4 – Thermogenesis and scaling of homeothermy). Megapodes are exceptional among terrestrial birds in that they are independent of their parents after they escape from their incubator mounds. Not surprisingly, these superprecocial neonates have strong thermogenic responses to cold and a juvenile plumage rather than natal down to enable effective thermoregulation (Booth 1984).

Among the taxa with precocial and semi-precocial modes of development, Galliformes (megapodes, pheasants, quail), Anseriformes (ducks and geese), Charadriiformes (auks, terns, gulls, shorebirds), and Sphenisciformes (penguins) there is considerable variability in the extent of thermogenic powers and the degree of homeothermy at hatching. Indices of homeothermy are highly correlated with neonate body mass (see Box 7.4 – Thermogenesis and scaling of homeothermy). Nevertheless, all develop thermogenic responses to cold early in the post hatching period. The timing of the increases in aerobic metabolic scope (the difference between resting and peak metabolic rates) appears to be related to the timing of nest departure (Chapter 5, G.H. Visser and Chapter 6, E. Hohtola and G.H. Visser in Deeming 2002b).

In most non-altricial species that attain homeothermy early in the post hatching period, parental brooding can provide a considerable energetic saving for chicks. However, the dependency of small precocial neonates on brooding can also constrain energy intake as they are self-feeding. In the King Quail (*Coturnix chinensis*), one of the smallest galliforms (hatchling and adult body mass 4 and 40–45 g respectively) with a subtropical and temperate distribution, it was established by respirometry that parental brooding at low ambient temperatures significantly reduces the costs to chick thermoregulation in the first week after hatching (Pearson 1994). Nevertheless, continual exposure to ambient temperatures < 20°C resulted in chicks failing to increase body mass. Food intake of the chick became limited by the necessity for prolonged brooding, and available energy reserves were probably exhausted by the energetic demands of thermoregulation. Time spent brooding at low ambient temperatures cannot be used for foraging by either the parent or chicks. At high latitudes small precocial chicks satisfy energy requirements during low ambient temperatures and inclement weather by taking advantage of continuous daylight for feeding.

It should be apparent that thermoregulation in cold climates presents a greater energetic burden for the precocial or semi-precocial chick than the adult bird due to allometric scaling effects and lack of mature plumage. The fraction of available energy allocated to growth in birds is dependent on the timing of the onset of endothermic responses, the chick's activity costs, and the thermal environment of the chick (Ricklefs 1979). Originally, Ricklefs suggested that there must be a trade-off between the maintenance of mature muscle function and the rate of cell proliferation to support further growth. However, whether post hatching growth in precocious chicks is limited by the rate of cell proliferation *per se* is not easily discernable from the data available. In the Arctic, where energy rich food supplies are abundant, small chicks might not be constrained by their ability to process enough food to support high growth rates, but rather how long their thermoregulatory abilities enables them to remain active to collect that quantity of food.

It remains unclear if the ancestral bird utilized a precocial or altricial mode of development. The phylogenies of Sibley and Ahlquist (1990) place the large birds with precocial young (Craciformes, Galliformes, Anseriformes, Tinamiformes, and Struthioformes) within Eoaves. All other taxa are included in Neoaves. Starck and Ricklefs (1998) reason that early in avian evolution there must have been both diversification of altricial forms and a change from altricial to precocial forms. For example, if the Turniciformes (button quail) are correctly placed at the base of Neoaves, then altriciality is the derived state in all extant altricial taxa. Also on the basis of distance analysis of DNA-DNA hybridization data, precociality must have evolved secondarily on four occasions within Gruiformes (grebes, rails) and Ciconiiformes (shorebirds). Many of those more recently evolved altricial taxa have delayed the development of thermoregulation and locomotion, and have benefited by remaining at the nest until fledging. The passerines have post hatching growth rates some 3–4 times higher than that of precocial young in Galliformes, which overlap in the range of neonatal body masses. In most passerines, the onset of substantial thermogenic responses is delayed until the middle of the nestling period. Effective homeothermy of individual chicks is achieved after growth is essentially complete.

It has long been believed that by delaying the development of thermoregulation, more energy is made available for rapid growth in altricial chicks (Ricklefs 1979). However, not all altricial taxa grow rapidly in the post hatching period, even though they are nest bound and relatively inactive. Parrots (Psittaciformes), kingfishers, bee-eaters, and hornbills (Coraciiformes) have post hatching growth rates intermediate between the fast growing altricial passerines and comparable sized precocial taxa. In parrots, thermogenic responses have been demonstrated in chicks less than one week old and adult levels of metabolic intensity are attained when 20–40% of the fledging period has elapsed (Bucher 1987; Pearson 1998). It is unlikely that parrots are exceptional in this regard. Other altricial taxa including pigeons, cormorants, and pelicans also attain homeothermy before chicks reach half of the adult body mass. Among these altricial birds there are large differences in growth rate, even within the same order. In Pelecaniformes, the Northern Gannet (*Sula bassanus*), and frigate birds (*Fregata* spp.) grow much slower than the pelicans. Pigeons and doves, also achieve fast growth early, but slow down after their early departure from the nest. On this basis, there is no simple inverse relation between the development of thermoregulation and post hatching growth rate. In addition to a possible trade-off between mature function and growth, other factors may act to constrain growth at either the level of the organism or individual tissues (Chapter 11, R.E. Ricklefs, J.M. Starck and M. Konarzewski in Deeming 2002b). Sibling competition favors rapid growth, but a limited or infrequent food supply favors slow growth. It is also possible that post hatching growth is more related to the development of the nervous system.

Box 7.4 Thermogenesis and scaling of homeothermy

Homeothermy is the ability of a chick to maintain a high constant body temperature (T_b) over a range of ambient temperatures (T_a). In young birds homeothermy indices are frequently used to refer to a chick's ability to produce heat equal to heat loss to the environment at low T_a. In order to determine a chick's thermogenic response (heat production, [mW or W]) to cold and thereby quantify the range of T_a over which a chick can remain active we need to know its rate of heat production, thermal conductance, and the range of T_b over which it can maintain function. Heat production can be determined by direct calorimetry, but more frequently aerobic metabolism determined by open-circuit respirometry is used to determine heat production indirectly.

Basal metabolic rate is routinely used to compare the minimum metabolism of adult animals under resting and fasted conditions within the thermoneutral zone (the range of T_a in which metabolism is minimal). However, young birds are rarely fasted in the wild and are continually growing, and therefore do not satisfy the definition of basal metabolism. Thus it is appropriate to determine the resting metabolic rate of chicks at a thermoneutral temperature (frequently not a T_a range in small chicks), which might be 35°C in large neonates and as high as 40°C in small ones. Below the thermoneutral temperature, metabolic rate must increase as T_a decreases to maintain a constant T_b, until at some point (peak metabolic rate, PMR) the chick is no longer able to sustain further increases in heat production, and metabolic rate and T_b decrease – this is the Newtonian Law of Cooling.

$$T_b - T_a = PMR / k \text{ (Eppley 1994)} \qquad (1)$$

According to Equation 1, a reduction in the minimum tolerable T_a and an increase in the maximal thermal gradient [°C] between a chick and the environment can be achieved either by an increase in PMR or by a reduction in minimal thermal conductance (k, [W.°C^{-1}]). Heat loss occurs through the insensible route of evaporative water loss from the respiratory tract and skin, as well as sensible radiation, convection, and conduction through the plumage and skin. The term k incorporates both sensible and insensible heat loss. Even though chick body temperature is not uniform during cooling, k is minimal at the minimum T_a where PMR occurs and therefore can be determined if PMR and thermal gradient are known.

Avian neonates are variable in their resting metabolic rates and their capacity for thermogenic heat production. The timing of the onset of endothermic responses varies among avian taxa and a major determinant of post hatching growth rates. A useful thermoregulatory index of homeothermy (H) compares the proportion of the adult thermal gradient maintained by a chick during cold exposure (Dunn 1975) [Equation 2], provided that cooling is measured under standardized conditions (T_a 20–25°C and 30 min duration); where T_c and T_A are the chick and adult body temperatures [°C] at the end of the cooling period.

$$H = 100 \times (T_c - T_a) / (T_A - T_a) \tag{2}$$

If a H index of 100% is considered to be physiological homeothermy, then effective homeothermy is accepted to be 75% of the adult ability. While these concepts are useful for investigating the relationship between growth rate and the development of mature function, they do not necessarily reflect the homeothermic capacity of a chick in the wild or its ability to tolerate typical environmental conditions. The cooling rates of chicks are slowed by exposure to strong incident radiation, nest insulation, and brood huddling behavior, and increased by wind and rain.

The body mass at which young birds achieve homeothermy is strongly related to neonatal body mass, surface area to volume ratio, and phylogeny. Independent of development mode, k allometrically scales with body mass$^{-0.17}$. Similarly, the surface area to volume ratio scales with body mass$^{0.67}$. Both relations infer that large neonates will achieve homeothermy earlier in development than smaller species of the same taxonomic group simply because they possess more favorable heat retention properties. Although these biophysical properties influence H, phylogeny (i.e., precocial vs. altricial status) is of overriding importance. H scales to mass$^{0.67}$ in precocial ducks, but to mass$^{0.82}$ in altricial passerines (Chapter 5, G.H. Visser in Deeming 2002b). A further consideration is the water vapour pressure in the environment. Evaporative water loss is elevated in the dry Polar environment, increasing insensible heat loss from small neonates. However, a high surface area to volume ratio is the major factor limiting homeothermy in small chicks. Consequently, the smallest shorebirds only attain homeothermy at about half the adult body mass, whereas larger shorebirds attain homeothermy shortly after hatching.

As previously mentioned, there can be considerable energetic savings for chicks from nest insulation and parental brooding. Although endothermic responses are considerably delayed in individual passerine chicks, as a brood in a well insulated nest chicks are able to reach effective homeothermy much earlier in the nestling period (Dunn 1976). In most altricial taxa, homeothermy is achieved before the completion of plumage growth towards the end of the nestling period. Efficient heat transfer during parental brooding and brood huddling requires that insulating feathers and down are absent or compressible (Webb 1993). A well insulated down serves to reduce heat loss to the environment, but it also hinders heat transfer by a parent through conduction. In a well insulated nest, young passerines can potentially achieve effective homeothermy before their plumage is complete. In the Cockatiel (*Nymphicus hollandicus*), a small parrot, endothermic responses develop well in advance of the end of parental brooding, which stops when the oldest chick is 12–13 d post hatching (Pearson 1998). Thereafter brood huddling enables these asynchronously hatching chicks to maintain high body temperatures during the daylight hours, when the parents are absent and the oldest chicks have attained homeothermy. It seems reasonable to speculate that the naked ventral surfaces of these young Cockatiel chicks enables more efficient heat flow between the siblings huddling in their nest hollows, which offer little resistance to cooling. In this parrot the early development of thermoregulation can be an energetic burden for the chicks at low temperatures. As the energy requirements of the brood peak earlier in development presumably more energy must also be spent by the parent in foraging, although the cumulative energy budget would be reduced as a result of a shorter growth period. On this basis, parental brooding is not the sole thermal strategy available to birds with altricial modes of development.

Altricial seabird chicks, such as some pelicans and penguins, form crèches during parental absences. This occurs at an age when the thermogenic powers of the chicks have developed and the parents are no longer capable of covering them during parental brooding. Down over the whole body acts to reduce heat loss from exposed chicks at low ambient temperatures. However, when crèches form chicks can make significant metabolic savings by huddling (Lawless *et al.* 2001). Contact between huddled chicks and reduced air flow around individual chicks act to reduce the surface area of chicks exposed to low temperatures, and thereby reduces the cost of thermoregulation in these large young birds.

In this section the focus has been on adaptations to cold in young birds. At the other extreme of environmental conditions, neonates in some coastal and desert breeding sites must deal with elevated body temperatures (hyperthermia) daily. At high ambient temperatures and low humidities the evaporation of water across the skin, legs, and respiratory passages in neonates is the primary means for heat dissipation. Neonatal respiratory water loss can be increased at high temperatures by a factor of 10 (Booth 1984). The altricial and semi-precocial neonates of pelicans, boobies, herons, terns, and gulls that breed in coastal regions are capable of gular fluttering (rapid vibration of the throat poach) to facilitate heat loss at high ambient temperatures. However, all neonates are thought to be able to increase respiration frequency as ambient temperature increases above the thermoneutral temperature.

The limited number of studies of evaporative water loss in neonatal birds suggests that neonates maintain lower maximal evaporative water loss rates than adults of the same body mass because they have lower resting metabolic rates at the same high ambient temperature (G.H. Visser in Deeming 2002b). A lower metabolic intensity in the neonate means a lower heat load (heat stress) and minimizes the need for heat dissipation. In the few species that have been tested, neonates appear to have better heat defenses than most adult birds. It has been suggested that a lower resting metabolic rate in neonatal birds in warmer climates conserves respiratory water loss (Klaassen and Drent 1991). Nevertheless, the limited metabolic water that can be generated from internal yolk stores of any neonate would not allow it to maintain water balance for long in a hot climate without water provision. Some well known adaptations of breeding birds that enable chicks to maintain water balance in desert environments include shading the young during the hottest part of the day and transporting water to the young (Cade and Maclean 1967; Hinsley and Ferns 1994).

Parental shading of neonates and young chicks reduces the heat load from direct solar radiation, and is employed by both small birds and birds as large as storks and ibis in their exposed stick nests. Plovers (*Pluvialis* and *Vanellus* spp.) and sandgrouse (*Pterocles* spp.) in the deserts and arid steppes not only soak their ventral feathers in distant water sources and bring it back to the nest to cool their eggs during incubation, but also so that the chicks can drink directly from their feathers (Hinsley and Ferns 1994).

For species with more altricial neonates, maintaining water balance is closely tied to maintaining energy balance, since all preformed and metabolic water is delivered in the food provided by the parents. Unlike precocial chicks, which can move independently to secure food and water, young birds of other developmental modes are entirely dependent on adult provisioning. Exceptionally, sandgrouse chicks are precocial, but nest long distances from water in some arid regions and the young are totally reliant on the male supplying water until the young are capable of making the long distance flights to water sources (Hinsley and Ferns 1994).

Ricklefs and Starck (1998, Chapter 16) speculated in their concluding remarks that the neonatal size differences between altricial and precocial taxa we see today reflect major evolutionary changes in avian life histories. Small active birds that utilize flight extensively for feeding dominate the altricial taxa in Neoaves, whereas large birds feeding on relatively immobile prey on the ground or water surface dominate Eoaves. These traits are also apparent in the taxa that have changed from precocial forms to semi-altricial taxa and altricial forms that secondarily became precocial taxa. From the discussion in this section it should be apparent that neonatal size and the development of metabolism are important to our understanding of evolution in birds.

Over the past 40 years the "growth rate – maturity hypothesis" developed by Robert Ricklefs has been accepted as the most plausible explanation for the diversity in post hatching growth rates across the altricial-precocial continuum. This model derives from the concept that mature function of tissues in general, and the skeletal muscles in particular, is incompatible with rapid cell proliferation (Ricklefs 1979). Supporting evidence has been largely centered on comparisons of growth rate and the development of thermoregulation in precocial and altricial chicks. However, as Klaassen and Drent (1991) demonstrate, precocial shorebird chicks in Arctic regions show more rapid growth associated with higher resting metabolic rates in comparison to conspecifics in temperate regions; in contradiction to the growth rate–maturity hypothesis.

Rigorous testing of the growth rate–maturity hypothesis has been difficult, as skeletal muscle from different body regions grows at different rates in different taxa, and the relative contributions of different muscle groups to whole animal thermogenic capacity is generally poorly known. Nevertheless, there is now evidence that in shorebird chicks early development of heat production might not preclude rapid skeletal muscle growth. Along with many other factors, catabolic enzymes such as citrate synthase and pyruvate kinase determine the capacity of muscle to produce adenosine triphosphate for sustained shivering. Measurements of resting and peak metabolic rates of shorebird chicks and their muscle catabolic enzyme activity suggest that faster growing pectoral muscles increases the metabolic intensity of some smaller species, enabling them to reach homeothermy early (Krijgsveld *et al.* 2001; Williams *et al.* 2007). Why other shorebird species grow more slowly is not clear. These studies suggest that the relationship between metabolism and growth rate is more complex than suggested in the basic growth rate–maturity hypothesis. Combining some of these biochemical approaches with studies of metabolism and thermoregulation in a range of avian taxa should provide new insights into avian evolution and how particular environments have influenced avian developmental strategies.

8

Approaches and Techniques

Although the emergence of new techniques, methods, and analytical approaches is hard to predict, research on avian biology has greatly benefited from technical advances (for example, tracking technology), and it should continue to do so in the coming decades. From that perspective, we briefly discuss several new areas of research that we believe will profoundly influence our understanding of avian biology in general and evolutionary physiology in particular. These range from molecular and genomic methods to new technical approaches addressing long-standing physiological questions.

8.1 Methods for Measuring Energy Expenditure and Movement

8.1.1 Patterns of Free-living Birds

One of the principal goals of ecophysiology is to characterize the physiological performance of animals in the context of their natural environment. This is typically done by capturing free-living animals and measuring their physiological response to variation in selected physical variables in a controlled laboratory setting. Such approaches are necessary to understand how particular environmental components affect the physiology of a species of interest. Additionally, the accumulation of data from numerous researchers using the same standards of environmental control has yielded substantial insight into such things as the effects of phylogeny and size on traits including basal metabolic rate, thermoregulatory response to cold, wind-chill effects, etc.

While these studies provide much insight into an animal's fundamental physiological capacity, they are usually difficult to extrapolate in a meaningful way to free-living conditions where abiotic features are very complex and, most importantly, where animals experience substantial biotic interactions. Thus questions concerning how free-living animals spend their time and energy require field-based methods to yield relevant answers. The two main approaches used currently rely either on isotopes or electronic surveillance. The choice in using these methods for measuring field metabolic rate (FMR) or for determining geographic locations and/or behavior over time depends on the temporal resolution required, animal size, likelihood of recapture, relative mobility during the measurement period – and of course, the nature of the question under investigation.

8.1.2 Free-Living Energy Expenditure

The most commonly used procedures for estimating FMR of free-living animals are the doubly labeled water (DLW) and heart rate (HR) methods (Butler *et al.* 2004). The DLW method requires injecting an animal with labeled O_2 (^{18}O) along with labeled hydrogen (deuterium or tritium) and then determining the rates of disappearance of these isotopes over time. Because the labeled O_2 contained in body water is in exchange equilibrium with respiratory CO_2, the labeled O_2 disappears from the animal as both water and CO_2, whereas labeled hydrogen is lost only as water (Fig. 8.1). Thus, the differential rate of elimination of these

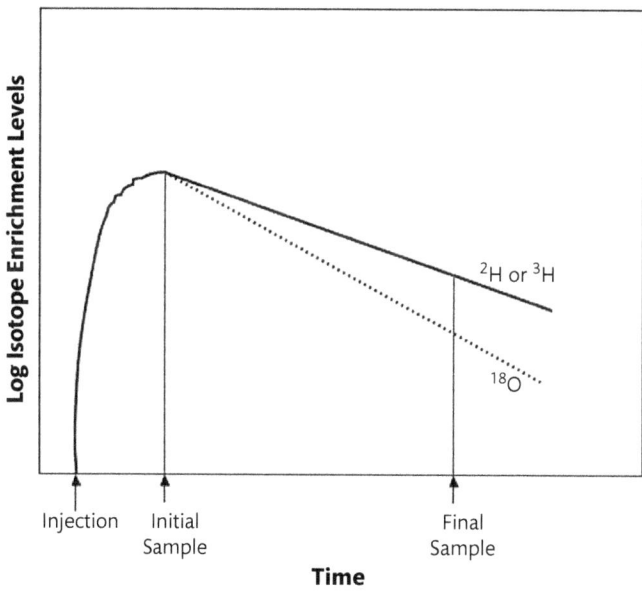

Fig. 8.1 Schematic representation of the time course of isotopic turnover in an animal following doubly labeled water injection. Body fluid content of the isotopes rises rapidly until reaching equilibrium, at which time an initial sample is taken. Subsequent samples are taken to compare elimination rates of labeled oxygen (^{18}O) to those for labeled hydrogen (deuterium or tritium, ^2H or ^3H, respectively). The difference in elimination rates of ^{18}O versus labeled H permits a quantifiable estimate of CO_2 production rate (redrawn from Butler *et al.* 2004).

two isotopes reflects an animal's CO_2 production, which, in turn, is related to its rate of aerobic metabolism (Lifson *et al.* 1955).

The HR method relies on conversion of bioelectrical signals accompanying the cardiac contractile cycle into transmitted or stored information on heart beat frequency. Transmitters usually send HR data via UHF or VHF signals to a receiver, whereas HR storage devices, sometimes called archival tags, digitally encode the HR data into onboard memory for later retrieval. Because archival tags are not burdened with the circuitry and power required for transmitting data, they consume less energy and can store substantial quantities of data in a smaller package than that required for transmitters designed for similar functions. The HR method presumes a strong correlation between heart rate and rate of O_2 consumption (VO_2), which has been demonstrated in laboratory measurements of graded activity in many vertebrate species. While both DLW and HR methods have the potential to provide good estimates of FMR, the decision to use either of them must consider the constraints associated with their use.

8.1.3 Temporal Considerations

Many questions posed by animal ecologists and evolutionary biologists require knowledge of the amount of energy or effort an animal expends for particular periods or activities. For example, those interested in examining life-history theory might be interested in contrasting the relative proportion of daily energy expenditure (DEE) that each parent allocates to offspring in species with short-life spans compared to long-living ones. Such a study would require quantification of energy expenditure for all daily activities related to chick rearing and thus would need an audit of daily activities along with measurements of their energy costs. By contrast, someone interested in altitudinal or latitudinal effects on a species daily energy needs might examine DEE across a species natural distribution over a range of seasons. These studies have very different sampling requirements and are, therefore, likely to favor one FMR method over the other.

Simply stated, the DLW method is not suitable for studies that need to resolve FMR values over periods of minutes to hours. It works best when the isotopes in labeled animals have reached at least one biological half life, which means usually from 1–5 days, depending on animal size and other variables. Thus, while the DLW method provides accurate estimates of total CO_2 expired over extended periods, the costs of component activities comprising the period of measurement can only be approximated when combined with concurrent time-activity budgets and knowledge of the thermal environment (Buttemer *et al.* 1986). If thermoregulatory requirements are not known, it is possible to estimate the costs of major activities using multiple regression or other statistical approaches if: (i) there is a large number of individuals in the study; (ii) time in different activities during the DLW measurement can be quantified; and (iii) there is substantial variation in activity levels among individuals (e.g., Chappell *et al.* 1993b. By contrast, the HR method dynamically reflects moment-to-moment changes in physical effort and thus has the potential to discriminate the energy costs of individual activities in a daily cycle. Consequently, those wishing to examine questions concerning allocation of effort to specific activities would need to validate this method for their study species.

8.1.4 Calibration Requirements

Heart rate: An examination of the Fick principle shows that predictions of VO_2 from HR may be complicated by variation in several associated variables:

$$VO_2 = f_h \cdot V_s (C_a O_2 - C_v O_2) \tag{1}$$

where, f_h is heart rate, V_s is cardiac stroke volume, and $C_a O_2$ and $C_v O_2$ represent arterial and mixed venous blood O_2 content, respectively. For f_h to accurately predict VO_2, the product of Vs and $(C_a O_2 - C_v O_2)$, also known as O_2 pulse (OP), must be invariant over the full range of animal activities, or the components of OP must covary in a predictable manner with VO_2. Measurement of these variables in resting and flying pigeons provides some insight into these interactions (Butler *et al.* 1977). Comparison of HR and VO_2 in resting birds before and after the fans of the wind tunnel were switched on show strong correspondence, rising 1.9 and 1.7 times, respectively, upon wind tunnel activation. By contrast, the VO_2 of birds flying at 36 kmh^{-1} increased nearly 10-fold above resting values, whereas HR was only 5.8 times higher. This disparity is explained completely by the 1.8-fold greater arteriovenous O_2 content difference in flying versus resting birds (Butler *et al.* 1977), implying that cardiac stroke volume changes little during exercise. There is also indirect evidence that OP may vary in relation to the type of physical activity irrespective of effort. For example, King Penguins (*Aptenodytes patagonicus*) showed a rise in HR with exposure to progressively lower temperatures and also showed a rise in HR with increased walking speed on a treadmill (Froget *et al.* 2002). The average HR at the lowest temperatures, however, was nearly double that of penguins walking at a speed provoking the same level of VO_2 (Fig. 8.2). Similarly, Barnacle Geese (*Branta leucopsis*) showed a far greater VO_2 at a commensurate HR while flying than while walking, thus implying a higher OP during flight (Ward *et al.* 2002).

Although activity-dependent variation in OP makes interpretation of FMR from HR more difficult, most studies have identified linear relations between HR and VO_2 with increased energy expenditure for particular activities. This means that establishment of HR and VO_2 relations for particular behaviors of interest can be confidently used to quantify the amount of effort a bird puts into such

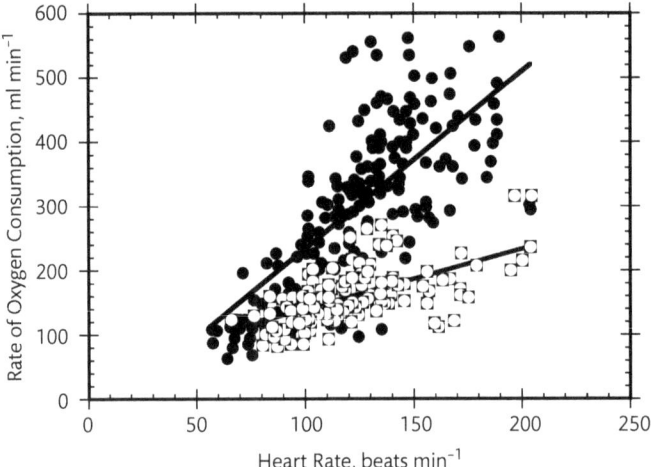

Fig. 8.2 The heart rate of 24 adult King Penuins (*Aptenodytes patagonicus*) during treadmill walking (filled symbols) and of 26 adult penguins during exposure to cold temperatures (open symbols). (Redrawn from Froget *et al*. 2002).

activities in the field. Among the constraints of this method is that it requires access to a respirometry system and a period of captivity to make the measurements, and that the bird can be tested in captivity while performing the same activity of interest in the field (i.e., flight, swimming, etc.). If captivity is likely to interfere with the bird's subsequent free-living activities, there are only two options. One is to establish the relation between HR and VO$_2$ using other birds from the same population and the other is to fit a HR recorder/transmitter onto focal birds and then capture these individuals upon conclusion of the period of interest to validate the relation between VO$_2$ and HR. It is important to note that the relation between HR and VO$_2$ for given activities has substantial inter-individual variation (e.g., Froget *et al*. 2002; Green *et al*. 2005). Thus studies needing accurate estimation of activity costs will require HR/VO$_2$ validation for each focal animal. If this is impractical, the alternative is to base the HR/VO$_2$ relation on a representative sample of birds collected at the same time and to apply this relation to HR recordings from free-living animals. This latter option will require a relatively large sample size to establish confidence limits for the HR/ VO$_2$ relations.

Another constraint of the HR method is the need to distinguish activities while recording HR to better predict associated VO$_2$. This could be done readily for animals under continuous surveillance, but most birds are not amenable to this level of scrutiny. However, variations in OP appear to be mainly associated with changes from low-level activities to those involving extended flight or walking. If this is so, birds could be fitted with movement sensors along with HR loggers/transmitters to identify the appropriate HR/ VO_2 relation to apply to the data. There have been significant advances in development of miniature triaxial accelerometers, which are able to quantify movement in three dimensions. These have been used successfully in Great Cormorants (*Phalacrocorax atriceps*) to distinguish periods of rest, flight, and walking (Wilson *et al.* 2006). The ability to record activity patterns and HR simultaneously should permit determination of time-energy budgets with much greater detail and precision that has previously been possible.

Doubly labeled water: There have been a number of studies validating the accuracy of DLW to predict CO_2 production in birds. This is done by either measuring CO_2 production in DLW-treated birds via respirometry or by passing effluent air from a flow-through chamber through a CO_2-absorbing material and determining total CO_2 production gravimetrically. Based on 18 validation studies for birds, CO_2 production predicted from DLW averaged 2.4% higher than measured values (Speakman 1997). While this might suggest that further validations are unnecessary, this is not the case if the intent of a study is to contrast the FMR between individuals. A validation study using 9 individual Budgerigars (*Melopsittacus undulatus*) had an average discrepancy of 0.04% between DLW-predicted and measured CO_2, but individual values ranged by up to 11.4% from one other (Buttemer *et al.* 1986). Similar ranges in variation are typical in validation studies, which reinforce the need to validate the accuracy of the technique in order to establish confidence limits when applying this method to interindividual comparisons. It is very important to recognize that valid estimates of FMR from DLW require both accurate predictions of CO_2 production and knowledge of the substrates being oxidized. The thermal (i.e., energy) equivalent of 1 litre of CO_2 production ranges from 20.9 kJ for carbohydrates to 28.0 kJ for fats (Schmidt-Nielsen 1997). Thus, improper substrate assignment could produce errors of up to 34%, despite accurate estimates of CO_2 production. This error can be minimized if the animal's diet composition is known and the animal maintains mass balance during the period of DLW measurement. If diet is not known or mass varies, a mid-range thermal equivalent for CO_2 production should be used. The reader is referred to the excellent review of DLW methodology by Speakman (1997) for further advice on this subject.

8.1.4 Size Considerations

Both methods are strongly affected by the size of the individual under study. The DLW method relies on comparison of three animal fluid samples: a pre-injection sample to determine the background isotopic content, a sample taken at the time of DLW equilibration with body fluids, and a sample collected upon recapturing the animal, usually a day or more later. Typically, the sampled fluid is blood, although other fluids are also applicable. For small birds, the extent of such bleeding and duration of capture may render them physiologically compromised and behaviorally distressed. This aspect is exacerbated by the fact that the biological half-life of DLW is inversely related to body size, which for small birds usually requires recapture within a day, or at most two days, after DLW injection. Consequently, FMR measurements in small birds are prone to handling effects, unless they quickly resume normal levels of daily activities upon release. This problem can be ameliorated by using the single-sample DLW technique (Ricklefs *et al.* 1986; Webster and Weathers 1989). Instead of getting the three fluid samples from the same individual, this method samples from three subsets of the population under study. Accordingly, uninjected birds are collected to establish background isotope levels, another group is injected with DLW to determine equilibrium fluid isotope enrichment levels, and a third is injected with an equivalent amount of DLW and released immediately after injection. In this way, birds examined for FMR are handled for at most 5 minutes before being released and then recaptured 24–48 h later. The general consensus is that this method reduces the effects of handling on subsequent behavior (Speakman 1997). At the other extreme of body size, the cost of adequately enriching the body fluids of large birds with stable isotopes can be very expensive and may limit the number of samples that are financially affordable and thus compromise the experimental design.

The HR method is also constrained by animal size, but for very different reasons. Recording or transmitting HR requires that the study animal carries the added mass of an electronic device. The consensus is that loads must be restricted to 5% or less of a volant species body mass to avoid compromising its health and behavior. There are HR transmitters as small as 0.6 g that have been used successfully to monitor HR of free-living *Catharus* thrushes that have ca. 30 g body masses (Bowlin *et al.* 2005). With the rapid gains in electronic and battery miniaturization, there is likely to be further downsizing of transmitters and archival tags to permit HR recording in birds as small as 10 g with less than 5% additional wing loading. Even when using appropriately sized HR devices, it is important to test the effects these have on the bird's activity patterns as the method of attachment and associated surgery is known to strongly influence consequent behavior and activity costs (Kenward 2001; Irvine *et al.* 2007).

8.1.5 Recapture and Mobility Considerations

Both HR and DLW methods require an initial capture to attach the HR logger/transmitter or to inject isotopes. This intervention may make birds wary and more difficult to recapture as soon as needed. Of the two methods, DLW is most constrained by the need to recapture animals soon after first handling, particularly for birds with fluid-rich diets or aquatic foragers due to their high rates of water turnover. Because this method estimates total CO_2 production from all activities, post-injection samples are usually taken at integral multiples of 24-h intervals to avoid biasing the FMR determination to particular phases of the daily cycle. In the case of small birds or those with high fluid intake, post-injection sampling is usually required 1–2 days after DLW treatment to optimize determination of isotope turnover. Delays in recapturing will lead to isotope enrichment levels being too close to background levels for accurate determination of FMR. Of the HR methods, most archival tags also require eventual recapture to retrieve stored data. The exceptions are hybrid tags that combine archival storage with interrogation via radio (VHF or UHF) or by mobile phone networks (GSM), however these are currently too large for most birds. Because data are stored in memory irrespective of subsequent battery failure in archival tags, retrieval can occur long after implantation. The use of HR transmitters is the only method for measuring FMR that does not require recapture of birds after placement of the device. This presumes, of course, that there is no need to calibrate the HR device against the individual's VO_2 following the field measures and that the method of attachment permits easy dislodgement from the bird if fitted externally or is sufficiently small to not adversely affect it if implanted internally.

For studies of FMR in highly mobile species, the use of HR loggers is the preferred method if recapture is possible. This is due to the temporal sampling limits of DLW and the need for receivers to be within range of transmitted signals from HR devices. An exception to this is the use of DLW to determine foraging costs in seabirds that return to breeding sites after extended foraging trips at substantial distances from their nests. Such studies have been successfully completed in Wandering Albatrosses (*Diomedea exulans*; Shaffer *et al.* 2003) and Blue Petrels (*Halobaena caerulea*; Weimerskirch *et al.* 2003). A more ambitious study overcame the need to have mobile animals return to the site of initial sampling by tracking animals after they were injected with DLW. This involved placing small radio transmitters on migrating *Catharus* after injecting them with DLW, and then following them in vehicles fitted with antennas and receivers to follow their movements. Once these birds landed, they were captured in mist nets to obtain a post-flight blood sample (Wikelski *et al.* 2003). Similarly, migratory costs have been estimated using HR transmitters by tracking them as they flew (Bowlin *et al.* 2005), but this required an incredible effort and much logistic support to follow their movements. Overall, FMR estimates in highly mobile birds are most easily obtained using archival HR devices, which can be retrieved from them at almost any time after attachment or implantation, provided that retrieval is technically and logistically feasible.

8.2 Measuring Movement Patterns

8.2.1 Electronic Methods

Much of the historic knowledge regarding bird movements has come from ringing studies, where many individuals are given unique leg bands and some of these are eventually captured or recovered at other locations and reported to the banding authorities. While these methods are still useful to get a snapshot of a species annual distribution, they are confounded by reporting rates being biased to heavily populated regions. This is especially problematic for pelagic marine species as they spend most of their lives over open oceans away from human contact. With the development of Global Positioning Systems (GPS) technology, it is now possible to track individual birds very accurately for extended periods. Some devices use a GPS platform combined with an electronic section that serves as a Platform Transmitter Terminal (PTT) that communicates to polar-orbiting ARGOS satellites. The satellites, in turn, can transmit this information to a ground station for communication to the researcher. The smallest commercially available GPS platform is only 2.5 g, but the addition of batteries, antenna, and optional UHF remote download attachment increase this load considerably. In addition to providing physical locations, other variables can potentially be measured (e.g., HR, temperature, etc.) and transmitted via PTT. These devices have been used to dramatic effect and have shown that a male Wandering Albatross flew over 15,000 km on a single foraging trip from its nest (Jouventin & Weimerskirch 1990). This method has gained popularity among those seeking to understand habitat requirements of vulnerable and highly mobile species (e.g, Higuchi *et al*. 2004) as well as to understand movement corridors of birds likely to spread avian influenza. The technology at present is limited to birds that can carry an additional 9.5 g, which conservatively would confine its use to birds weighing 300 g or more.

Movements of smaller species can still be estimated using electronic devices, but the accuracy of these is far less than that possible using GPS-based telemetry. This is most commonly done using archival light-intensity loggers. The basis for determining bird locations from daylight and time stems from the relations between day length and latitude and between time of sunrise and longitude. This works well if the calendar date is also recorded and if the bird stays in one location sufficiently long to get consecutive readings of similar value. Accuracy is lost when birds are continuously on the move, as going eastward will shorten perceived daylength, going westward will lengthen it, going north or south may increase or decrease daylength, depending on season, and moving on a diagonal will complicate determination of both latitude and longitude. These effects can be substantial, as large birds can move hundreds of kilometers per day.

Because of its reliance on detecting light quality, this method is most amenable to birds using shade-free environments such as open seas. Even in seabirds staying in locations for suitably long periods, however, the biggest challenge is getting an accurate appraisal of latitude from times of sunrise and sunset. This is particularly true when the dates of concern are close to a solar equinox, or at high latitudes in summer or winter when the onset of sunrise or sunset is very gradual. This problem can be largely overcome if birds fitted with light loggers routinely sit on the sea surface, as this permits sea surface temperature (SST) to also be recorded. By comparing measurements of the data logger identified as being water temperature with contemporaneous recordings of satellite-based remote sensing of global SST, further resolution of location from the light loggers is achieved. This was validated in Laysan and Black footed Albatrosses (*Phoebastria immutabilis* and *P. nigripes*, respectively) that were fitted with satellite transmitters to determine GPS-based locations as well as light and temperature loggers (Shaffer *et al.* 2005). They found that the average error between satellite-based and light-based locations was reduced from 400 km to 200 km by incorporating SST measurements to resolve latitude. Similar calibrations to determine the effectiveness of light-based estimates of geolocation in terrestrial birds have not been attempted.

8.2.2 Stable Isotopic Methods

Some of the major elements comprising animal structure occur naturally in multiple forms as stable isotopes. Thus the ubiquitous carbon atom is found in tissues as both ^{12}C and ^{13}C, hydrogen as ^{1}H (protium) and ^{2}H (deuterium or D), and nitrogen as ^{14}N and ^{15}N. The superscripted value in each case designates the atomic mass and the higher values for each element are a consequence of an extra neutron. The difference in atomic mass has functional consequences in their distribution. Generally speaking, the biological products formed from a mix of isotopes will contain proportionately more of the lighter isotopic form than found in the reactants. This is because it is easier to make or break bonds of lighter isotopes of a given element than heavier ones. This is the basis for isotope fractionation and it underlies much of the forensic potential that isotopic analysis affords in animal studies.

The process of isotope fractionation also occurs in phase changes of water. When vapor forms from water, the lighter isotopes are more readily vaporized than the heavier ones. Thus the ratio of D:^{1}H in ocean water will exceed the ratio of these isotopes in vapor derived from it. When vapor from this source reaches saturation and condenses, the situation is reversed, with D condensing more readily than ^{1}H. As a consequence, rainfall from a given airmass will contain proportionately less D as the condensation process continues. Because the Earth's atmospheric moisture is mainly derived from warm, low-latitude oceanic regions, the ensuing precipitation from this vapor source contains proportionately less and less D as it travels over higher latitudes and altitudes and dissipates. This can has been verified by measuring the D and ^{1}H content of precipitation at a global scale.

Before discussing these results, it is pertinent to describe the convention in reporting isotope content of samples. The very small amount of heavy relative to light isotopes makes use of ratios very awkward when comparing isotopic contents. Accordingly, the convention is to express the relative abundance of the isotopes of concern as their ratio in a sample normalized to a standard in units of parts per thousand (o/oo; Karasov & Martinez del Rio 2007). The calculation for this determination has the form:

$$\partial D = \left(\frac{R_{samples} - R_{standard}}{R_{standard}} \right) \times 1000 \qquad (5.2)$$

where, R_{sample} and $R_{standard}$ are the ratios of D: ^{1}H in the sample and standard sample, respectively, and, in the case of deuterium, the standard is typically Vienna Standard Mean Ocean Water (VSMOW). When the sample has a lower ratio of heavy to light isotopes than the standard, the δ value is negative and the heavy isotope content of the sample is said to be depleted. Conversely, when the δ value is positive, the sample is said to have an enriched level of heavy isotope.

In the case of global patterns of D contained in rainfall, there is strong evidence that rainwater is progressively depleted of D along a gradient from low latitude sources to polar locations (Bowen and Revenaugh 2003; Fig. 8.3). This

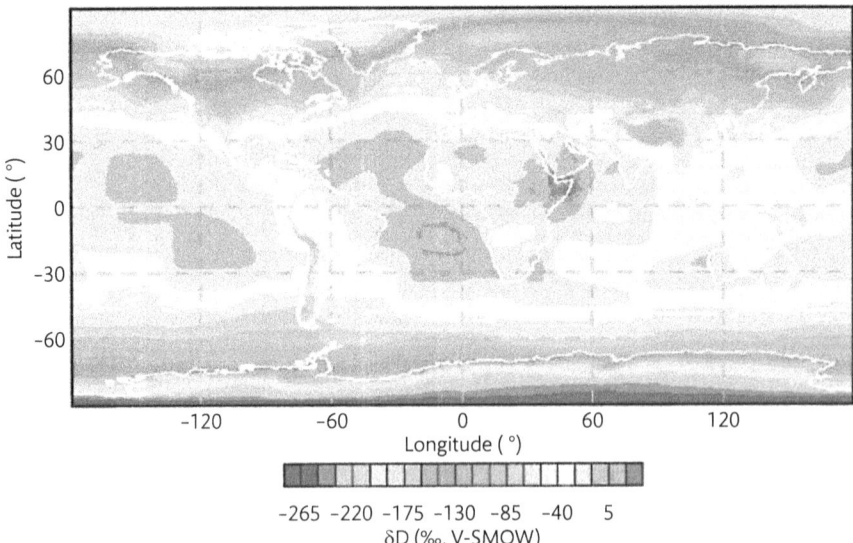

Fig. 8.3 Mean annual precipitation content of deuterium, δD, in parts per thousand, in relation to Vienna Standard Mean Ocean Water. The maps are based on interpolation of data from 340 sites distributed globally. (From Bowen & Revenaugh 2003).

figure also shows that the gradients of δD in precipitation (δDp) to be more delineated in northern than in southern continents. Because of continuous turnover of materials in most living tissues, the ratios of D: ¹H in plants and animals will be strongly influenced by the δD$_p$ of their immediate environment. This can be exploited to determine the regions where individual birds have resided, based on isotope signatures of their tissues. Feathers are metabolically inert once grown, consequently their δD should mirror δD$_p$ at the time (and place) of their production. For many north-temperate birds that have distinct molt phenologies, this provides an opportunity to sample feathers that are grown at breeding sites as well as those formed at wintering grounds. This method has been validated in feathers of known origin in both North America (Wassenaar & Hobson 2000) and in Europe (Hobson *et al.* 2004). Although both studies show a significant relation between feather δD values and annual as well as growing season δD$_p$, the variance in the regressions for European birds was much higher.

Variance between feather δD and δD$_p$ is likely when birds rely on ephemeral sources of water, consume food that is reliant on groundwater, or associate with lakes or wetlands (Hobson 2005). There is also evidence that birds residing in hot arid environments have higher tissue δD levels than expected due to fractionation of D in body water during evaporative cooling (McKechnie *et al.* 2004). Surprisingly, there is also evidence that δD can vary among feathers grown at the same time, with primaries having higher δD values than coverts in Northern Goshawks (*Accipter gentilis*; Smith & Dufty 2005). At a broad scale, however, this method has the potential to quickly identify latitudinal migratory limits among populations of species that have previously been unstudied (e.g., Hobson *et al.* 2001). Resolution of habitat associations will be improved by inclusion of other isotopes or trace elements known to vary in a predictable manner across continents, as well as from the further development of analytical procedures based on feathers of known geographical origin (e.g., Wunder *et al.* 2005).

8.3 Flight Energy Requirements

Perhaps the most fundamental behavioral, ecological, and physiological property of birds is their ability to fly. We know that flight is essential to the biology of the vast majority of birds that are volant, and that the physiological demands and energy costs of flight are very high. However, the magnitude and variation of flight costs according to behavior and environmental conditions is poorly understood in most species.

In principle, energy costs incurred during bird flight can be deduced from the amount of fuel combusted or from the energy that is being transferred during combustion. Energy expenditure can also be derived from the exchange of O_2 and CO_2, or even from the amount of water produced. However, in most cases we do not know precisely which substrates animals are using, and how much energy is being transferred from the combustion processes (see Videler, 2005).

There are a variety of approaches to determine the energy costs of bird flight. Laboratory measurements introduce biases that are quite different from the uncertainties associated with field approaches.

8.3.1 Mass Loss in the Field

The first attempts to estimate the metabolic costs of flight date back to the early 1950s (Videler, 2005). These investigators used the rate of mass change during extended flights as an indicator of energy use, with assumptions about the composition of the mass loss (fuel versus water, etc.). The use of mass loss figures seemed a promising approach back then. Nisbet (1963) summarized the information collected before 1963. However, these estimates suffered from large standard deviations because there was no information on the mass before departure and on the conditions faced during flight.

Dolnik and Gavrilov (1973) made flight cost estimates of small passerines during flights over a distance of 50 km in the Neringa Spit, a narrow strip of sand dunes, about 100 km long and 0.7–3.5 km wide, covered with trees and shrubs, separating the Gourlandic Haff from the Baltic Sea. This location is a route of very intense daytime migration. At two sites, birds were trapped, measured and released to estimate the energy expenditure during the 50 km flight. Dolnik and Gavrilov were able to calculate the average flight speed under windless conditions, and measure the mass and fatness of a large number of birds before and after their flights. Wind speeds and directions were measured midway between the traps, and were assumed to be constant along the spit and similar to that at the monitoring point. Wind data are used to calculate an equivalent flight distance in still air.

The study conducted by Dolnik and Gavrilov, in which field conditions could be robustly monitored, show that mass loss measurements in the field to estimate flight costs in birds can provide accurate flight cost estimates. Because it is necessary to have good knowledge of environmental conditions, especially wind speed, this technique may be restricted to short flight distances only.

8.3.2 Wind Tunnel Mass Loss, Metabolic, and Heart Rate Measurements

Wind tunnels offer the opportunity to fly birds at different speeds under carefully controlled conditions. It appears that Greenwalt (1961) was the first to apply wind tunnels to the study of bird flight, by placing a hummingbird feeder directly downstream of an electric fan. He was able to control the bird's air speed by adjusting the speed of the fan, while maintaining its ground speed at zero. This enabled him to make measurements of wing kinematics while the bird remained stationary relative to the camera. Zero ground speed is not the same as "hovering", which means flight at zero air speed. The bird's exertions depend on its air speed, not its ground speed, and these two speeds are only the same if the wind speed is zero. If the wind speed in the wind tunnel is uniform and turbulence is low, then flight at zero ground speed is mechanically and physiologically the same as flight through still air, at an air speed equal and opposite to the wind speed. The difference from the free-flight situation is that the bird's air speed can be determined and held constant by the experimenter. In addition to the modulation of wind speed, tunnels can permit control of temperature. If the wind tunnel can be tilted, the angle of climb and descent is also controllable (Pennycuick *et al.* 1997).

While it is not difficult to generate a flow of wind in which a bird can fly, it is another matter altogether to do this in a way that would allow one to obtain reliable measurements of fuel consumption, optimum flight speeds, and so on. It is not always easy to train birds to fly in a wind tunnel. Space is always confined, the airflow can be turbulent and uneven across the vertical and horizontal cross-section, and the engines and fans or propellers usually make considerable noise. However, significant improvements have been achieved with the construction of a new low-turbulence wind tunnel for bird flight experiments at Lund University, Sweden (see Pennycuick *et al.* 1997).

Klaassen *et al.* (2000), using the Lund wind tunnel, estimated the flight costs of a Thrush Nightingale (*Luscinia luscinia*) during eight experimental flights; seven of these lasted 12 h non-stop. Flight speed was kept constant at $10\,\mathrm{ms}^{-1}$. Fuel combustion was used as a measure of flight costs. The diet of this individual consisted of mealworms containing 44% fat and 56% protein. It took the bird three days to completely recover from each 12 h flight bout and regain its starting body mass. The energy consumed minus the energy lost during these three days was considered equivalent to the costs of the experimental flight.

Metabolic measurements are possible by analyzing the composition of the air directly in closed-circuit tunnels or by using lightweight, carefully fitted masks and tubes, through which gases in the expired air are collected and analyzed, in systems that are open. Theoretically, one could perform such measurements without masks in an open-circuit tunnel if there were sensitive enough O_2 and CO_2 analyzers. Early work with these methods shows that flight costs may vary considerably with wind speed (see Tucker 1968b, 1972). Nevertheless, gas exchange measurements are difficult or impossible with many species that are intolerant of masks or will not fly reliably in tunnels, and wind tunnels large enough for big birds are generally unavailable. In these cases, other approaches are necessary.

One increasingly popular approach, especially for field studies, is measurement of heart rate. Heart rates can reflect O_2 consumption but the relationship has to be established in each case. This technique has been used in studies of free-ranging

Black-browed Albatrosses (*Thalassarche melanophrys*; Bevan *et al.* 1995) and Barnacle Geese (Butler *et al.* 1998). In a wind tunnel study, Ward *et al.* (2002) found a linear relationship between O_2 consumption, measured through a mask, and heart rates of Barnacle and Bar-headed Geese (*Anser indicus*) flying at velocities between 14 and 20 ms^{-1}, and 16 and 21 ms^{-1}, respectively. Interestingly, and conveniently for field measurements, heart rates and O_2 consumption did not vary with flight speed. These relationships made it possible to calculate O_2 consumption of birds flying without a mask using measured heart rates.

Liechti and Bruderer (2002) compared the effective wing beat frequency (Feff) of Barn Swallows (*Hirundo rustica*) and House Martins (*Delichon urbica*) during free flight and wind tunnel experiments. They found that under comparable flight conditions, Barn Swallows and House Martins in free flight had significantly lower values of Feff (by 40% and 32%, respectively) than individuals in wind tunnel experiments, and they suggest that wind tunnel experiments might overestimate birds' flight costs compared with free flight.

Maximal energy requirements during hovering flight were cleverly determined by Chai and Dudley (1995) while Ruby-throated Hummingbirds (*Archilochus colubris*) hovered inside a Plexiglas chamber, the contents of which were varied by means of gradual replacement of air (20% O_2, 70% N_2, 1% trace gases, density 1.2 kgm^{-3} at sea level) with 79% He/21% O_2 (heliox, density 0.4 kgm^{-3}). Since lift generation is a function of gas density, this reduction means that wingbeat kinematics must be altered to match the increased requirements for mechanical power. Using this technique, Chai and Dudley measured mechanical performance, metabolic expenditure, and efficiency of hummingbird flight muscles during hovering, and they claim that maximum performance in these experiments was unequivocally elicited because the birds failed to sustain hovering when gas density fell below a certain threshold.

8.4 Evolutionary and Ecological Functional Genomics

The emerging field of evolutionary and ecological functional genomics seeks to understand how gene function and genomic changes underlie the ability of diverse organisms to flourish despite challenges from their biotic and abiotic environments (Feder and Mitchell-Olds, 2003). To accomplish this, the genetic mechanisms that affect ecologically important traits must be investigated (Chown *et al.*, 2004). Ecological knowledge is crucial for the interpretation of many types of genomic data (Baldwin, 2001; Weinig *et al.*, 2002), particularly in establishing the consequences of genetic variation. Transcription profiles (e.g., cDNA microarrays; see description below) are extremely sensitive to the interactions and kinetics of environmental factors, which necessitates great care in establishing the physiological and ecological relevance of experimental conditions (Churchill, 2002). Similarly, the fitness consequences of genetic variants can differ substantially between laboratory and field experiments, with phenotypes that are absent in one venue being present in the other. For example, some quantitative trait loci (QTLs) for salt-tolerance traits in sunflowers differ substantially when measured in model soil in greenhouses and in the wild (Lexer *et al.*, 2003).

A meaningful integration of ecology and genomics promises important new insights into the processes by which individuals and populations respond and adapt to their environment. Ecological genomic studies of adaptation seek to link organism and population level processes through an understanding of genome organization and function. Although still restricted to only a few organisms (often, the very small number of so-called "model" organisms), these studies are quite promising. Given the rapid advances in molecular and genomic techniques (particularly high-speed sequencing and gene expression methods), they will almost certainly be extended to a wide range of other species, including birds. For example, the planktonic crustacean *Daphnia*, which has been an important model system for ecology, is now also being used as a genomic model. In this species production of parthenogenetic male offspring occurs through environmental cues (temperature, photoperiod, and crowding), which involves endocrine regulation. Recent progress has uncovered a putative juvenoid *cis*-response element, which together with microarray analysis will stimulate further research into nuclear hormone receptors and their associated transcriptional regulatory networks (Eads *et al.*, 2008). These studies are thus providing opportunities to extend laboratory research from the molecular basis of morphological, physiological, and behavioral differences to questions addressing the degree and importance of genetic variation within natural populations.

8.4.1 Studies on Birds

In a recent and thorough review, Ellegren and Sheldon (2008) discuss the importance of genomic studies to understand the relationships between genotype and phenotype in natural populations. They also argue that the accumulating body of knowledge, which has been acquired from such studies in recent years, indicates that a "new understanding of the quantitative genetics of fitness variation in the wild may result in a synthesis of ecological and molecular approaches in evolutionary biology." In their review, Ellegren and Sheldon give us an elegant example of the use of these approaches from studies in which Darwin's finches (*Geospiza spp.*) have been used as model systems. It is widely known that adaptive radiation among Darwin's finches has resulted in more than ten different species exploiting different ecological niches on the Galapagos Islands. Classic studies have shown that adaptation to different niches is reflected in the distinct beak morphologies and behaviors observed in Darwin's finches. The large-billed ground finch, for example, which eats seeds on the ground, possess a deep and wide beak. On the other hand, finches with pointed and long beaks usually feed on cactus flowers, and finches possessing pointed and slender beaks are mainly insect eaters. Ellegren and Sheldon remark that the existing differences observed in external beak morphology are consistent with corresponding differences in craniofacial skeletons, and the experimental work of Schneider and Helms (2003), using other bird species in their investigation, have shown that the cellular origin of beak development is located in the neural crest-derived mesenchyme.

Ellegren and Sheldon (2008) report also on a recent study, conducted by Abzhanov et al. (2004), where the expression pattern of growth factors which are involved in avian craniofacial development was closely examined. In this study, a strong correlation between beak morphology of Darwin's finches and the level of expression of bone morphogenetic protein 4 (Bmp4) in the mesenchyme of the upper beak was found. Abzhanov and colleagues therefore suggest that regulation of Bmp4 expression is a key variable directly influencing quantitative variation in beak morphology of Darwin's finches. Their results, however, could not distinguish between a role of cis-regulatory elements or variation in the induction/transduction of upstream factors. In another related study, Abzhanov and colleagues (2006) used microarray hybridization techniques in order to identify, among Darwin's finches possessing different beak morphology, differentially expressed genes. These studies led to the discovery that calmodulin (CaM), a known key component of a Ca^{++}-dependent signal transduction pathway, which in turn is a determinant factor for the control of bone differentiation and growth, was expressed at much higher levels in cactus-feeding finches than in other species of Darwin's finches.

According to Ellegren and Sheldon (2008), such studies provide new mechanistic insight into "this textbook example of evolution by natural selection." The experimentally produced manipulation by Abzhanov and colleagues (2006), in which misexpression of Bmp4 and CaM in developing chicken embryos has similar effects on beak morphology as those observed in Galapagos finches, is a typical case of experimental evidence of the functional significance of the proposed genetic mechanism (Ellegren and Sheldon, 2008).

Ellegren and Sheldon (2008) also suggest that with the completion of the zebra finch (*Taeniopygia guttata*) genome sequence (Warren et al., 2010), and taking into account that bird genomes show a high degree of conservation with regards to their organization, many more functional genomics-based studies of wild birds will become feasible in the near future.

8.5 Molecular Biology

8.5.1 MicroRNAs

Changes in the patterns of gene expression are widely believed to underlie many of the phenotypic differences within and between species. The emergence of multicellular organisms was accompanied, and perhaps facilitated, by dramatic increases in the complexity of gene regulatory mechanisms (Levine and Tijan, 2003; Moore, 2005). At the level of transcriptional regulation, this can be clearly verified in the massive expansions of transcription-factor families and the pervasive combinatorial control of genes by multiple transcription factors in higher organisms (Levine and Tijan, 2003; Davidson, 2006). Transcription factors are proteins that either activate or repress transcription of other genes by binding to short *cis*-regulatory elements called transcription-factor binding sites that lie in the vicinity of the target genes. Transcription factors are usually grouped into families on the basis of shared DNA-binding domains, which are an important determinant of transcription-factor binding specificity (Chen and Rajewsky, 2007).

At the level of post-transcriptional control, entirely new mechanisms arose, typified by a large and growing class of approximately 22-nucleotide-long non-coding RNAs, known as microRNAs (miRNAs), which function as repressors in all known animal and plant genomes (Ambros, 2004; Bartel, 2005). MicroRNAs have diverse expression patterns and regulate various developmental and physiological processes (e.g., suppression of apoptosis and regulation of fat metabolism in the fruit fly *Drosophila melanogaster* (Xu *et al.*, 2003). The emerging picture is that miRNAs have the potential to regulate almost all aspects of cellular physiology (He and Hannon, 2004). Although transcription factors and miRNA are two of the best-studied gene regulatory mechanisms, there are many other layers of gene regulation, including cell signaling, mRNA modifications (splicing, polyadenylation, and localization), chromatin modifications, and mechanisms of protein localization, modification and degradation (Chen and Rajewsky, 2007).

8.5.2 Microarrays

The genomes of a rapidly growing number of organisms have been sequenced completely, and millions of additional genetic sequences have been deposited in international repositories. However, the biological functions of most of these genes remain unknown or have been predicted only through homology to genes with known functions. One way to determine the functions of such genes is by repeated measurements of their RNA transcripts. For example, knowing that a particular gene is expressed only in cardiac muscle, and only under particular conditions, gives us implicit functional knowledge about that gene. Functional genomics is the study of gene function through parallel expression measurements of a genome (Butte, 2002). The most common tools to carry out these measurements include complementary DNA (cDNA) microarrays (Schena *et al.* 1995), oligonucleotide microarrays (Lockhart *et al.* 1996) or serial analysis of gene expression (SAGE) (Velculescu *et al.* 1995).

Microarrays can be broadly defined as tools for massively parallel ligand binding assays where very large sets of features (e.g., oligonucleotides) are placed on a solid support (e.g., a glass slide) at high density for recognizing a complex mixture of target molecules (Ekins and Chu 1999). For biological applications the features on the arrays can be DNA, RNA, proteins, polyssacharides, lipids, small organic compounds or even whole cells (Hoheisel, 2006). Therefore, microarray technology in principle allows the estimation of target abundance and the detection of biological interactions at the molecular or cellular level (Shiu and Borevitz, 2008).

Although there are many protocols and systems available, the basic technique involves extraction of RNA from biological samples in either control or interventional states. The RNA, or in some protocols isolated messenger RNA, is then copied while incorporating either fluorescent nucleotides or a tag that is later stained with fluorescence. The labeled DNA is then hybridized to a microarray for a period of time, after which the excess is washed off and the microarray is scanned under laser light. With oligonucleotide microarrays, for which all probes have been designed to be theoretically similar with regard to hybridization temperature and binding affinity, each microarray measures a single sample and provides an absolute measurement level for each RNA molecule, although this absolute measurement might not correlate exactly with *in vivo* concentration. With cDNA microarrays, for which each probe has its own hybridization characteristic, each microarray measures two samples, and provides a relative measurement level for each different RNA species in the sample. Regardless of the technique, the end result is 4,000–50,000 measurements of gene expression per sample. As a complete experiment might involve anywhere up to hundreds of microarrays, the resultant RNA-expression data sets can be huge, which poses daunting analytical challenges.

Many free and commercial software packages are available to analyze microarray data, although it is still difficult to find a single off-the-shelf software package that answers all functional-genomics questions. Before multiple microarray measurements can be integrated into a single analysis, the measurements need to be normalized, or adjusted to make them comparable. When microarrays are used in an experiment in which the measurements from different treatments or test groups are made simultaneously, with homogenous populations of similar cells and using a single type of microarray, normalization might be a matter of equalizing the overall brightness of each microarray image, assuming that the quantity of RNA is equal (Wu, 2001). Other normalization methods include comparing expression levels of "housekeeping" genes (Eickhoff *et al* 1999), which assumes that such genes do not change across experiments (Zien *et al.*, 2001), and using "splines" (Li and Hung Wong, 2001), or other nonlinear techniques (Ramdas *et al.*, 2001; Tseng *et al.*, 2001). The spline approach is based on fitting data segments with a series of small, simple curves instead of attempting to fit a single complex polynomial to the combined data. Normalization techniques for one microarray technology might not apply to another, due to differences in assumptions and the distributions of the results. For instance, if we assume that in a given experiment, expression of most genes in a cell does not change and an equal number of genes are up- and down-regulated (not always a valid assumption), then differential-expression measurements from spotted arrays (microarrays based on pen tip deposition on glass slides) might be normally distributed, whereas measurements from oligunucleotide microarrays will have a different, perhaps skewed, distribution (Butte, 2002). Thus, expression measurements made across different microarray technologies are not directly comparable.

Current methods for analyzing RNA-expression can be divided into two categories: (i) supervised approaches (analysis to determine genes that fit a predetermined pattern); and (ii) unsupervised approaches (analysis to characterize the components of a data set without *a priori* input; reviewed by Butte (2002).

Although restricted to a few model systems (e.g., *Daphnia*, *Drosophila*), these techniques, as they become more and more consolidated, could be applied to address still unanswered questions with regards to avian biology. For example: (i) what is the change in neurohormonal gene expression in social species relative to "classic" territorial pairs?; (ii) what genomic changes occur in preparation for migration, including regulation of fat metabolism (miRNAs)?; (iii) what genes are involved in bird orientation to migratory flights?; and so on.

8.5.3 Phenotypic Plasticity

Behavioral and phenotypic plasticity are crucial determinants of fitness as they allow individuals to respond appropriately to environmental variation such as seasonal changes and reproductive opportunities, at much shorter time scales than genetic evolution (which can only produce change across generations). As complex phenotypes are the consequences of the interaction between genes and environment, an essential step to understand complex traits and their evolution in general is to identify the proximal mechanism of behavioral and phenotypic plasticity (Gibson, 2002).

One interesting example of studies attempting to identify expression changes in this context is a comparison between nursing and foraging bees using cDNA microarrays (Whitfield *et al.*, 2003). As the transition to foraging in the honey bee involves environmentally modulated behavioral changes that are associated with changes in brain structure and neurochemistry, the expression profile differences between brains of nursing and foraging bees were compared. Of the nearly 5000 genes tested, 39% showed significant change in transcript abundance between these two types of bees. Although the nurse-to-forager transition is age-related, they were able to compare nurse and precocious forager bees of the same age and found that most gene expression differences between these types of bees are due to behavior instead of age. In addition, the brain messenger RNA profiles of individual bees were found to be a good predictor of whether it is a nurse or a forager.

This study exemplifies the use of microarrays to identify gene expression changes that can be attributed to behavioral or phenotypic differences between genetically similar individuals. Although major questions remain unanswered, such as how these changes are triggered by the environment and if the behavioral and phenotypic differences are due to some or all of the expression changes, microarrays can be more broadly used to dissect expression differences between individuals and populations in different environments (Shiu and Borevitz, 2008). Such approaches can also be used in future studies to address problems involving bird phenotypic plasticity, as we have seen in earlier chapters. For example, complex social behaviors such as maintaining territorial spaces or nest keeping are crucial to birds, and within this context what are the gene expression changes that may occur when an individual shifts from being a nest helper to being a territory holder?

8.6 Variability Analysis

8.6.1 Lévy Flight (or Walk) Searching Patterns

Foraging is a vital aspect of the life of essentially all animals, and particularly birds, which often search for small, cryptic food items over very wide areas. The general problem of how to search for randomly-located "target sites" (e.g., food) efficiently can be quantitatively described as Brownian random flights or Lévy flights (walks). When food is abundantly available, foraging can be most effectively found by Brownian motion with a Gaussian distribution. However, when food is scarce, a Lévy flight could be more efficient than the usual Gaussian motion (Bartumeus *et al.*, 2003).

Lévy flights are special class of random flights whose steps lengths are not constant, but instead are chosen from a probability distribution with a power-law tail, such that clusters of short steps are connected by long rare steps. Lévy flights display fractal properties, have no typical scale, and occur in physical (Shlesinger *et al.* 1995; ben-Avraham and Havlin, 2000; Bardou *et al*, 2002) and chemical systems (Ott *et al.* 1990). Shlesinger (1986) proposed that Lévy flights may be observed in the behavior of foraging ants. Cole (1995) argued that *Drosophila* might also perform Lévy flights, and Lévy flights were also observed in bumblebees (Heinrich 1979). Viswanathan *et al.* (1996) tested the hypothesis that foraging Wandering Albatrosses show Lévy flights. They collected data over a three–month summer period as part of a study of Wandering Albatross foraging biology in the South Atlantic. Electronic recording devices were attached to the legs of the birds on 19 separate foraging trips. The devices took measurements every three seconds, and recorded the number, $u(t)$, of 15 seconds intervals in each hour (t) for which the animal was wet for nine seconds or more (i.e., when it alighted on the ocean, presumably to catch prey or rest). Nineteen time series were obtained, one for each of the foraging trips studied. The series ranged from 77 hours to 416 hours with a mean of 175 hours.

It has been speculated that scale invariance may reflect the exploitation of highly complex environments which might themselves have fractal properties (Russell *et al.* 1992; Johnson *et al.* 1992; Drossel *et al.* 1992). Viswanathan and colleagues investigated the origin of scale invariance in their albatross data and found that the distribution of flight-time intervals approximately followed a power law. Conventional random walks have been used to model the foraging behavior of bacteria and other organisms (Berg 1983; Edelstein-Keshet 1988). However, according to Viswanathan and colleagues, such models predict not a scale-invariant power law but rather a Poisson distribution, and in their view these cannot explain the observed foraging behavior of Wandering Albatrosses, i.e., temporal scale invariance and Lévy flight behavior. The landing points (i.e., geographical locations) of Wandering Albatrosses have spatially scale-invariant properties, which may indicate that the distribution of food on the ocean surface is also scale invariant (McClatchie *et al.* 1994), and birds that fail to produce a scale invariant distribution of flight time intervals would face a greater difficulty finding food, and hence surviving. Lévy flight search patterns in albatrosses foraging behavior may reflect the solution of the biological search problem in complex environments (Shlesinger *et al.* 1993).

This empirical evidence has been challenged recently by Edwards *et al.* (2007). However, their arguments are based on mathematical technicalities, which do not invalidate the general conclusions derived, and Lévy flight as a potential real-time monitoring tool (Bertrand *et al.*, 2005). The questions that remain to be answered are whether albatrosses follow a Lévy-like distribution as a simple consequence of the abundance of food sources, whether such behavior is a hard-wired genetic trait, or whether it is dependent on the "terrain geography" associated with the availability of food. Statistical tools for analyzing search or other movement patterns can be further refined to address these questions.

9

Conclusions and Future Directions

The conspicuous nature of birds and their high-energy lifestyles has made them an attractive research model for many behavioral and physiological studies. Previous chapters give a sense of the scope of physiological research undertaken to date, but also reveal a growing interest in avian physiological studies. This stems in part from their utility in addressing questions fundamental to evolutionary and ecological theories, but also from the growing concern of the consequences of rapidly deteriorating habitat and resource availability on the world's fauna. This presents an urgency to better understand how animals respond to rapidly changing environments from a conservation perspective, but it also creates natural experiments for testing life-history and evolutionary theories. We strongly believe that physiological studies of birds are very important in addressing these questions and highlight a few research areas we feel are particularly pertinent.

9.1 Responses to Habitat Variation

Predictions of future global climate variation foreshadow rapid and broad-scale changes in important features of major habitats. While the mobility of most avian species makes them better suited to seek alternate sites than non-volant species, this presumes that such sites are available and provide the necessary food and shelter to accommodate their life-history needs. Presumably, generalist species will be better at occupying new habitats than specialist ones, and there are ample opportunities to test this idea and refine conservation techniques in advance of major habitat change. There are many areas of the world where humans have inoculated areas they have colonized with non-endemic species, including birds. This presents an opportunity to compare selected physiological and immunological traits of species that colonize readily and show rapid range expansion with those of species that display restricted distribution. There is evidence that good colonizing species differ from poorer ones in how they respond to novel immunological challenges (Lee *et al.* 2005). It would also be interesting to examine stress indicators in species differing in colonizing ability (corticosterone,

corticosterone-binding globulins, oxidative stress measures, anti-oxidants, etc.) to see whether they differ in a consistent manner. Birds could be captured at different locations in their current distribution to see if birds comprising leading edges of expansion differ from those at the core, but it would also be important to compare the physiological and immunological traits of introduced birds with their counterparts in source populations in Europe and Asia. Predictive models arising from such studies could then be applied to native species that show increases or decreases in their ranges in response to climate or habitat variation. This research would be nicely complemented by quantitative genetic assessment of species showing contracting and expanding distributions.

9.2 Changes in Movement Patterns and Food Use

The predicted change in climatic patterns will likely give rise to altered migration timing as well as changes in destinations. The broad scale of such events requires an equally broad assessment of the origins of migrant birds followed by more careful tracking of species identified as being vulnerable. With the refinement of stable isotope assessment, it should be possible to track shifts in patterns of migration among species breeding in particular areas. An added benefit of this approach is that changes in major food types could also be surmised by examining carbon and nitrogen isotopes in tissue and feather samples (Caut *et al.* 2006). Precise delineation of habitat use by larger species could be tracked using GPS-based transmitters, but there are also plans for developing a global tracking system that would permit monitoring movement of small species fitted with light-weight transmitters (Wikelski *et al.* 2007; Hobson and Wassenaar 2008). Once the migratory paths and staging areas of a particular population are identified, ground-level studies could be used to establish their diets.

There is still a critical need to elucidate the basic processes of migration physiology to better understand the transport capacity of a given species. More field studies are necessary to determine the influence of weather patterns and altitude on energy and water demands during long-distance flight. The few studies of free-living birds conducted to date provide conflicting evidence regarding altitudinal effects on energy and water requirements. The question as to whether birds normally dehydrate during migratory flight is still unresolved and invites more field studies. Modeling of migration costs will also require a better understanding of the relative importance of wind-assisted flight among species differing in morphology.

Physiological, morphological, and sensory differences between species will also influence their migratory ability, particularly in relation to detecting geographic location. Attempts to identify traits unique to migrating species have concluded that most traits that are useful for migration are shared by both sedentary and highly mobile species. Furthermore, phylogenetic analysis of species displaying nocturnal migration suggests that this ability is either evolutionarily ancient, or has evolved convergently many times (Piersma *et al.* 2005). Insight into the evolution of migratory behavior would follow from closer examination of the physiological and morphological traits of species showing adaptive modification of their movement patterns in response to environmental change. In this regard, a better understanding of magnetoreception among birds, particularly the phylogenetic distribution of retinal cryptochromes and endogenous magnetites, is critically needed (Mouritsen and Ritz 2005). Future studies should examine these and other navigational traits among closely related species that differ in migratory tendencies.

9.3 Temperature Effects on Development

Avian growth rates have been best studied in taxa breeding in cold and temperate environments, where the costs of temperature regulation are potentially the highest for young birds. However, we poorly understand how environmental temperature during development affects growth efficiency in taxa other than domesticated species. Before we can predict the impact of rapidly changing climate on avian reproduction, we need to better understand how species body size and location on the altricial-precocial continuum affects the growth requirements of their offspring and the amount of brooding by their parents. Parental behavior during incubation and post-hatching periods will largely determine the range of environmental temperatures to which the young are exposed during development, but time at the nest comes at the expense of parental self-maintenance. For species breeding at sites with abundant food resources, parents can afford to devote much time and energy maintaining optimal conditions for chick growth and fledging. If food were to become more limiting in the future, however, time away from the nest will increase and bring with it suboptimal thermal conditions coupled with reduced feeding rates.

At a more fundamental level, our understanding of the consequences on nestling growth of developing endothermy early (precocial species) versus later (altricial) has been based largely on distantly related species. Within family comparisons in Charadriidae now offer evidence to suggest that the "growth rate-maturity hypothesis" cannot explain how some shorebirds achieve early homeothermy concurrent with fast growth (Williams *et al.* 2007). It is clear that molecular approaches are needed to understand the basis of the present diversity of developmental strategies in birds. For example, insight into ontogenetic differences in the age of establishing endothermy might follow from monitoring the signaling pathways that regulate aerobic metabolism over the course of development (Walter and Seebacher 2007).

9.4 Maternal Effects on Development

Apart from the obvious effect of differential provisioning of eggs on chick size, there is accumulating evidence that a range of other non-genetic maternal effects have profound consequences on the phenotypes of offspring. These include equipping eggs with a complement of maternal antibodies, providing them with antioxidants, and maintaining optimal incubation temperatures. Because all of these factors can be modified by maternal physical condition and health, there is an interplay between the phenotype of the mother and that of her progeny. Maternal antibody allocation to yolk and albumen, for example, is directly related to her circulating levels at the time of egg laying (Hasselquist and Nilsson 2009). Thus, immunosuppressed females will produce slower-growing immune-deficient progeny than their healthier counterparts. There is also evidence that the eggs of females with elevated plasma corticosterone levels produce eggs with higher content of this hormone. Embryos exposed to elevated corticosterone have exaggerated stress responses when mature and show a passive personality (Cockrem 2007). What is unclear, however, is whether this phenotypic lability represents mechanisms favoring chick versus maternal survivorship and future fecundity (Marshall and Uller 2007). Resolution of these questions will require an understanding of the species life-history, and an integrated use of physiological, immunological, and endocrinological techniques.

A more dramatic maternal effect on progeny relates to the effects of female nutrition on the gender of her progeny. Female Seychelle Warblers (*Acrocephalus sechellensis*) produce mainly males in habitats with poor food supply, and mainly females when food is plentiful (Komdeur *et al.* 1997). By contrast, Kakapo (*Strigops habroptila*) females produce mainly sons when food is abundant and their bodies are replete with energy reserves, but mainly females when their body condition is poorer (Robertson *et al.* 2006). The profound life-history differences between these species explains why different genders are selected when food is abundant, but it is unclear what mechanisms underlie the prezygotic selection of gender by females. Understanding these processes has profound implications for evolutionary and life-history theories, but is equally important for conservation concerns. Indeed, the supplemental feeding program for the threatened Kakapo resulted in an insufficient number of females being recruited into the population, to the detriment of population recovery.

9.5 Establishing Physiological Health Indices

Population declines in avifauna are occurring globally and are likely to get worse. With increasing numbers of bird populations coming under threat, species recovery programs are being rapidly implemented, but their success depends on how well and how quickly remedial actions are taken. In this regard, it is important to be able to dynamically track individual health in relation to conservation measures to gauge their effectiveness. Traditional measures of animal quality usually resort to body condition measures (see review by Stevenson and Woods 2006), which are mainly based on ratios of body mass to an estimate or surrogate of body volume. These presume that to be relatively heavier is better, a situation that we are regularly reminded of by the medical profession to be patently false. While it is true that animals with low condition indices are unlikely to cope with periods of food deprivation or energy imbalance as well as more replete animals, being fat brings with it requirements for greater periods of feeding and exposure to predators, as well as reduced mobility. There is therefore a need to place condition indices in perspective (Walsberg 2003), but also to establish and validate a suite of physiological, endocrinological, and immunological measures that distinguish vulnerable from healthy animals. Suggested measurements pertinent to this goal were highlighted by Wikelski and Cooke (2006), but more effort should be directed to extending and refining this approach.

9.6 Physiological Performance Measures

The expected changes in climate across many of the world's continents will likely result in a redistribution of species from current locations. It would be intriguing to see whether species movements were correlated with their capacity to deal with changed thermal conditions and the relative contribution of physiological and behavioral adjustments to this process. For example, species with limited scope for thermal adjustment would be expected to move to locations with thermal regimes that are similar to their previous habitat, whereas those more tolerant to temperature variation might reside in very different thermal environments. This could be tested by examining thermoregulatory responses and thermogenic capacities of individuals collected throughout their range. Their responses could then be compared to measurements made after acclimation to various temperature ranges to identify their acclimation potential. Species showing limited thermal acclimation potential, but living over a greater range of thermal environments would be predicted to be behaviorally compensating for these excursions, which could be validated by comparing the activities and habitat associations at the extremes of the range with conspecifics living in more benign environments. On the other hand, directional selection of thermoregulatory performance is suggested if individuals at the extremes of the range differ in their thermoregulatory capabilities, but show limited acclimation potential.

Physiological performance measures also have tremendous potential to test the assumptions of many evolutionary and life-history theories. Many of these assume that particular activities (e.g., advertisement to attract mates; signals to dissuade rivals; etc.) are costly tradeoffs and therefore are best afforded by the most fit and capable individuals. Such signals are therefore considered as honest indicators of the displayer's health and genetic quality. While costs are easy to imagine, they can often be difficult to quantify. Estimating them within a physiological framework can at least validate the more obvious costs in terms of energy, physical performance, immunological functions, and endocrine correlates and test whether costs in any form are evident. For example, Medium Ground Finches (*Geospiza fortis*) with deeper bills can eat harder seeds and survive periods of drought more readily than birds with shallower bills. Female finches, however, prefer the qualities of song that are produced by males having shallower bills. Thus, there is a tradeoff in males of this species between being an effective breeder and being an effective feeder. Applying similar functional approaches to evaluate the costs of mating and maintenance activities has been found useful in distinguishing true handicaps from indices of male quality (Vanhooydonck *et al.* 2007).

There are many more areas of biology that will find physiological approaches to be of fundamental importance in the years to come. We emphasize that the topics we identify in this chapter are not meant to be prescriptive or comprehensive, but to serve as examples of how physiological understanding adds tremendous value to many traditionally non-mechanistic areas of research gaining popularity. It is also fundamentally important to recognize that the major breakthroughs in comparative physiology have been driven by curiosity-based fascination in wanting to know how animals work. This motivation will sustain ecophysiology and the many disciplines it influences well into the future.

Bibliography

Abzhanov, A., Protas, M., Grant, B. R., Grant, P. R. and Tabin, C. J. (2004). Bmp4 and morphological variation of beaks in Darwin's finches. *Science* **305**, 1462–1465.

Abzhanov, A., Kuo, W. P., Hatmann, C., Grant, R. B., Grant, P. R. and Tabin, C. J. (2006). The calmodulin pathway and evolution of elongated beak morphology in Darwin's finches. *Nature* **442**, 563–567.

Ackerman, R. A., Whittow, G. C., Paganelli, C. V. and Petit, T. N. (1980). Oxygen consumption, gas exchange, and growth of the embryonic wedge-tailed shearwaters (*Puffinus pacificus chlororhynchus*). *Physiol Zool* **53**, 210–221.

Adkins-Regan, E. (2008). Do hormonal control systems produce evolutionary inertia? *Phil Trans R Soc B* **363**, 1599–1609.

Afik, D. and Karasov, W. H. (1995). The trade-offs between digestion rate and efficiency in warblers and their ecological implications. *Ecology* **76**, 2247–2257.

Afik, D., Caviedes-Vidal, E., Martínez del Rio, C. and Karsov, W. H. (1995). Dietary modulation of intestinal hydrolytic enzymes in yellow-rumped warblers. *Am J Physiol* **269**, R413–R420.

Afik, D., Darken, B. W. and Karasov, W. H. (1997a). Is diet-shifting facilitated by modulation of intestinal nutrient uptake? Test of an adaptation hypothesis in Yellow-rumped warblers. *Physiol Zool* **70**, 213–221.

Afik, D., McWilliams, S. R. and Karasov, W. H. (1997b). Test for passive absorption of glucose in yellow-rumped warblers and its ecological implications. *Physiol Zool* **70(3)**, 370–377.

Ainley, D. G., O'Connor, E. F. and Boekelheide, R. J. (1984). *The Marine Ecology of Birds in the Ross Sea, Antarctica*. American Ornithologists' Union, Washington, D.C.

Åkesson, S. and Hedenström, A. (2000). Wind selectivity of migratory flight departures in birds. *Behav Ecol Sociobiol* **47**, 140–144.

Alerstam, T. (1990). *Bird Migration*. Cambridge University Press, Cambridge, UK.

Alerstam, T. and Lindstrom, A. (1990). Optimal bird migration: The relative importance of time, energy and safety. In E. Gwinner, ed. *Bird Migration: Physiology and Ecophysiology*, pp. 331–351. Springer-Verlag, Berlin, Germany.

Alexander, J. and Stimson, W. H. (1988). Sex-hormones and the course of parasitic infection. *Parasitol Today* **4**, 189–193.

Alexander, D. A. (2004). Nature's Flyers: Birds, Insects, and the Biomechanics of Flight. Johns Hopkins University Press, Baltimore, MD.

Allman, J. (2000). *Evolving Brains*. Scientific American Library, New York, NY.

Alonso-Alvarez, C., Bertrand, S., Devevey, G., Prost, J., Faivre, B. and Sorci, G. (2004). Increased susceptibility to oxidative stress as a proximate cost of reproduction. *Ecol Lett* **7**, 363–368.

Alonso-Alvarez, C., Bertrand, S., Devevey, G., Prost, J., Faivre, B., Chastel, O. and Sorci, G. (2006). An experimental manipulation of life-history trajectories and resistance to oxidative stress. *Evolution* **60**, 1913–1924.

Alscher, R. G., Erturk, N. and Heath, L. S. (2002). Role of superoxide dismutases (SODs). in controlling oxidative stress in plants. *J Exp Bot* **53**, 1331–1341.

Altshuler, D. L. (2006). Flight performance and competitive displacement of hummingbirds across elevational gradients. *Am Nat* **167**, 216–229.

Altshuler, D. L. and Dudley, R. (2006). The physiology and biomechanics of avian flight at high altitude. *Integr Comp Biol* **46(1)**, 62–71.

Altshuler D. L., Dudley, R., Ellington, C. P. (2004a). Aerodynamic forces of revolving hummingbird wings and wing models. *J Zool* **264**, 327–332.

Altshuler, D. L., Stiles, F. G. and Dudley, R. (2004b). Of hummingbirds and helicopters: Hovering costs, competitive ability, and foraging strategies. *Am Nat* **163**, 16–25.

Ambros, V. (2004). The functions of animal microRNAs. *Nature* **431**, 350–355.

Amo, L., Galvan, I., Tomás, G. and Sanz, J. J. (2008). Predator odour recognition and avoidance in a songbird. *Funct Ecol* **22**, 289–293.

Amundsen, T. and Slagsvold, T. (1991). Asynchronous hatching in the Pied Flycatcher: an experiment. *Ecology* **92**, 797–804.

Amundsen, T. and Slagsvold, T. (1998). Hatching asynchrony in great tits: a bet-hedging strategy? *Ecology* **79**, 295–304.

Ancel, A., Starke, L. N., Ponganis, P. J., Van Dam, R. and Kooyman, G. L. (2000). Energetics of surface swimming in Brandt's cormorant (*Phalacrocorax penicillatus* Brandt). *J Exp Biol* **203**, 3727–3731.

Anderson, K. J. and Jetz, W. (2005). The broad-scale ecology of energy expenditure of endotherms. *Ecol Letters* **8**, 310–318.

Ar, A. and Rahn, H. (1980). Water in the avian egg: overall budget of incubation. *Am Zool* **20**, 373–384.

Ar, A., Paganelli, C. V., Reeves, R. B., Greene, D. G. and Rahn, H. (1974). The avian egg: water vapor conductance, shell thickness, and functional pore area. *Condor* **76**, 153–158.

Ar, A., Arieli, B., Belinsky, A. and Yom-Tov, Y. (1987). Energy in avian eggs and hatchlings: utilization and transfer. *J Exp Zool Suppl* **1**, 151–164.

Ar, A., Barnea, A., Yom-Tov, Y. and Mersten-Katz, C. (2004). Woodpecker cavity aeration: a predictive model. *Respir Physiol Neurobiol* **144**, 237–249.

Ardia, D. R., Schat, K. A. and Winkler, D. W. (2003). Reproductive effort reduces long-term immune function in breeding tree swallows (*Tachycineta bicolor*). *Proc R Soc B* **270**, 1679–1683.

Arieli, Y., Peltonen, L. and Marder, J. (1988). Reproduction of rock pigeon exposed to extreme ambient temperatures. *Comp Biochem Physiol* **90A**, 497–500.

Arnold, S. J. (1983). Morphology, performance and fitness. *Am Zool* **23**, 347–361.

Aschoff, J. and Pohl, H. (1970a). Der ruheumsatz von vogelen als funktion der tageszeit under der korpergrosse. *J Ornithol* **111**, 38–47.

Aschoff, J. and Pohl, H. (1970b). Rhythmic variations in the energy metabolism. *Fed Proc* **29**, 1541–1552.

Ashmole, N. P. (1971). Sea bird ecology and the marine environment. In D. S. Farner, J. R. King and K. C. Parke, eds. *Avian Biology*, vol. 1, pp. 224–286. Academic Press, New York.

Askenmo, C. (1979). Reproductive effort and return rate of male pied flycatchers. *Am Nat* **114**, 748–753.

Askin, R. A. and Spicer, R. A. (1995). The late Cretaceous and Cenozoic history of vegetation and climate at Northern and Southern latitudes. In *Effects of Past Global Change on Life*, pp. 156–173, National Academy Press, Washington, DC.

Astheimer, L. B. and Grau, C. R. (1985). A comparison of yolk growth rates in seabird eggs. *Ibis* **132**, 380–394.

Atkins, P. W. (1992). The Elements of Physical Chemistry. Oxford University Press, Oxford, UK.

Austin, G. T. (1978). Daily time budget of the postnesting verdin. *Auk* **95**, 247–251.

Axelrod, D. I. and Raven, P. H. (1978). Late Cretaceous and Tertiary vegetation history of Africa. In M. J. A. Werger, ed. *Biogeography and Ecology of Southern Africa*, pp. 77–130, Junk, The Hague.

Backman, J. and Alerstam, T. (2001). Confronting the winds: Orientation and flight behaviour of roosting swifts, *Apus apus. Proc Roy Soc B* **268**, 1081–1087.

Balaban, E. (1988). Bird song syntax: Learned intraspecific variation is meaningful (cultural variation/vocal communication/population differences). *PNAS* **85**, 3657–3660.

Baldwin, I. T. (2001). An ecologically motivated analysis of plant—herbivore interactions in native tobacco. *Plant Physiol* **127**, 1449–1458.

Baldwin, J. (1988). Predicting the swimming speed and diving behaviour of penguins from muscle biochemistry. In J. M. Ferris, H. R. Burton, G. W. Johnstone and I. A. E. Bayly, eds. *Biology of the Vestford Hills, Antartica. Hidrobiologia* **165**, 255–261.

Ball, G. F. and Balthazart, J. (2008). Individual variation and the endocrine regulation of behaviour and physiology in birds: A cellular/molecular perspective. *Phil Trans R Soc B* **363**, 1699–1710.

Ball, G. F. and Wingfield, J. C. (1987). Changes in plasma levels of luteinizing hormone and sex steroid hormones in relation to multiple broodedness and nest-site density in male starlings. *Physiol Zool* **60**, 191–199.

Balthazart, J., Baillien, M., Cornil, C. A. and Ball, G. F. (2004). Preoptic aromatase modulates male sexual behavior: Slow and fast mechanisms of action. *Physiol Behav* **83**, 247–270.

Bang, B. G. (1966). The olfactory apparatus of tubenosed birds (Procellariiformes). *Acta Anat* **65**, 391–415.

Banzett, R. B., Butler, J. P., Nations, C. S., Barnas, G. M., Lehr, G. L. and Jones, J. H. (1987). Inspiratory aerodynamic valving in goose lungs depends on gas density and velocity. *Respir Physiol* **70**, 287–300.

Banzett, R. B., Nations, C. S., Wang, N., Fredberg, J. J. and Butler, J. P. (1991). Pressure profiles show features essential to aerodynamic valving in geese. *Respir Physiol* **84**, 295–309.

Barbosa, A. (1995). Foraging strategies and their influence on scanning and flocking behaviour of waders. *J Avian Biol* **26**, 182–186.

Barbosa, A. and Moreno, E. (1999). Evolution of foraging strategies in shorebirds: an ecomorphological approach. *Auk* **116**, 712–725.

Bardou, F., Bouchaud, J.-P., Aspect, A. and Cohen-Tannoudji, C. (2002). *Lévy Statistics and Laser Cooling: How Rare Events Bring Atoms to Rest*. Cambridge University Press, Cambridge, UK.

Barja, G. (2004). Aging in vertebrates, and the effect of caloric restriction: A mitochondrial free radical production-DNA damage mechanism? *Biol Rev* **79**, 235–251.

Barja, G., Cadenas, S., Rojas, C., Perezcampo, R. and Lopeztorres, M. (1994). Low mitochondrial free-radical production per unit O_2 consumption can explain the simultaneous presence of high longevity and high aerobic metabolic rates in birds. *Free Rad Res* **21**, 317–327.

Barker, F. K., Cibois, A., Schikler, P., Feinstein, J. and Cracraft, J. (2004). Phylogeny and diversification of the largest avian radiation. *PNAS* **101(30)**, 11040–11045.

Barnes, G. G. and Thomas, V. G. (1987). Digestive organ morphology, diet, and guild structure of North American Anatidae. *Can J Zool* **65**, 1812–1817.

Barré, H., Nedergaard, J. and Cannon, B. (1986). Increased respiration in skeletal muscle mitochondria from cold-acclimated ducklings: uncoupling effects of free fatty acids. *Comp Biochem Physiol B* **85**, 343–348.

Bartel, D. P. (2005). MicroRNAs: genomics, biogenesis, mechanism and function. *Cell* **116**, 281–297.

Bartholomew, G. A. (1972). The water economy of seed-eating birds that survive without drinking. pp. 237–254. *Proceedings 15th International Ornithological Congress,*. The Hague, Netherlands.

Bartholomew, G. A. and Cade, T. J. (1963). The water economy of land birds. *Auk* **80**, 504–539.

Barton, R. A. (1996). Neocortex size and behavioural ecology in primates. *Proc R Soc B* **263**, 173–177.

Bartumeus, F., Peters, F., Pueyo, S., Marrasé, C. and Catalan, J. (2003). Helical Lévy walks: Adjusting searching statistics to resource availability in microzooplankton. *PNAS* **100**, 12771–12775.

Battley, P. F., Piersma, T., Dietz, M. W., Tang, S., Dekinga, A. and Hulsman, K. (2000). Empirical evidence for differential organ reductions during trans-oceanic bird flight. *Proc R Soc B* **267**, 191–195.

Battley, P. F., Dekinga, A., Dietz, M. W., Piersma, T., Tang, S. and Hulsman, K. (2001). Basal metabolic rate declines during long-distance migratory flight in Great Knots. *Condor* **103**, 838–845.

Bauchinger, U. and Biebach, H. (2006). Transition between moult and migration in a long-distance migratory passerine: organ flexibility in the African wintering area. *J Ornithol* **147**, 266–273.

Beauchamp, A. J. (1989). Panbiogeography and rails of the genus *Gallirallus*. *New Zealand J Zool* **16**, 763–772.

Bech, C. and Præsteng, K. E. (2004). Thermoregulatory use of heat increment of feeding in the Tawny owl (*Strix aluco*). *J Thermal Biol* **29**, 649–654.

Beckman, K. B. and Ames, B. N. (1998). The free radical theory of aging matures. *Physiol Rev* **78**, 547–581.

Belovsky, G. E. (1984). Herbivore optimal foraging: A comparative test of three models. *Am Nat* **124**, 97–115.

ben-Avraham, D. and Havlin, S. (2000). *Diffusion and Reactions in Fractals and Disordered Systems*. Cambridge University Press, Cambridge,UK.

Bennett, P. M. and Harvey, P. H. (1987). Active and resting metabolism in birds: allometry, phylogeny and ecology. *J Zool* **213**, 327–363.

Berg, H. C. (1983). *Random Walks in Biology*. Princeton University Press, Princeton, NJ.

Berger, J., Swenson, J. E. and Persson, I.-L. (2001). Recolonizing carnivores and naïve prey: conservation lessons from Pleistocene extinctions. *Science* **291**, 1036–1039.

Bernstein, M. H. (1991). Hypoxic birds: temperature and respiration. In S. Lahiri, N. S. Cherniack and R. S. Fitzgerald, eds. *Response Adpatation to Hypoxia: Organ to Organelle*, pp. 223–234, Oxford University Press, Oxford, UK.

Berresheim, H., Andreae, M. O., Ayers, G. P. and Gillett, R. W. (1989). Distribution of biogenic sulfur-compounds in the remote southern-hemisphere. In E. J. Saltzman and W. J. Cooper, eds. *Biogenic Sulfur in the Envrionment*, pp. 352–356. American Chemical Society Washington, D.C.

Berteaux, D., Réale, D., McAdam, A.G. and Boutin, S. (2004). Keeping pace with fast climate change: can arctic life count on evolution? *Integr Comp Biol* **44**, 140–151.

Berthold, A. A. (1849). Transplantation of testes. English translation by D. P. Quiring. *Bull Hist Med* **16**, 399–401.

Berthold, P. (1975). Migration: control and metabolic physiology. In D. S. Farner and J. R. King, eds. *Avian Biology*, pp. 77–128. Academic Press, New York, NY.

Berthold, P. (1993). *Bird Migration: a General Survey*. Oxford Ornithological Series. Oxford University Press, Oxford, UK.

Berthold, P. and Querner, U. (1981). Genetic basis of migratory behavior in European warblers. *Science* **212**, 77–79.

Berthold, P., Helbig, A. J., Mohr, G. and Querner, U. (1992). Rapid microevolution of migratory behaviour in a wild bird species. *Nature* **360**, 668–670.

Bertrand, S., Burgos, J. M., Gerlotto, F. and Atiquipa, J. (2005). Lévy trajectories of Peruvian purse-seiners as an indicator of the spatial distribution of anchovy (*Engraulis ringens*). *ICES J Mar Sci* **62**, 477–482.

Bevan, R. M., Butler, P. J., Woakes, A. J. and Prince, P. A. (1995). The energy expenditure of free-ranging black-browed albatrosses. *Phil Trans R Soc* **350**, 119–131.

Bicudo, J. E. P. W., Bianco, A. C. and Vianna, C. R. (2002). Adaptive thermogenesis in hummingbirds. *J Exp Biol* **205**, 2267–2273.

Biebach, H. (1996). Energetics of winter and migratory fattening. In C. Carey, ed. *Avian, Energetics and Nutritional Ecology*, pp. 280–323. Chapman and Hall, New York, NY.

Bishcoff, K. B. (1966). Optimal continuous fermentation reaction design. *Can J Chem Eng* **44**, 281–284.

Biviano, A. B., Martínez del Rio, C. and Phillips, D. L. (1993). Ontogenesis of intestine morphology and intestinal disaccharidases in chickens (*Gallus gallus*) fed contrasting purified diets. *J Comp Physiol B* **163**, 508–518.

Bjorklund, M. and Westman, B. (1986). Adaptive advantages of monogamy in the great tit (*Parus major*) – an experimental test of the polygyny threshold-model. *Anim Behav* **34**, 1436–1440.

Black, C. P. and Tenney, S. M. (1980). Oxygen transport during progressive hypoxia in high-altitude and sea-level waterfowl. *Respir Physiol* **39**, 217–239.

Blake, J. G. and Loiselle, B. A. (1991). Variation in resource abundance affects capture rates of birds in three lowland habitats in Costa Rica. *Auk* **108**, 114–130.

Blake, R. W. (1981). Mechanics of drag-based mechanisms of propulsion in aquatic vertebrates. *Symp Zool Soc Lond* **48**, 29–52.

Blaxter, K. M. (1989). *Energy Metabolism in Animals and Man*. Cambridge University Press, Cambridge, UK.

Bleiweiss, R. (1988). Origin of hummingbird faunas. *Biol J Linn Soc* **65**, 77–97.

Blem, C. R. (1980). The energetics of migration. In S. A. Gauthreaux Jr., ed. *Animal Migration, Orientation, and Navigation*, pp. 175–224. Academic Press, New York, NY.

Blount, J. D., Houston, D. C. and Moller, A. P. (2000). Why egg yolk is yellow. *TREE* **15**, 47–49.

Boersma, P. D. (1986). Body temperature, torpor, and growth of chicks of fork-tailed storm petrels (*Oceanodroma furcata*). *Physiol Zool* **59**, 10–19.

Boersma, P. D. and Wheelwright, N. T. (1979). Egg neglect in the Procellariiformes: reproductive adaptations in the fork-tailed storm-petrel. *Condor* **81**, 157–165.

Boles, W. E. (1995). The world's oldest songbird. *Nature* **374**, 21–22.

Bolton, M., Houston, D. and Monaghan, P. (1992). Nutritional constraints on egg formation in the lesser black-backed gull – an experimental study. *J Anim Ecol* **61**, 521–532.

Bonadonna, F., Caro, S., Jouventin, P. and Nevitt, G. A. (2006). Evidence that blue petrel, *Halobaena caerulea*, fledglings can detect and orient to dimethyl sulfide. *J Exp Biol* **209**, 2165–2169.

Bonadonna, F., Miguel, E., Grosbois, V., Jouventin, P. and Bessiere, J.-M. (2007). Individual odor recognition in birds: an endogenous olfactory signature on petrels' feathers? *J Chem Ecol* **33**, 1819–1829.

Booth, D. T. (1984). Thermoregulation in neonate mallee fowl *Leipoa ocellata*. *Physiol Zool* **57**, 251–260.

Bottjer, S. W., Miesner, E. A. and Arnold, A. P. (1984). Forebrain lesions disrupt development but not maintenance of song in passerine birds. *Science* **224**, 901–903.

Boulinier, T. and Staszewski, V. (2008). Maternal transfer of antibodies: Raising immuno-ecology issues. *TREE* **23**, 282–288.

Bouverot, P. (1985). *Adaptation to Altitude-Hypoxia in Vertebrates*. Springer-Verlag, Berlin, Germany.

Bouverot, P. and Dejours, P. (1971). Pathway of respired gas in the air-sac-lung apparatus of fowls and duck. *Respir Physiol* **12**, 330–342.

Bouwstra, J. A., Honeywell-Nguyen, P. L., Gooris, G. S. and Ponec, M. (2003). Structure of the skin barrier and its modulation by vesicular formulations. *Prog Lipid Res* **42**, 1–36.

Bowen, G. J. and Revenaugh, J. (2003). Interpolating the isotopic composition of modern meteoric precipitation. *Water Resources Res* **39**, 9–14.

Bowlin, M. S., Cochran, W. W. and Wikelski, M. C. (2005). Biotelemetry of New World thrushes during migration: Physiology, energetics and orientation in the wild. *Integr Comp Biol* **45**, 295–304.

Brackenbury, J. H. (1971). Air flow dynamics in the avian lung as determined by direct and indirect methods. *Respir Physiol* **13**, 319–329.

Braude, S., Tang-Martinez, Z. and Taylor, G. T. (1999). Stress, testosterone, and the immunoredistribution hypothesis. *Behav Ecol* **10**, 345–350.

Braun, E. J. and Duke, G. E. (1989). Function of the avian cecum. *J Exp Biol* **3**, 1–130.

Braun, J. and Reif, W.-E. (1985). A survey of aquatic locomotion in fishes and tetrapods. *N Jahrb Geol Paläont Abh* **169**, 307–332.

Brenner, R. R. (1984). Effect of unsaturated-acids on membrane-structure and enzyme-kinetics. *Progr Lipid Res* **23**, 69–96.

Brenowitz, E. (1991). Evolution of the vocal control system in the avian brain. *Neurosciences* **3**, 399–407.

Bretz, W. L. and Schmidt-Nieslen, K. (1971). Bird respiration: flow patterns in the duck lung. *J Exp Biol* **54**, 103–118.

Breuner, C. W. and Orchinik, M. (2002). Plasma binding proteins as mediators of corticosteroid action in vertebrates. *J Endocrinol* **175**, 99–112.

Brice, A. T. and Grau, C. R. (1991). Protein requirements of Costa's hummingbirds *Calypte costae*. *Physiol Zool* **64(2)**, 611–626.

Briggs, G. E. and Haldane, J. B. S. (1925). A note on the kinetics of enzyme action. *Biochem J* **19**, 338–339.

Brigham, R. M., Woods, C. P., Lane, J. E., Fletcher, Q. E. and Geiser, F. (2006). Ecological correlates of torpor use among five caprimulgiform birds. *Acta Zool Sinica* **52(suppl.)**, 401–404.

Britto, L.R.G., Keyser, K.T., Hamassaki, D. E. and Karten, H.J. (2004). Catecholaminergic subpopulation of retinal displaced ganglion cells projects to the accessory optic nucleus in the pigeon (*Columba livia*). *J Comp Neurol* **269**, 109–117.

Brocklehurst, B. (2002). Magnetic fields and radical reactions: Recent developments and their role in nature. *Chem Soc Rev* **31**, 301–311.

Brody, S. (1945). *Bioenergetics and Growth*. Reinhold, Baltimore, MD.

Broggi, J., Hohtola, E., Orell, M. and Nilsson, J.-Å. (2005). Local adaptation to winter condition in a passerine spreading north: A common-garden approach. *Evolution* **59**, 1600–1603.

Brooker, M. L., Davies, N. B. and Noble, D. G. (1998). Rapid decline of host defences in response to reduced Cuckoo parasitism: behavioural flexibility of Reed Warblers in a changing world. *Proc R Soc B* **265**, 1277–1282.

Brown, J. A., Balment, R. J. and Rankin, J. C. (1993). *New Insights in Vertebrate Kidney Function*. Cambridge University Press, Cambridge, UK.

Brown, J. H. and West, G. B. (2000). *Scaling in Biology*. Oxford University Press, Oxford, UK.

Brown, J. H., Calder, W. A. and Kodric-Brown, A. (1978). Correlates and consequences of body size in nectar-feeding birds. *Am Zool* **18**, 687–700.

Bruderer, B., Underhill, L.G. and Liechti, F. (1995). Altitude choice by night migrants in a desert area predicted by meteorological factors. *Ibis* **137**, 44–55.

Brummermann, M. and Reinertsen, R. E. (1991). Adaptation of homeostatic thermoregulation: comparison of incubating and non-incubating bantam hens. *J Comp Physiol B: Biochem Syst Environ Physiol* **161**, 133–140.

Brummermann, M. and Reinertsen, R. E. (1992). Cardiovascular responses to thoracic skin cooling: comparison of incubating and non-incubating bantam hens. *J Comp Physiol B: Biochem Syst EnvironPhysiol* **162**, 16–22.

Buchanan, K. L., Evans, M. R. and Goldsmith, A. R. (2003). Testosterone, dominance signalling and immunosuppression in the house sparrow, *Passer domesticus*. *Behav Ecol Sociobiol* **55**, 50–59.

Bucher, T. L. (1987). Patterns in the mass-independent energetics of avian development. *J Exp Zool Suppl* **1**, 139–50.

Buddington, R. K., Chen, J. W. and Diamond, J. M. (1991). Dietary regulation of intestinal brush-border sugar and amino acid transport in carnivores. *Am J Physiol, Reg Integr Comp Physiol* **261**, R793–R801.

Buehler, D. M., Baker, A. J. and Piersma, T. (2006). Reconstructing palaeoflyways of the late Pleistocene and early Holocene Red Knot (*Calidris canutus*). *Ardea* **94**, 485–498.

Bunnell, F. L. and Harestad, A. S. (1990). Activity budgets and body weight in mammals. How sloppy can mammals be? *Curr Mammal* **2**, 245–305.

Burish, M. J., Kueh, H. Y. and Wang, S. S. H. (2004). Brain architecture and social complexity in modern and ancient birds. *Brain Behav Evol* **63**, 107–124.

Burns, K. J. (1997). Molecular systematics of tanagers (Thraupinae): evolution and biogeography of a diverse radiation of neotropical birds. *Mol Phylogenet Evol* **8**, 334–348.

Burton, A. C. (1934). The application of the theory of heat flow to the study of energy metabolism. *J Nutr* **7**, 497–533.

Butler, P. J. and Woakes, A. J. (2001). Seasonal hypothermia in a large migrating bird: saving energy for fat deposition? *J Exp Biol* **204**, 1361–1367.

Butler, P. J., West, N. H. and Jones, D. R. (1977). Respiratory and cardiovascular responses of the pigeon to level flight in a wind-tunnel. *J Exp Bio* **71**, 7–26.

Butler, P. J., Woakes, A. J. and Bishop, C. M. (1998). Behaviour and physiology of Svalbard barnacle geese, *Branta leucopsis*, during their autumn migration. *J Avian Biol* **29**, 536–545.

Butler, P. J., Green, J. A., Boyd, I. L. and Speakman, J. R. (2004). Measuring metabolic rate in the field: the pros and cons of the doubly labeled water and heart rate methods. *Funct Ecol* **18**, 168–183.

Butler, R. W., Williams, T. D., Warnock, N. and Bishop, M. A. (1997). Wind assistance: A requirement for migration of shorebirds? *Auk* **114**, 456–466.

Butte, A. (2002). The use and analysis of microarray data. *Nature Rev* **1**, 951–960.

Buttemer, W. A. and Astheimer, L. B. (2000). Testosterone does not affect basal metabolic rate or blood parasite load in captive male white-plumed honeyeaters *Lichenostomus penicillatus*. *J Avian Biol* **31**, 479–488.

Buttemer, W. A. and Dawson, T. J. (1989). Body temperature, water flux and estimated energy-expenditure of incubating emus (*Dromaius novaehollandiae*). *Comp Biochem Physiol A* **94**, 21–24.

Buttemer, W. A., Hayworth, A. M., Weathers, W. W. and Nagy, K. A. (1986). Time-budget estimates of avian energy-expenditure: Physiological and meterological considerations. *Physiol Zool* **59**, 131–149.

Buttemer, W. A., Astheimer, L. B., Weathers, W. W. and Hayworth, A. M. (1987). Energy savings attending winter-nest use by verdins (*Auriparus flaviceps*). *Auk* **104**, 531–535.

Buttemer, W. A., Warne, S., Bech, C. and Astheimer, L. B. (2008a). Testosterone effects on avian basal metabolic rate and aerobic performance: Facts and artifacts. *Comp Biochem Physiol A, Mol Integr Physiol* **150**, 204–210.

Buttemer, W. A., Battam, H. and Hulbert, A. J. (2008b). Fowl play and the price of petrel: Long-living procellariiformes have peroxidation-resistant membrane composition compared with short-living galliformes. *Biol Lett* **4**, 351–354.

Buxton, P. A. (1923). *Animal Life in Deserts: A Study of the Fauna in Relation to the Environment.* Arnold, London, UK. Reprinted, 1955.

Byrne, R. W. and Corp, N. (2004). Neocortex size predicts deception in primates. *Proc R Soc B* **271**, 1693–1699.

Byrne, R. W. and Whiten, A. (1988). *Machiavellian Intelligence: Social Expertise and the Evolution of Intellect in Monkeys, Apes, and Humans.* Oxford University Press, Oxford, UK.

Cade, T. J. and Greenwald, L. (1966). Nasal salt excretion in falconiform birds. *Condor* **68**, 338–350.

Cade, T. J. and Maclean, G. I. (1967). Transport of water by adult sandgrouse to their young. *Condor* **69**, 323–343.

Calder, W. A. (1973). An estimate of the heat balance of a nesting hummingbird in a chilling climate. *Comp Biochem Physiol A* **46**, 291–300.

Calder, W. A. (1974). The thermal and radiant environment of a winter hummingbirds nest. *Condor* **76**, 268–273.

Calder, W. A. (1984). *Size, Function and Life History.* Harvard University Press, Cambridge, MA.

Calder, W. A. (1994). When do hummingbirds use torpor in nature? *Physiol Zool* **67**, 1051–1076.

Calder, W. A. and King, J. R. (1974). Thermal and caloric relationships of birds. In D. S. Farner and J. R. King, eds. *Avian Biology*, pp. 259–413. Academic Press, New York, NY.

Canoine, V., Fusani, L., Schlinger, B. and Hau, M. (2007). Low sex steroids, high steroid receptors: Increasing the sensitivity of the nonreproductive brain. *Develop Neurobiol* **67**, 57–67.

Caraco, T. (1979). Time budgeting and group size: a test of theory. *Ecology* **60**, 618–627.

Carey, C. (1980). Adaptation of the avian egg to high altitude. *Am Zool* **20**, 449–459.

Carey, C. (1993). Does nonshivering thermogenesis exist in birds? Overview. In C Carey, G. L. Florant, B. A. Wunder and B. Horwitz, eds. *Life in the Cold. Ecological, Physiological and Molecular Mechanisms*, pp. 527–528. Westview Press, Boulder, CO.

Carey, C. (1996). Female reproductive energetics. In C. Carey, ed. *Avian Energetics and Nutritional Ecology.* Chapman and Hall, New York, NY.

Carey, J. R. and Judge, D. S. (2000). *Longevity Records.* Odense University Press, Odense, Denmark.

Carmi, N., Pinshow, B. Porter, W. P. and Jaeger, J. (1992). Water and energy limitations on flight duration in small migrating birds. *Auk* **109**, 268–276.

Carpenter, F. L. and Hixon, M. A. (1988). A new function for torpor: fat conservation in a wild migrant hummingbird. *Condor* **90**, 373–378.

Carpenter, F. L., Paton, D. C. and Hixon, M. A. (1983). Weight gain and adjustment of feeding territory size in migrant hummingbirds. *PNAS* **80**, 7259–7263.

Carpenter, F. L., Hixon, M. A., Beuchat, C. A., Russell, R. W. and Paton, D. C. (1993). Biphasic mass gain in migrant hummingbirds: body composition changes, torpor, and ecological significance. *Ecology* **74**, 1173–1182.

Case, T. J. (1978). On the evolution and adaptive significance of postnatal growth rates in terrestrial vertebrates. *Quart Rev Biol* **53**, 243–282.

Cashmore, A., Jarillo, J. A., Wu, Y. J. and Liu, D. (1999). Cryptochromes: blue light receptors for plants and animals. *Science* **284**, 760–765.

Casler, C. L. (1973). The air-sac systems and buoyancy of the anhinga and double-crested cormorant. *Auk* **90**, 324–340.

Casotti, G. and Ricardson, K. C. (1992). A stereological analysis of kidney structure of honeyeater birds (*Meliphagidae*). inhabiting either arid or wet environments. *J Anat* **180**, 281–288.

Casotti, G., Waldron, T., Misquith, G., Powers, D. and Slusher, L. (2007). Expression and localization of an aquaporin-1 homologue in the avian kidney and lower intestinal tract. *Comp Biochem Physiol A, Mol Integr Physiol* **147(2)**, 355–362.

Cassey, P., Blackburn, T. M., Sol, D., Duncan, R. P. and Lockwood, J. L. (2004). Global patterns of introduction effort and establishment success in birds. *Proc R Soc B* **271**, S405–S408.

Casto, J. M., Nolan, V. and Ketterson, E. D. (2001). Steroid hormones and immune function: Experimental studies in wild and captive dark-eyed juncos (*Junco hyemalis*). *Am Nat* **157**, 408–420.

Caut, S., Roemer, G.W., Donlan, C.J. and Courchamp, C. (2006). Coupling stable isotopes with bioenergetics to estimate interspecific interactions. *Ecol Applications* **16**, 1893–1900.

Caviedes-Vidal, E. and Karasov, W. H. (1995). Influences of diet composition on pancreatic enzyme activities in house sparrows. *Am Zool* **35**, 78A.

Caviedes-Vidal, E. and Karasov, W. H. (1996). Glucose and amino acid absorption in house sparrow intestine and its dietary modulation. *Am J Physiol, Reg Integr Comp Physiol* **271(40)**, R561–R568.

Caviedes-Vidal, E., Afik, D., Martínez del Rio, C. and Karasov, W. H. (2000). Dietary modulation of intestinal enzymes of the house sparrow (*Passer domesticus*): Testing an adaptive hypothesis. *Comp Biochem Physiol A* **125**, 11–24.

Caviedes-Vidal, E. McWhorter, T. J., Lavin, S. R., Chediack, J. G., Tracy, C. R. and Karasov, W. H. (2007). The digestive adaptation of flying vertebrates: high intestinal paracellular absorption compensates for smaller guts. *PNAS* **104(48)**, 19132–19137.

Cavieres, G. and Sabat, P. (2008). Geographic variation in the response to thermal acclimation in rufous-collared sparrows: are physiological flexibility and environmental heterogeneity correlated? *Funct Ecol* **22**, 509–515.

Chai, P. and Dudley, R. (1995). Limits to vertebrate locomotor energetics suggested by hummingbirds hovering in heliox. *Nature* **377**, 722–725.

Challenger, W. O., Williams, T. D., Christians, J. K. and Vezina, F. (2001). Follicular development and plasma yolk precursor dynamics through the laying cycle in the European starling (*Sturnus vulgaris*). *Physiol Biochem Zool* **74**, 356–365.

Chandler, C. R., Ketterson, E. D., Nolan, V. and Ziegenfus, C. (1994). Effects of testosterone on spatial activity in free-ranging male dark-eyed juncos, *Junco hyemalis. Animal Behav* **47**, 1445–1455.

Chandler, C. R., Ketterson, E. D. and Nolan, V. (1997). Effects of testosterone on use of space by male dark-eyed juncos when their mates are fertile. *Animal Behav* **54**, 543–549.

Chappell, M. A., Morgan, K. R., Souza, S. L. and Bucher, T. L. (1989). Convection and thermoregulation in two Antarctic seabirds. *J Comp Physiol* **159**, 313–322.

Chappell, M. A., Morgan, K. R. and Bucher, T. L. (1990). Weather, microclimate, and energy costs of thermoregulation for breeding Adélie penguins. *Oecologia* **83**, 420–426.

Chappell, M. A., Shoemaker, V. H., Janes, D. N., Bucher, T. L. and Maloney, S. K. (1993a). Diving behavior during foraging in breeding Adélie Penguins. *Ecology* **74(4)**, 1204–1215.

Chappell, M. A., Shoemaker, V. H., Janes, D. N., Maloney, S. K. and Bucher, T. L. (1993b). Energetics of foraging in breeding Adélie Penguins. *Ecology* **74(8)**, 2450–2461.

Chappell, M. A., Bech, C. and Buttemer, W. A. (1999). The relationship of central and peripheral organ masses to aerobic performance variation in house sparrows. *J Exp Biol* **202**, 2269–2279. ˙

Chatterjee, S. (1997). *The Rise of Birds*. Johns Hopkins University Press, Baltimore, MD.

Chatterjee, S. and Templin, R. J. (2007). Biplane wing planform and flight performance of the feathered dinosaur *Microraptor gui. PNAS* **104**, 1576–1580.

Chen, K. and Rajewsky, N. (2007). The evolution of gene regulation by transcription factors and microRNAs. *Nat Rev Genet* **8**, 93–103.

Cherel, Y., Robin, J. -P. and Le Maho, Y. (1988). Physiology and biochemistry of long-term fasting birds. *Can J Zool* **66**, 159–166.

Chesser, R. T. and Levey, D. J. (1998). Austral migrants and the evolution of migration in New World birds: diet, habitat and migration revisited. *Am Nat* **152**, 311–319.

Chew, S. J., Mello, C., Nottebohm, F., Jarvis, E. and Vicario, D. S. (1995). Decrements in auditory responses to a repeated conspecific song are long-lasting and require two periods of protein synthesis in the songbird forebrain. *PNAS* **92**, 3406–3410.

Chown, S. L., Gaston, K. J. and Robinson, D. (2004). Macrophysiology: large-scale patterns in physiological traits and their ecological implications. *Funct Ecol* **18**, 159–167.

Churchill, G. A. (2002). Fundamentals of experimental design for cDNA microarrays. *Nature Genet* **32**, 490–495.

Cichon, M. (2000). Costs of incubation and immunocompetence in the collared flycatcher. *Oecologia* **125**, 453–457.

Ciminari, M. E., Afik, D., Karasov, W. H. and Caviedes-Vidal, E. (2001). Is diet-shifting facilitated by modulation of pancreatic enzymes? Test of an adaptation hypothesis in Yellow-rumped warblers. *The Auk* **118(4)**, 1101–1107.

Cippolini, M. L. (2000). Secondary metabolites of vertebrate-dispersed fruits: evidence for adaptive functions. *Rev Chilena Hist Nat* **73**, 421–440.

Clark, L. and Ricklefs, R. E. (1988). A model for evaluating time constraints on short-term reproductive success in altricial birds. *Am Zool* **28**, 853–862.

Clarke, J. A., Zhou, Z. and Zhang, F. (2006). Insight into the evolution of avian flight from a new clade of Early Creataceous ornithurines from China and the morphology of *Yixianornis grabaui. J Anat* **208**, 287–308.

Cochran, W. W., Mouritsen, H. and Wikelski, M. (2004). Migrating songbirds recalibrate their magnetic compass daily from twilight cues. *Science* **304**, 405–408.

Cockburn, A. (2006). Prevalence of different modes of parental care in birds. *Proc R Soc B* **273**, 1375–1383.

Cockrem, J.F. (2007). Stress, corticosterone responses and avian personalities. *J Ornithol* **148**, 169–178.

Cohen, A., Klasing, K. and Ricklefs, R. (2007). Measuring circulating antioxidants in wild birds. *Comp Biochem Physiol B, Biochem Mol Biol* **147**, 110–121.

Cohen, A. A., Mcgraw, K. J., Wiersma, P., Williams, J. B., Robinson, W. D., Robinson, T. R., Brawn, J.D. and R. E. Ricklefs, R.E. (2008). Interspecific associations between circulating antioxidant levels and life-history variation in birds. *Am Nat* **172**, 178–193.

Cole, B. J. (1995). Fractal time in animal behaviour: the movement activity of *Drosophila*. *J Anim Behav* **50**, 1317–1324.

Coleman, J.D. (1974). Breakdown rates of foods ingested by starlings. *J Clin Invest* **44**, 69–172.

Collin, A., Buyse, J., Van as, P., Darras, V. M., Malheiros, R. D., Moraes, V. M. B., *et al.* (2003). Cold-induced enhancement of avian uncoupling protein expression, heat production, and triiodothyronine concentrations in broiler chicks. *Gen Comp Endocrinol* **130**, 70–77.

Collins, S. (2004). Vocal fighting and flirting: the functions of birdsong. In P. Marler and H. Slabbekoorn, eds. *Nature's Music: the Science of Birdsong*, pp. 39–79. Elsevier/Academic Press, San Diego, CA.

Conway, C. J. and Martin, T. E. (2000). Evolution of passerine incubation behavior: Influence of food, temperature, and nest predation. *Evolution* **54**, 670–685.

Cooper, A. and Penny, D. (1997). Mass survival of birds across the Cretaceous–Tertiary boundary: Molecular evidence. *Science* **275**, 1109–1113.

Cooper, B. A. and Ritchie, R. J. (1995). The altitude of bird migration in east-central Alaska: a radar and visual study. *J Field Ornithol* **66**, 590–608.

Cooper, J. (1972). Sexing the Jackass Penguin. *Safring News* **1**, 23–25.

Cooper, J. (1981). Pelagic birds and mammals of the southern Benguela region. *Trans R Soc S Afr* **44**, 373–378.

Cooper, J. (1985). Biology of the Bank Cormorant, part 2: morphometrics, plumage, bare parts and moult. *Ostrich* **56**, 79–85.

Costantini, D. and Moller, A. P. (2008). Carotenoids are minor antioxidants for birds. *Funct Ecol* **22**, 367–370.

Cottam, M., Houston, D., Lobley, G. and Hamilton, I. (2002). The use of muscle protein for egg production in the zebra finch, *Taeniopygia guttata*. *Ibis* **144**, 210–217.

Cracraft, J. (1973). Continental drift, palaeoclimatology, and the evolution and biogeography of birds. *J Zool* **169**, 455–545.

Cracraft, J. (2001). Avian evolution, Gondwana biogeography and the Cretaceous-Tertiary mass extinction event. *Proc R Soc Lond B* **268**, 459–469.

Cramp, S. and Simmons, K. E. L. (1983). *Handbook of the Birds of Europe, the Middle East and North Africa: the Birds of the Western Palearctic*, vol. 3, *Waders to Gulls*. Oxford University Press, Oxford, UK.

Crawford, E. C. and Schmidt-Nielsen, K. (1967). The temperature regulation and evaporative cooling in the ostrich. *Am J Physiol* **212**, 347–353.

Criscuolo, F., Gonzalez-Barroso, M. D., Le Maho, Y., Ricquier, D. and Bouillaud, F. (2005). Avian uncoupling protein expressed in yeast mitochondria prevents endogenous free radical damage. *Proc R Soc B* **272**, 803–810.

Croat, T. B. (1975). Phenological behavior of habit and habitat classes on Barro Colorado Island (Panama Canal Zone). *Biotropica* **7**, 270–277.

Croxall, J. P. and Linshman, G. S. (1987). The food and feeding ecology of penguins. In J. P. Croxall, ed. *Seabirds: Feeding Ecology and Role in Marine Ecosystems*, pp. 101–133. Cambridge University Press, Cambridge, UK.

Croxall, J. P. and Prince, P. A. (1994). Dead or alive, night or day – how do albatrosses catch squid. *Antarct Sci* **6**, 155–162.

Culik, B. (2001). Finding food in the open ocean: foraging strategies in Humboldt penguins. *Zool* **104**, 327–338.

Custer, T. W. and Pitelka, F. A. (1975). Correction factors for digestion rates for prey taken by Snow Bunting (*Plectrophenax nivalis*). *Condor* **77**, 210–212.

Daan, S., Deerenberg, C. and Dijkstra, C. (1996). Increased daily work precipitates natural death in the kestrel. *J Anim Ecol* **65**, 539–544.

Dabelow, A. (1925). Die Schwimmanpassnung der Vögel. Ein Beitrag zur biologischen Anatomie der Fortbewegung. *Jb Morph Mikroskop Anat* **54**, 288–321.

Dacey, J. W. H. and Wakeham, S. G. (1986). Oceanic dimethylsulfide: production during zooplankton grazing on phytoplankton. *Science* **233**, 1314–1316.

Daly, K. L. and DiTullio, G. R. (1996). Particulate dimethylsulfoniopropionate removal and dimethyl sulfide production by zooplankton in the Southern Ocean. In R. P. Kiene, P. T. Visscher, M. D. Kellor and G. O. Kirst, eds. *Biological and Environmental Chemistry of DMSP and Related Sulfonium Compounds*, pp.223–238. Plenum Press, New York, NY.

Dantzler, W. H. (1989). *Comparative Physiology of the Vertebrate Kidney*. Springer-Verlag, New York, NY.

Davidson, E. H. (2006). *The Regulatory Genome: Gene Regulatory Networks in Development and Evolution*. Academic Press, New York, NY.

Dawson, W. R. (1982). Evaporative losses of water by birds. *Comp Biochem Physiol A* **71**, 495–509.

Dawson, W. R. (1984). Physiological studies of desert birds: present and future considerations. *J Arid Environ* **7**, 133–155.

Dawson, W. R. and Bartholomew, G. A. (1968). Temperature regulation and water economy of desert birds. In G. W. Brown Jr, ed. *Desert Biology*, pp. 357–394. Academic Press, New York, NY.

Dawson, W. R. and Whittow, G. C. (2000). Regulation of body temperature. In G. C. Whittow, ed. *Sturkie's Avian Physiology*, pp. 343–390. Academic Press, New York, NY.

De Heij, M. E., Van Den Hout, P. J. and Tinbergen, J. M. (2006). Fitness cost of incubation in great tits (*Parus major*) is related to clutch size. *Proc R Soc B* **273**, 2353–2361.

De Ridder, E., Pinxten, R. and Eens, M. (2000). Experimental evidence of a testosterone-induced shift from paternal to mating behaviour in a facultatively polygynous songbird. *Behav Ecol Sociobiol* **49**, 24–30.

De Vries, J. and Van Eerden, M. R. (1995). Thermal conductance in aquatic birds in relation to the degree of water contact, body mass and body fat: energetics implications of living in a strong cooling environment. *Physiol Zool* **68**, 1143–1163.

DeBose, J. L. and Nevitt, G. A. (2008). The use of odors at different spatial scales: comparing birds with fish. *J Chem Ecol* **34**(7), 867–881.

DeBose, J. L., Lema, S. C. and Nevitt, G. A. (2008). Dimethylsulfoniopropionate as a foraging cue for reef fishes. *Science* **319**, 1356.

Deeming, D. C. (2002). Behaviour patterns during incubation. In D. C. Deeming, ed. *Avian Incubation: Behaviour, Environment, and Evolution*. Oxford University Press, Oxford, UK.

Dekinga, A., Dietz, M. W., Koolhaas, A. and Piersma, T. (2001). Time course and reversibility of changes in the gizzards of red knots alternately eating hard and soft food. *J Exp Biol* **204**, 2167–2173.

del Hoyo, J., Elliott, A. and Sargatal, J. (1996). *Handbook of the Birds of the World*, Vol. 3. Lynx Edicions, Barcelona, Spain.

Denny, M. W. (1993). *Air and Water: the Biology and Physics of Life's Media*. Princeton University Press, Princeton, NJ.

Desagher, S., Glowinski, J. and Premont, J. (1996). Astrocytes protect neurons from hydrogen peroxide toxicity. *J Neurosci* **16**, 2553–2562.

Deviche, P. and Parris, J. (2006). Testosterone treatment to free-ranging male dark-eyed juncos (*Junco hyemalis*). exacerbates hemoparasitic infection. *Auk* **123**, 548–562.

Deviche, P., Breuner, C. and Orchinik, M. (2001). Testosterone, corticosterone, and photoperiod interact to regulate plasma levels of binding globulin and free steroid hormone in dark-eyed juncos, *Junco hyemalis*. *Gen Comp Endocrinol* **122**, 67–77.

Dhabhar, F. S. and McEwen, B. S. (1997). Acute stress enhances while chronic stress suppresses cell-mediated immunity in vivo: A potential role for leukocyte trafficking. *Brain Behav Immun* **11**, 286–306.

Dial, K. P. (2003a). Wing-assisted incline running and evolution of flight. *Science* **299**, 402–404.

Dial, K. P. (2003b). Evolution of avian locomotion: correlates of flight style, locomotor modules, nesting biology, body size, and the origin of flapping flight. *Auk* **120(4)**, 941–952.

Dial, K. P., Biewener, A. A., Tobalske, B. W. and Warrick, D. R. (1997). Mechanical power output of bird flight. *Nature* **390**, 67–70.

Dial, K. P., Randall, R.J. and Dial, T. R. (2006). What use is half wing in the ecology and evolution of birds? *Bioscience* **56(5)**, 437–445.

Diamond, J. M. (1981). Flightlessness and fear of flying in island species. *Nature* **293**, 507–508.

Diamond, J. M. (1991). Evolutionary design of intestinal nutrient absorption: enough but not too much. *News Physiol Sci* **6**, 92–96.

Diamond, J. M. and Hammond, K. A. (1992). The matches, achieved by natural selection, between biological capacities and their natural loads. *Experientia* **48**, 551–557.

Diamond, J. M., Karasov, W. H., Phan, D. and Carpenter, F. L. (1986). Digestive physiology is a determinant of foraging bout frequency in hummingbirds. *Nature* **320**, 62–63.

Diamond, J. M., Bishop, K. D. and Gilardi, J. D. (1999).Geophagy in New Guidea birds. *Ibis* **141**, 181–193.

Díaz, L. and Carrascal, L. M. (2006). Influence of habitat structure and nest site features on predation pressure of artificial nests in Mediterranean oak forests. *Ardeola* **53**, 69–81.

Dijkstra, C., Bult, A., Bijlsma, S., Daan, S., Meijer, T. and Zijlstra, M. (1990). Brood size manipulations in the kestrel (*Falco tinnunculus*) – effects on offspring and parent survival. *J Anim Ecol* **59**, 269–285.

Dingle, H. (1996). The Biology of Life on the Move. Oxford University Press, Oxford, UK.

Dittami, J., Hoi, H. and Sageder, G. (1991). Parental investment and territorial sexual-behaviour in male and female reed warblers – are they mutually exclusive. *Ethology* **88**, 249–255.

Dobbs, R. C., Styrsky, J. D. and Thompson, C. F. (2006). Clutch size and the costs of incubation in the house wren. *Behav Ecol* **17**, 849–856.

Dolnik, V. R. and Gavrilov, V. M. (1973). Energy metabolism during flight of some passerines. In B. E. Byikhovskii, ed. *Bird Migrations: Ecological and Physiological Factors*, pp. 288–296. John Wiley & Sons, New York, NY.

Dow, D. D. (1965). The role of saliva in food storage by the Gray Jay. *Auk* **82**, 139–154.

Drent, R. H., Fox, A. D. and Stahl, J. (2006). Traveling to breed. *J Ornithol* **147**, 122–134.

Drinnan, R. E. (1957). The winter feeding of the oystercatcher (*Haematopus ostralegus*). on the edible cockle. *J Anim Ecol* **26**, 439–469.

Drossel, B., Clar, S. and Schwabl, F. (1992). Self-organized critical forest-fire model. *Phys Rev Lett* **69**, 1629–1632.

Duchamp, C., Marmonier, F., Denjean, F., Lachuer, J., Eldershaw, T. P. D., Rouanet J.-L., Morales, A., Meister, R., Bénistant, C., Roussel, D. and Barré, H. (1999). Regulatory, cellular and molecular aspects of avian muscle nonshivering thermogenesis. *Ornis Fennica* **76**, 151–165.

Dudley, R. (1998). Atmospheric oxygen, giant Paleozoic insects and the evolution of aerial locomotor performance. *J Exp Biol* **201**, 1043–1050.

Duffy, D. L., Bentley, G. E., Drazen, D. L. and Ball, G. F. (2000). Effects of testosterone on cell-mediated and humoral immunity in non-breeding adult European starlings. *Behav Ecol* **11**, 654–662.

Dugas-Ford, J. and Ragsdale, C. (2003). 23rd Annual J. B. Johnston Club Meeting and 15th Annual Karger Workshop 2003. *Brain Behav Evol* **62**, 168–174.

Dukas, R. (1998). Evolutionary ecology of learning. In R. Dukas, ed. *Cognitive Ecology: The Evolutionary Ecology of Information Processing and Decision Making*, pp. 129–174. University of Chicago Press, Chicago, IL.

Dukas, R. and Bernays, E. A. (2000). Learning improves growth rate in grasshoppers. *PNAS* **97**, 2637–2640.

Duke, G. E. (1989). Gastrointestinal motility and its regulation. *Poult Sci* **61**, 1245–1256.

Dumonteil, E., Barré, H. and Meissner, G. (1995). Expression of sarcoplasmic reticulum Ca^{2+} transport proteins in cold-acclimating ducklings. *Am J Physiol* **269**, C955–C960.

Dunn, E. H. (1975). The timing of endothermy in the development of altricial birds. *Condor* **77**, 288–293.

Dunn, E. H. (1976). The relationship between brood size and age of effective homeothermy in nestling house wrens. *Wilson Bull* **88**, 478–482.

Dykstra, C. R. and Karasov, W. H. (1992). Changes in gut structure and function of House Wrens (*Troglodytes aedon*). in response to increased energy demands. *Physiol Zool* **65**, 422–442.

Dykstra, C. R. and Karasov, W. H. (1993). Nesting energetics of House Wrens (*Troglodytes aedon*). in relation to maximal rates of energy flow. *Auk* **110**, 481–491.

Eads, B. D., Andrews, J. and Colbourne, J. K. (2008). Ecological genomics in *Daphnia*: stress responses and environmental sex determination. *Heredity* **100**, 184–190.

Echtay, K. S. (2007). Mitochondrial uncoupling proteins – what is their physiological role? *Free Rad Biol Med* **43**, 1351–1371.

Eckman, V. W. (1905). On the influence of the earth's rotation on ocean-currents. *Ark Math Astrom Fys* **2**, 1–53.

Edelstein-Keshet, L. (1988). *Mathematical Models in Biology*. Random House, New York, NY.

Edwards, A. M, Phillips, R. A., Watkins, N. W., Freeman, M. P., Murphy, E. J., Afanasyev, V., Buldyrev, S. V., da Luz, M. G. E., Raposo, E. P., Stanley, H. E. and Viswanathan, G. M. (2007). Revisiting Lévy flight search patterns of wandering albatrosses, bumblebees and deer. *Nature* **449**, 1044–1049.

Eens, M., Van Duyse, E., Berghman, L. and Pinxten, R. (2000). Shield characteristics are testosterone-dependent in both male and female moorhens. *Hormones Behav* **37**, 126–134.

Eickhoff, B., Korn, B., Schick, M., Poustka, A. and van der Bosch, J. (1999). Normalization of array hybridization experiments in differential gene expression analysis. *Nucleic Acids Res* **27**, 33.

Ekins, R. and Chu, F. W. (1999). Microarrays: their origins and applications. *Trends Biotechnol* **17**, 217–218.

Elbrønd, V. S., Dantzer, V., Mayhew, T. M. and Skadhauge, E. (1993). Dietary and aldosterone effects on the morphology and electrophysiology of the chicken coprodeum. In P. J. Sharp. Ed. Avian Endocrinology, pp. 217–226. *Journal of Endocrinology* Ltd., Bristol, UK.

Elbrønd, V. S., Laverty, G., Dantzer, V., Grøndahl, C. and Skadhauge, E. (2009). Ultrastructure and electrolyte transport of the epithelium of coprodeum, colon and the proctodeal diverticula of *Rhea Americana*. *Comp Biochem Physiol A* **152**, 357–365.

Elgar, M. A. and Harvey, P. H. (1987). Basal metabolic rates in mammals: allometry, phylogeny and ecology. *J Funct Ecol* **1**, 25–36.

Ellegren, H. and Sheldon, B. C. (2008). Genetic basis of fitness differences in natural populations. *Nature Rev* **452**, 169–175.

Ellis, H. E. and Gabrielsen, G. W. (2002). Energetics of free-ranging seabirds. In E. A. Schreiber and J. Burger, eds. *Biology of Marine Birds*, pp. 359–407. CRC Press, Boca Raton, CA.

Ellington, C. P. (1990). Limitations on animal flight performance. *J Exp Biol* **160**, 71–91.

Else, P. L. and Wu, B. J. (1999). What role for membranes in determining the higher sodium pump molecular activity of mammals compared to ectotherms? *Journal of Comp Physiol B, Biochem Syst Environ Physiol* **169**, 296–302.

Emery, N. J. (2006). Cognitive ornithology: The evolution of avian intelligence. *Phil Trans R Soc B* **361**, 23–43.

Emery, N. J. and Clayton, N. S. (2004). The mentality of crows: convergent evolution of intelligence in corvids and apes. *Science* **306**, 1903–1907.

Emery, N. J. and Clayton, N. S. (2005). Evolution of the avian brain and intelligence. *Curr Biol* **15(23)**, R946–R950.

Emlen, S. T. (1967). Migratory orientation in the Indigo bunting, *Passerina cyanea*. Part I. *Auk* **84**, 309–342.

Engstrand, S. M. and Bryant, D. M. (2002). A trade-off between clutch size and incubation efficiency in the barn swallow, *Hirundo rustica*. *Funct Ecol* **16**, 782–791.

Enstipp, M. R., Jones, D. R., Lorentsen, S.-H. and Grémillet, D. (2007). Energetic costs of diving and prey-capture capabilities in cormorants and shags (Phalacrocoracidae). underline the unique adaptation to the aquatic environment. *J Ornithol* **148**, S593–S600.

Enstipp, M.R., Grémillet, D. and Jones, D. R. (2008). Heat increment of feeding in double-crested cormorants (*Phalacrocorax auritus*). and its potential for thermal substitution. *J Exp Biol* **211**, 49–57.

Enstrom, D. A., Ketterson, E. D. and Nolan, V. (1997). Testosterone and mate choice in the dark-eyed junco. *Anim Behav* **54**, 1135–1146.

Eppley, Z. A. (1984). Development of thermoregulatory abilities in Xantus's murrelet chicks, *Synthiliboramphus hypoleucus*. *Physiol Zool* **57**, 307–317.

Eppley, Z. A. (1994). A mathematical model of heat flux applied to developing endotherms. *Physiol Zool* **67**, 829–854.

Estes, J. A., Tinker, M. T., Williams, T. M. and Doak, D. F. (1998). Killer whale predation on sea otters liking oceanic and nearshore ecosystems. *Science* **282**, 473–476.

Evans, M. R., Goldsmith, A. R. and Norris, S. R. A. (2000). The effects of testosterone on antibody production and plumage coloration in male house sparrows (*Passer domesticus*). *Behav Ecol Sociobiol* **47**, 156–163.

Evans, R. M. (1984). Some causal and functional correlates of crèching in young white pelicans. *Can J Zool* **62**, 814–819.

Evock-Clover, C. M., Poch, S. M., Richards, M. P., Ashwell, C. M. and Mcmurtry, J. P. (2002). Expression of an uncoupling protein gene homolog in chickens. *Comp Biochem Physiol A – Mol Integr Physiol* **133**, 345–358.

Fain, M. G. and Houde, P. (2004). Parallel radiations in the primary clades of birds. *Int J Org Evolution* **58**, 2558–2573.

Faivre, B., Preault, M., Salvadori, F., Thery, M., Gaillard, M. and Cezilly, F. (2003). Bill colour and immunocompetence in the European blackbird. *Anim Behav* **65**, 1125–1131.

Falkowski, P. G., Katz, M. E., Milligan, A. J., Fennel, K., Cramer, B. S., Aubry, M. P., Berner, R. A., Novacek, M. J. and Zapol, W. M. (2005). The rise of oxygen over the past 205 million years and the evolution of large placental mammals. *Science* **309**, 2202–2204.

Faraci, F. M. 1991. Adaptations to hypoxia in birds: how to fly high. *Ann Rev Physiol* **53**, 59–70.

Fedde, M. R. 1990. High-altitude bird flight: exercise in a hostile environment. *News Physiol Sci* **5**, 191–193.

Feder, M. E. and Mitchell-Olds, T. (2003). Evolutionary and ecological functional genomics. *Nat Rev Gen* **4**, 649–655.

Feder, M. E., Bennett, A. F., Burggren, W. W. and Huey, R. B. (1987). *New Directions in Ecological Physiology*. Cambridge University Press, New York, NY.

Feduccia, A. (1980). *The Age of Birds*. Harvard University Press, Cambridge, Massachusetts.

Feduccia, A. (1995). Explosive evolution in Tertiary birds and mammals. *Science* **267**, 637.

Feduccia A. (1996). *The Origin and Evolution of Birds*. Yale University Press, New Haven, CT.

Feduccia, A., Lingham-Soliar, T. and Hinchcliffe, J. R. (2005). Do feathered dinosaurs exist? Testing the hypothesis on neontological and paleontological evidence. *J Morphol* **266**, 125–166.

Feenders, G., Liedvogel, M., Rivas, M., Zapka, M., Horita, H., Hara, E., Wada, K., Mouritsen, H. and Jarvis, E. D. (2008). Molecular mapping of movement-associated areas in the avian brain: A Motor Theory for Vocal Learning Origin. *PloS One* **3(3)**, e1768.

Feinsinger, P. and Colwell, R. K. (1978). Community organization among Neotropical nectar-feeding birds. *Am Zool* **18**, 779–795.

Felsenstein, J. (1985). Phylogenies and the comparative method. *Am Nat* **125**, 1–15.

Ferreira, A. R. J., Smulders, T. V., Sameshima, K., Mello, C. V. and Jarvis, E. D. (2006). Vocalizations and associated behaviors of the somber hummingbird (*Aphantochroa cirrhochloris*). and the rufous-breasted hermit (*Glaucis hirsutus*). *Auk* **123(4)**, 1129–1148.

Feuerbacher, I. and Prinzinger, R. (1981). The effects of the male sex-hormone testosterone on body temperature and energy metabolism in male Japanese quail (*Coturnix coturnix japonica*). *Comp Biochem Physiology A* **70**, 247–250.

Fish, F. E. (1996). Transitions from drag-based to lift-based propulsion in mammalian swimming. *Am Zool* **36**, 628–641.

Fisher, R. A. (1930). *The Genetical Theory of Natural Selection*. Clarendon Press, Oxford, UK.

Fisher, R. J., Poulin, R. G., Todd, L. D. and Brigham, R. M. (2004). Nest stage, wind speed, and air temperature affect the nest defense behaviours of burrowing owls. *Can J Zool* **82**, 707–713.

Fite, K., Brecha, M., Karten, H. & Hunt, S. (1981). Displaced ganglion cells and the accessory optic system of pigeons. *J Comp Neurol* **195**, 278–288.

Fleming, P. A., Hartman Bakken, B., Lotz, C. N. and Nicolson, S. W. (2004). Concentration and temperature effects on sugar intake and preferences in a sunbird and a hummingbird. *Funct Ecol* **18**, 223–232.

Foerster, K. and Kempenaers, B. (2005). Effects of testosterone on male-male competition and male-female interactions in blue tits. *Behav Ecol Sociobiol* **57**, 215–223.

Fogden, M. P. L. and Fogden, P. M. (1979). Role of fat and protein reserves in the annual cycle of the grey-backed camaroptera in Uganda (Aves, Sylvidae). *J Zool* **189**, 233–258.

Foley, W. J., McLean, S. and Cork, S. J. (1995). Consequences of biotransformation of plant secondary metabolites in acid-base metabolism in mammals – a final common pathway? *J Chem Ecol* **21**, 721–743.

Folstad, I. and Karter, A. J. (1992). Parasites, bright males, and the immunocompetence handicap. *Am Nat* **139**, 603–622.

Freckleton, R. P., Harvey, P. H. and Pagel, M. D. (2002). Phylogenetic analysis and comparative data: a test and review of evidence. *Am Nat* **160**, 712–726.

French, A. R. (1993). Hibernation in birds: comparisons with mammals. In C. Carey, G. L. Florant, B. A. Wunder and B. Horwitz, eds. *Life in the Cold: Ecological, Physiological and Molecular Mechanisms*, pp. 43–53. Westview Press, Boulder, CO.

Fridovich, I. (1989). Superoxide dismutases. An adaptation to a paramagnetic gas. *J Biol Chem* **264**, 7761–7764.

Froget, G., Handrich, Y., Le Maho, Y., Rouanet, J.-L., Woakes, A. J and Butler, P. J. (2002). The heart/oxygen consumption relationship during cold exposure of the king penguin: a comparison with that during exercise. *J Exp Biol* **205**, 2511–2517.

Fu, Z., Inaba, M., Noguchi, T. and Kato, H. (2002). Molecular Cloning and Circadian Regulation of Cryptochrome Genes in Japanese Quail (*Coturnix coturnix japonica*). *J Biol Rhythms* **17**, 14–27.

Furness, R. W. and Bryant, D. (1996). Effect of wind on field metabolic rates of breeding northern fulmars. *Ecology* **77**, 1181–1188.

Fyhn, M., Gabrielsen, G. W., Nordoy, E. S., Moe, B., Langseth, I. and Bech, C. (2001). Individual variation in field metabolic rate of kittiwakes (*Rissa tridactyla*). during the chick-rearing period. *Physiol Biochem Zool* **74**, 343–355.

Gabrielsen, G. W. and Steen, J. B. (1979). Tachycardia during egg-hypothermia in incubating ptarmigan (*Lagopus lagopus*). *Acta Physiol Scand* **107**, 273–277.

Garland, T. and Ives, A. R. (2000). Using the past to predict the present: confidence intervals for regression equations in phylogenetic comparative methods. *Am Nat* **155**, 346–364.

Garland, T., Harvey, P. H. and Ives, A. R. (1992). Procedures for the analysis of comparative data using phylogenetically independent contrasts. *Syst Biol* **41**, 18–32.

Gause, G. F. (1934). *The Struggle for Existence*. Williams and Wilkins, Baltimore, MD.

Gauthier-Clerc, M., LeMaho, Y., Clerquin, Y., Drault, S. and Handrich, Y. (2000). Penguin fathers preserve food for their chicks. *Nature* **408**, 928–929.

Gertsberger, R. and Gray, D. A. (1993). Fine structure, innervation, and functional control of avian salt glands. *Int Rev Cytol* **144**, 129–215.

Gibson, G. (2002). Microarrays in ecology and evolution: a preview. *Mol Ecol* **11**, 17–24.

Gil, D. and Gahr, M. (2002). The honesty of bird song: multiple constraints for multiple traits. *TREE* **17**, 133–141.

Giladi, I. and Pinshow, B. (1999). Evaporative and excretory water loss during free flight in pigeons. *J Comp Physiol B* **169**, 311–318.

Giovani, B., Byrdin, M., Ahmad, M. and Brettel, K. (2003). Light-induced electron transfer in a cryptochrome blue-light photoreceptor. *Nat Struct Biol* **10**, 489–490.

Goldberg, T. L. and Ewald, P. W. (1991). Territorial song in the Anna's Hummingbird, *Calypte anna*: costs of attraction and benefits of deterrence. *Anim Behav* **42**, 221–226.

Goldstein, D. L. and Braun, E. J. (1989). Structure and concentrating ability in the avian kidney. *Am J Physiol* **256**, R501–R509.

Goldstein, D. L. and Nagy, K. A. (1985). Resource utilization by desert quail: time and energy, food and water. *Ecology* **66**, 378–387.

Goldstein, D. L. and Shadhauge, E. (2000). Renal and extrarenal regulation of body fluid composition. In G. C. Whittow, ed. *Sturkie's Avian Physiology*, pp. 265–298. Academic Press, New York, NY.

Gorman, H. E. and Nager, R. G. (2004). Prenatal developmental conditions have long-term effects on offspring fecundity. *Proc Roy Soc B* **271**, 1923–1928.

Goth, A. and Evans, C. S. (2004). Egg size predicts motor performance and postnatal weight gain of Australian brush-turkey (*Alectura lathami*) hatchlings. *Can J Zool* **82**, 972–979.

Gowaty, P. A. (1996). Field studies of parental care in birds: New data focus questions on variation among females. *Adv Stud Behav* **25**, 477–531.

Graham, J. B., Dudley, R., Aguilar, N. M. and Gans, C. (1995). Implications of the late Palaeozoic oxygen pulse for physiology and evolution. *Nature* **375**, 117–120.

Green, J., Woakes, A., Boyd, I. and Butler, P. (2005). Cardiovascular adjustments during loco-motion in penguins. *Can J Zool* **83**, 445–454.

Green, J. A., Frappell, P. B., Clark, T. D. and Butler, P. J. (2006). Physiological response to feeding in little penguins. *Physiol Biochem Zool* **79**, 1088–1097.

Green, M., Alerstam, T., Gudmundsson, G. A., Hedenström, A. and Piersma, T. (2004). Do Arctic waders use adaptive wind drift? *J Avian Biol* **35**, 305–315.

Greenwalt, C. H. (1961). *Hummingbirds*. Doubleday, New York, NY.

Grémillet, D. (1995). "Wing-drying" in cormorants. *J Avian Biol* **26**, 176.

Grémillet, D., Chauvin, C., Wilson, R. P., Le Maho, Y. and Wanless, S. (2005a). Unusual feather structure allows partial plumage wettability in diving great cormorants *Phalacrocorax carbo*. *J Avian Biol* **36**, 57–63.

Grémillet, D., Kuntz, G., Woakes, A. J., Gilbert, C., Robin, J.-P., Le Maho, Y. and Butler P. J. (2005b). Year-round recordings of behavioural and physiological parameters reveal the sur-vival strategy of a poorly insulated diving endotherm during the Arctic winter. *J Exp Biol* **208**, 4231–4241.

Grindstaff, J. L., Brodie, E. D. and Ketterson, E. D. (2003). Immune function across genera-tions: Integrating mechanism and evolutionary process in maternal antibody transmission. *Proc Roy Soc B* **270**, 2309–2319.

Grossman, C. J. (1984). Regulation of the immune system by sex steroids. *Endocr Rev* **5**, 435–455.

Gryj, E., Martínez del Rio, C. and Baker, I. (1990). Avian pollination and nectar use in *Combretum fruticosum* (Loefl.). *Biotropica* **22**, 266–271.

Guglielmo, C. G., Karasov, W. H. and Jakubas, W. J. (1996). Nutritional costs of a plant sec-ondary metabolite explain selective foraging by ruffed grouse. *Ecology* **77**, 1103–115.

Gupta, B. B. P. and Thapliyal, J. P. (1984). Role of thyroid and testicular hormones in the regu-lation of basal metabolic rate, gonad development, and body weight of spotted munia, *Lonchura punctulata*. *Gen Comp Endocrinol* **56**, 66–69.

Hackett, S. J., Kimball, R. T., Reddy, S., Bowie, R. C. K., Braun, E. L, Braun, M. J., Chojnowski, J. L., Cox, W. A., Han, K.-L., Harshman, J., Huddleston, C. J., Marks, B. D., Miglia, K. J., Moore, W. S., Sheldon, F. H., Steadman, D. W., Witt, C. C. and Yuri, T. (2008). A Phylogenomic study of birds reveals their evolutionary history. *Science* **320**, 1763–1767.

Haesler, S., Wada, K., Nshdejan, A., Morrisey, E. E. and Lints, T. (2004). FoxP2 expression in avian vocal learners and non-learners. *J Neurosci* **24**, 3164–3175.

Haftorn, S. (1959). The proportion of spruce seeds removed by the tits in a Norwegian spruce forest in 1954–55. *Kongelige Norske Videnskabers Selskab Forhandlinger* **32**, 121–125.

Haftorn, S. (1988). Survival strategies of small birds during winter. In H Ouellet, ed. *Acta XIX Congressus Internationalis Ornithologici*, Vol. II, pp. 1973–1980.

Haftorn, S. and Reinertsen, R. E. (1985). The effect of temperature and clutch size on the energetic cost of incubation in a free-living blue tit (*Parus caeruleus*). *Auk* **102**, 470–478.

Hagelin, J. C. (2007). Odors and chemical signaling. In B. G. M. Jamieson, ed. *Reproductive Behavior and Phylogeny of Aves*, Vol. 6B, pp. 76–119. Science Publishers, Enfield, NH.

Halliwell, B. and Gutteridge, J. M. C. (1999). *Free Radicals in Biology and Medicine*. Oxford University Press, Oxford, UK.

Halsey, L. G., White, C. R., Fahlman, A., Handrich, Y. and Butler, P. J. (2007). Onshore energetics in penguins: Theory, estimation and ecological implications. *Comp Biochem Physiol A* **147**, 1009–1014.

Hammond, K. A. and Diamond, J. (1997). Maximal sustained energy budgets in humans and animals. *Nature* **386**, 457–462.

Hanssler, I. and Prinzinger, R. (1979). Influences of the sex-hormone testosterone on body temperature and metabolism of the male Japanese quail (*Coturnix coturnix japonica*). *Experientia* **35**, 509–510.

Haque, R., Chaurasia, S. S., Wessel, J. H. and Iuvone, P. M. (2002). Dual regulation of cryptochrome 1 mRNA expression in chicken retina by light and circadian oscillators. *NeuroReport* **13**, 2247–2251.

Hargitai, R., Prechl, J. and Torok, J. (2006). Maternal immunoglobulin concentration in collared flycatcher (*Ficedula albicollis*) eggs in relation to parental quality and laying order. *Funct Ecol* **20**, 829–838.

Harman, D. (1956). Aging: a theory based on free radical and radiation chemistry. *J Gerontol* **11**, 298–300.

Harper, D. G. C. (1988) Robin *Erithacus rubecula* species account. In S. Cramp, ed. *Handbook of Birds of Europe, the Middle East and North Africa*, Vol. V. *Tyrant Flycatchers to Thrushes*, pp. 605. Oxford University Press, Oxford, UK.

Harper, P. C., Croxall, J. P. and Cooper, J. (1985). A guide to foraging methods used by marine birds in Antarctic and Subantarctic seas. *Biomass Handbook* **24**, 1–22.

Harrisson, T. H. and Hollom, P. A. D. (1932). The great crested grebe enquiry. *Brit Birds* **26**, 142–155.

Harvey, P. H. and Pagel, M. D. (1991). *The Comparative Method in Evolutionary Biology*. Oxford University Press, Oxford, UK.

Hassan, H. M. (1989). Microbial superoxide dismutases. *Adv Gen* **26**, 65–97.

Hasselquist, D. and Nilsson, J.-A. (2009). Maternal transfer of antibodies in vertebrates: transgenerational effects on offspring immunity. *Phil Trans Roy Soc B* 364(1513): 51–60.

Hasselquist, D., Marsh, J. A., Sherman, P. W. and Wingfield, J. C. (1999). Is avian humoral immunocompetence suppressed by testosterone? *Behav Ecol Sociobiol* **45**, 167–175.

Hasselquist, D., Wasson, M. F. and Winkler, D. W. (2001). Humoral immunocompetence correlates with date of egg-laying and reflects work load in female tree swallows. *Behav Ecol* **12**, 93–97.

Hau, M. (2007). Regulation of male traits by testosterone: Implications for the evolution of vertebrate life histories. *Bioessays* **29**, 133–144.

Hau, M., Wikelski, M., Soma, K. K. and Wingfield, J. C. (2000). Testosterone and year-round territorial aggression in a tropical bird. *Gen Comp Endocrinol* **117**, 20–33.

Haury, L. and Weihs, D. (1976). Energetically efficient swimming behavior of negatively buoyant zooplankton. *Limnol Oceanogr* **21**, 797–803.

Hawkins, P. A. J., Butler, P. J., Woakes, A. J. and Gabrielsen, G. W. (1997). Heat increment of feeding in Brünnich's guillemot *Uria lomvia*. *J Exp Biol* **200**, 1757–1763.

Hay, M. E. and Kubanek, J. (2002). Community and ecosystem level consequences of chemical cues in the plankton. *J Chem Ecol* **28**, 2001–2016.

He, L. and Hannon, G. J. (2004). MicroRNAs: small RNAs with a big role in gene regulation. *Nat Rev Genet* **5**, 522–531.

Hedrick, P. W. (1986). Genetic polymorphism in heterogeneous environments: a decade later. *Ann Rev Ecol and Syst* **17**, 535–566.

Hedrick, T. L., Tobalske, B. W. and Biewener, A. A. (2002). Estimates of circulation and gait change based on a three-dimensional kinematic analysis in cockatiels (*Nymphicus hollandicus*). and ringed turtle-doves (*Streptopelia risoria*). *J Exp Biol* **205**, 1389–1409.

Hedrick, T. L., Usherwood, J. R. and Biewener, A. A. (2004). Wing inertia and whole-body acceleration: an analysis of instantaneous aerodynamic force production in cockatiels (*Nymphicus hollandicus*). flying across a range of speeds. *J Exp Biol* **207**, 1689–1702.

Hegner, R. E. & Wingfield, J. C. (1986). Behavioural and endocrine correlates of multiple brooding in the semicolonial house sparrow, *Passer domesticus* 1. Males. *Hormones and Behavior* **20**, 294–312.

Hegner, R. E. and Wingfield, J. C. (1987a). Effects of experimental manipulation of testosterone levels on parental investment and breeding success in male house sparrows. *Auk* **104**, 462–469.

Hegner, R. E. and Wingfield, J. C. (1987b). Effects of brood-size manipulations on parental investment, breeding success, and reproductive endocrinology of house sparrows. *Auk* **104**, 470–480.

Heimhofer, U., Hochuli, P. A., Burla, S., Dinis, J. and Weissert, H. (2005). Timing of Early Cretaceous angiosperm diversification and possible links to major paleoenvironmental change. *Geology* **33**, 141–144.

Heinrich, B. (1979). Resource heterogeneity and patterns of movement in foraging bumblebees. *Oecologia* **40**, 235–245.

Heinrich, B. (2003). Overnighting of golden-crowned kinglets during winter. *The Wilson Bulletin*, **115**, 113–114.

Hepple R. T., Agey, P. J. Hazelwood, L. Szewczak, J. M. MacMillen, R. E. and Mathieu-Costello, O. (1998). Increased capillarity in leg muscle of finches living at altitude. *J Appl Physiol* **85**, 1871–1876.

Herrero, A. and Barja, G. (1997). ADP-regulation of mitochondrial free radical production is different with complex I- or complex II-linked substrates: Implications for the exercise paradox and brain hypermetabolism. *J Bioenerg Biomembr* **29**, 241–249.

Herrero, A. and Barja, G. (1998). H_2O_2 production of heart mitochondria and aging rate are slower in canaries and parakeets than in mice: Sites of free radical generation and mechanisms involved. *Mechanisms of Ageing and Development* **103**, 133–146.

Hiebert, S. M. (1993). Seasonality of daily torpor in a migratory hummingbird. *Auk* **103**, 453–464.

Higuchi, H., Pierre, J. P., Krever, V., Andronov, V., Fujita, G., Ozaki, K., Goroshko, O., Ueta, M., Smirensky, S. and Mita, N. (2004). Using a remote technology in conservation: Satellite tracking White-naped Cranes in Russia and Asia. *Cons Biol* **18**, 136–147.

Hilton, G. M., Houston, D. C., Barton, N. W. H., Furness, R. W. and Ruxton, G. D. (1999). Ecological constraints on digestive physiology in carnivorous and piscivorous birds. *J Exp Zool* **283**, 365–376.

Hilton, G. M., Ruxton, G. D., Furness, R. W. and Houston, D. C. (2000). Optimal digestion strategies in seabirds: a modeling approach. *Evol Ecol Res* **2**, 207–230.

Hinds, D. S., Baudinette, R. V., MacMillen, R. E. and Halpern, E. A. (1993). Maximum metabolism and the aerobic factorial scope of endotherms. *J Exp Biol* **182**, 41–56.

Hinsley, S. A. and Ferns, P. N. (1994). Time and energy budgets of breeding males and females in sandgrouse *Pterocles* species. *Ibis* **136**, 261–270.

Hixon, M. A., Carpenter, F. L. and Paton, D. C. (1983). Territory area, flower density, and time budgeting in hummingbirds: An experimental and theoretical analysis. *Am Nat* **122**, 366–391.

Hobson, K. A. (2005). Stable isotopes and the determination of avian migratory connectivity and seasonal interactions. *Auk* **122**, 1037–1048.

Hobson, K. A. and Wassenaar, L. I. (Eds). (2008). *Tracking Animal Migration with Stable Isotopes*. Academic Press, New York, NY.

Hobson, K. A., McFarland, K. P, Wassenaar, L. J., Rimmer, C. C. and Goetz, J. E. (2001). Linking breeding and wintering ground of Bicknell's thrushes using stable isotope analyses of feathers. *Auk* **16**, 16–23.

Hobson, K. A., Bowen, G. J., Wassenaar, L. I., Ferrand, Y. and Lormee, H. (2004). Using stable hydrogen and oxygen isotope measurements of feathers to infer geographical origins of migrating European birds. *Oecologia* **141**, 477–488.

Hoffman, T. C. M. and Walsberg, G. E. (1999). Inhibiting ventilatory evaporation produces an adaptive increase in cutaneous evaporation in mourning doves *Zenaida macroura*. *J Exp Biol* **202**, 3021–3028.

Hoffman, T. C. M., Walsberg, G. E. and DeNardo, D. F. (2007). Cloacal evaporation: an important and previously undescribed mechanism for avian thermoregulation. *J Exp Biol* **210**, 741–749.

Hoheisel, J. D. (2006). Microarray technology: beyond transcript profiling and genotype analysis. *Nat Rev Genet* **7**, 200–210.

Hohtola, E. (2004). Shivering thermogenesis in birds and mammals. In B. M. Barnes and H. V. Carey, eds. *Life in the Cold: Evolution, Mechanisms, Adaptation, and Application. Twelfth International Hibernation Symposium*, pp. 241–252, Biological Papers of the University of Alaska, Institute of Arctic Biology, number 27. Fairbanks, AK.

Holling, C. S. (1959). The components of predation as revealed by a study of small mammal predation of European pine sawfly. *Can Entomol* **91**, 293–320.

Holman, R. T. (1954). Autoxidation of fats and related substances. In R. T. Holman, W. O. Lundberg and T. Malkin, eds. *Progress in Chemistry of Fats and Other Lipids*. Pergamon Press, London, UK.

Homberger, D. G. and de Silva, K. N. (2000). Functional microanatomy of the feather-bearing integument: implications for the evolution of birds and avian flight. *Amer Zool* **40**, 553–574.

Houston, A. I., Mcnamara, J. M., Barta, Z. and Klasing, K. C. (2007). The effect of energy reserves and food availability on optimal immune defence. *Proc Roy Soc B* **274**, 2835–2842.

Houston, D. C., Donnan, D. and Jones, P. J. (1995a). The source of the nutrients required for egg-production in zebra finches, *Poephila guttata*. *J Zool* **235**, 469–483.

Houston, D. C., Donnan, D. and Jones, P. J. (1995b). Use of labeled methionine to investigate the contribution of muscle proteins to egg-production in zebra finches. *J Comp Physiol B* **165**, 161–164.

Houston, D. C., Donnan, D., Jones, P., Hamilton, I. and Osborne, D. (1995c). Changes in the muscle condition of female zebra finches, *Poephila guttata* during egg-laying and the role of protein storage in bird skeletal-muscle. *Ibis* **137**, 322–328.

Hove, J. R., O'Bryan, L. M., Gordon, M. S., Webb, P. W. and Weihs, D. (2001). Boxfishes (Teleostei: Ostraciidae). as a model system for fishes swimming with many fins: kinematics. *J Exp Biol* **204**, 1459–1471.

Hudson, D. M. and Jones, D. R. (1986). The influence of body mass on the endurance to restrained submersion in the Pekin duck. *J Exp Biol* **120**, 351–367.

Hudson, J. W. and Kimsey, S. L. (1966). Temperature regulation and metabolic rhythms in populations of the house sparrow, *Passer domesticus*. *Comp Biochem Physiol* **17**, 203–217.

Hughes, A. (1977). The topography of vision in mammals of contrasting life style: comparative optics and retinal organisation. In F. Crescitelli, ed. *Handbook of Sensory Physiology*, Vol. VII/5, pp.613–756. Springer, Berlin, Germany.

Hughes, J. and Criscuolo, F. (2008). Evolutionary history of the UCP gene family: gene duplication and selection. *BMC Evol Biol* **8**, 306.

Hughes, M. R. (1970). Relative kidney size in nonpasserine birds with functional salt glands. *Condor* **72**, 164–168.

Hughes, M. R. and Chadwick, A. (1989). *Progress in Avian Osmoregulation*. Leeds Philosophical and Literary Society, Leeds, UK.

Hulbert, A. J. (2005). On the importance of fatty acid composition of membranes for aging. *J Theor Biol* **234**, 277–288.

Hulbert, A. J. (2008). The links between membrane composition, metabolic rate and lifespan. *Comp Biochem Physiology A, Mol Integr Physiol* **150**, 196–203.

Hulbert, A. J. and Else, P. L. (1999). Membranes as possible pacemakers of metabolism. *J Theor Biol* **199**, 257–274.

Hulbert, A. J. and Else, P. L. (2004). Basal metabolic rate: history, composition, regulation, and usefulness. *Physiol Biochem Zool* **77**, 869–876.

Hulbert, A. J., Faulks, S., Buttemer, W. A. and Else, P. L. (2002a). Acyl composition of muscle membranes varies with body size in birds. *J Exp Biol* **205**, 3561–3569.

Hulbert, A. J., Rana, T. and Couture, P. (2002b). The acyl composition of mammalian phospholipids: an allometric analysis. *Comp Biochem Physiol B, Biochem Mol Biol* **132**, 515–527.

Hulbert, A. J., Pamplona, R., Buffenstein, R. and Buttemer, W. A. (2007). Life and death: Metabolic rate, membrane composition, and life span of animals. *Physiol Rev* **87**, 1175–1213.

Hume, I. and Biebach, H. (1996). Digestive tract function in the long-distance migratory garden warbler, *Sylvia borin*. *J Comp Physiol B* **166**, 388–395.

Hunt, K. E., Hahn, T. P. and Wingfield, J. C. (1999). Endocrine influences on parental care during a short breeding season: Testosterone and male parental care in lapland longspurs (*Calcarius lapponicus*). *Behav Ecol Sociobiol* **45**, 360–369.

Hutchinson, J. C. D. (1955). Evaporative cooling in fowls. *J Agric Sci* **45**, 48–59.

Huxley, J. (1932). *On Relative Growth*. Methuen, London, UK.

Illius, A. W. and Jessop, N. S. (1995). Modeling metabolic costs of allelochemical ingestion by foraging herbivores. *J Chem Ecol* **21**, 693–719.

Ilmonen, P., Taarna, T. and Hasselquist, D. (2002). Are incubation costs in female pied flycatchers expressed in humoral immune responsiveness or breeding success? *Oecologia* **130**, 199–204.

Ilmonen, P., Hasselquist, D., Langefors, A. and Wiehn, J. (2003). Stress, immunocompetence and leukocyte profiles of pied flycatchers in relation to brood size manipulation. *Oecologia* **136**, 148–154.

IPCC (2007). *Climate Change 2007: The Physical Science Basis. Contribution of Working Group I to the Fourth Assessment Report of the Intergovernmental Panel on Climate Change*. S. Solomon, D. Qin, M. Manning, Z. Chen, M. Marquis, K. B. Averyt, M. Tignor and H.L. Miller, eds. 996 pp. Cambridge University Press, Cambridge, UK.

Irvine, R. J., Leckie, F. and Redpath, S. M. (2007). Cost of carrying radio transmitters: a test with racing pigeons *Columba livia*. *Wildlife Biol* **13**, 238–243.

Iverson, G. C., Warnock, S. E., Butler, R. W., Bishop, M. A. and Warnock, N. (1996). Spring migration of western sandpipers along the pacific coast of North America: a telemetry study. *Condor* **98**, 10–21.

Ivlev, V. S. (1961). *Experimental Ecology of the Feeding of Fishes*. Yale University Press, New Haven, CT.

Iwaniuk, A., Nelson, J. E. and Pellis, S. M. (2001). Do big-brained animals play more? Comparative analyses of play and relative brain size in mammals. *J Comp Psychol* **115**, 29–41.

Jackson, S., Nicolson, S. W. and Lotz, C. N. (1998). Sugar preferences and "side bias" in cape sugarbirds and lesser double-collared sunbirds. *Auk* **115(1)**, 156–165.

Jacobson, J. D. and Ansari, M. A. (2004). Immunomodulatory actions of gonadal steroids may be mediated by gonadotropin-releasing hormone. *Endocrinol* **145**, 330–336.

Jaeger, E. C. (1948). Does the poorwill "hibernate"? *Condor* **50**, 45–46.

Jaeger, E. C. (1949). Further observations on the hibernation of the poorwill. *Condor* **51**, 105–109.

Jakubas, W. J. and Mason, J. R. (1991). Role of avian trigeminal sensory system in detecting conyferal bonzoate, a plant allelochemical. *J Chem Ecol* **17**, 2213–2221.

Jakubas, W. J., Karasov, W. H. and Guglielmo, C. G. (1993). Ruffed grouse tolerance and biotransformation of the plant secondary metabolite coniferyl benzoate. *Condor* **95**, 625–640.

Janes, D. N. and Chappell, M. A. (1995). The effect of ration size and body size on specific dynamic action in Adélie penguin chicks, *Pygoscelis adeliae*. *Physiol Zool* **68**, 1029–1044.

Janik, V. M. and Slater, P. J. B. (1997). Vocal learning in mammals. *Adv Study Behav* **26**, 59–99.

Jarvis, E. D. (2004). Learned birdsong and the neurobiology of human language. *Ann N Y Acad Sci* **1016**, 749–777.

Jarvis, E. D. (2009). Bird Brain: Evolution. *Encyclopedia of Neuroscience*. **2**, 209–215.

Jarvis, E. D. and Mello, C. V, (2000). Molecular mapping of brain areas involved in parrot vocal communication. *J Comp Neurol* **419**, 1–31.

Jarvis, E. D., Mello, C. V. and Nottebohm, F. (1995). Associative learning and stimulus novelty influence the song-induced expression of an immediate early gene in the canary forebrain. *Learn Mem* **2**, 62–80.

Jarvis, E. D., Scharff, C., Grossman, M. R., Ramos, J. A. and Nottebohm, F. (1998). For whom the bird sings: context-dependent gene expression. *Neuron* **21**, 775–788.

Jarvis, E. D., Ribeiro, S., da Silva, M. L., Ventura, D. and Vielliard, J. (2000). Behaviourally driven gene expression reveals song nuclei in hummingbird brain. *Nature* **406**, 628–632.

Jarvis, E. D., Gunturkun, O., Bruce, L., Csillag, A., Karten, H., Kuenzel, W., Medina, L., Paxinos, G., Perkel, D.J., Shimizu, T., Striedter, G., Wild, J. M., Ball, G. F., Dugas-Ford, J., Durand, S. E., Hough, G. E., Husband, S., Kubikova, L., Lee, D. W., Mello, C. V., Powers, A., Siang, C., Smulders, T. V., Wada, K., White, S. A., Yamamoto, K., Yu, J., Reiner, A. and Butler, A. B. (2005). The Avian Brain Nomenclature Consortium. Avian brains and a new understanding of vertebrate brain evolution. *Nature Rev Neurosci* **6**, 151–159.

Javed, S., Takekawa, J. Y., Douglas, D. C., Rahmani, A. R., Kanai, Y., Nagendran, M., Choudhury, B. C., Sharma, S. (2000). Tracking the spring migration of a bar-headed goose (*Anser indicus*). across the Himalaya with satellite telemetry. *Global Environ Res* **2**, 195–205.

Jehl, R. J. (1990). Aspects of the molt migration. In E. Gwinner, ed. *Bird Migration: The Physiology and Ecophysiology*, pp.102–113. Springer-Verlag, Berlin, Germany.

Jerison, H. J. (1973). *Evolution of the Brain and Intelligence*. Academic Press, New York, NY.

Jeschke, J. M., Koop, M. and Tollrian, R. (2002). Predator functional responses: discriminating between handling and digesting prey. *Ecol Monogr* **72**, 95–112.

Johansson, L. C. and Norberg, U. M. L. (2000). Asymmetric toes aid underwater swimming. *Nature* **407**, 582–583.

Johansson, L. C. and Norberg, U. M. L. (2001). Lift-based paddling in diving grebe. *J Exp Biol* **204**, 1687–1696.

Johnsen, S. and Lohmann, K. J. (2005). The physics and neurobiology of magnetoreception. *Nature Rev Neurosci* **6**, 703–712.

Johnsen, T. S. (1998). Behavioural correlates of testosterone and seasonal changes of steroids in red-winged blackbirds. *Anim Behav* **55**, 957–965.

Johnson, A. R., Wiens, J. A., Milne, B. T. and Crist T. O. (1992). Animal movements and population dynamics in heterogeneous landscapes. *Landscape Ecol* **7**, 63–75.

Jones, J. H., Effmenn, E. L. and Schmidt-Nielsen, K. (1981). Control of air flow in bird lungs: Radiographic studies. *Respir Physiol* **45**, 121–131.

Jones, P. J. and Ward, P. (1976). The level of reserve protein as the proximate factor controlling the timing of breeding and clutch-size in the red-billed Quelea *Quelea quelea*. *Ibis* **118**, 547–574.

Jouventin, P. and Weimerskirch, H. (1990). Satellite tracking of wandering albatrosses. *Nature* **343**, 746–748.

Jump, D. B. (2002). Dietary polyunsaturated fatty acids and regulation of gene transcription. *Curr Opin Lipidol* **13**, 155–164.

Jump, D. B. and Clarke, S. D. (1999). Regulation of gene expression by dietary fat. *Annu Rev Nutr* **19**, 63–90.

Kaas, J. H. (2000). Why is brain size so important: design problems and solutions as neocortex gets bigger or smaller. *Brain Mind* **1**, 7–23.

Karasov, W. H. (1988). Nutrient transport across vertebrate intestine. In R. Gilles, ed. *Advances in Comparative and Environmental Physiology*, pp. 131–172. Spring-Verlag, Berlin, Germany.

Karasov, W. H. (1990). Digestions in birds: chemical and physiological determinants and ecological implications. In M. L. Morrison, C. J. Ralph, J. Verner and J. R. Jehl, eds. *Avian Foraging: Theory, Methodology and Applications. Studies in Avian Biology* **13**, 391–415. Cooper Ornithological Society, Lawrence, KS.

Karasov, W. H. (1996). Digestive plasticity in avian energetic and feeding ecology. In C. Carrey, ed. *Avian Energetics and Nutritional Ecology*, pp. 61–84. Chapman & Hall, New York, NY.

Karasov, W. H. and Cork, S. J. (1994). Glucose absorption by a nectarivorous bird: the passive pathway is paramount. *Am J Physiol, Gastrointest Liver Physiol* **267**(*30*), G18–G26.

Karasov, W. H. and Cork, S. J. (1996). Test of a reactor-based digestion optimization model for nectar-eating rainbow lorikeets. *Physiol Zool* **69**, 117–138.

Karasov, W. H. and Diamond, J. M. (1983). Adaptive regulation of sugar and amino acid transport by vertebrate intestine. *Am J Physiol, Gastrointest Liver Physiol* **245**(*8*), G443–G462.

Karasov, W. H. and Hume, I. D. (1997). Vertebrate gastrointestinal system. In W. H. Dantzler, ed. *Comparative Physiology* 1: *Handbook of Physiology*, Section 13, pp. 409–480. Oxford University Press, New York, NY.

Karasov, W. H. and Levey, D. J. (1990). Digestive system trade-offs and adaptations of frugivorous passerine birds. *Physiol Zool* **63**, 1248–1270.

Karasov, W. H. and Martínez del Rio, C. (2007). *Physiological Ecology: How Animals Process Energy, Nutrients, and Toxins*. Princeton University Press, Princeton, NJ.

Karasov, W. H. and McWilliams, S. R. (2005). Digestive constraints in mammalian and avian ecology. In J. M.Starck and T. Wang, eds. *Physiological and Ecological Adaptations to Feeding in Vertebrates*, pp.87–112. Science Publishers, Inc., Enfield, NH.

Karasov, W. H. and Pinshow, B. (1998). Changes in lean mass and organs of nutrient assimilation in a long-distance passerine migrant at a springtime stopover site. *Physiol Zool* **71**(**4**), 435–448.

Karasov, W. H. and Pinshow, B. (2000). Test for physiological limitation to nutrient assimilation in a long-distance passerine migrant at a springtime stopover site. *Physiol Biochem Zool* **73**, 335–343.

Karasov, W. H., Phan, J. M., Diamond, J. M. and Carpenter, F. L. (1986). Food passage and intestinal nutrient absorption in hummingbirds. *Auk* **103**, 453–464.

Karasov, W. H., Meyer, M. W. and Darken, B. W. (1992). Tannic acid inhibition of amino acid and sugar absorption by mouse and vole intestine: Tests following acute and subchronic exposure. *J Chem Ecol* **18**, 719–736.

Karasov, W. H., Pinshow, B., Starck, J. M. and Afik, D. (2004). Anatomical and histological changes in the alimentary tract of migrating blackcaps (*Sylvia atricapilla*): A comparision among fed, fasted, food-restricted and refed birds. *Physiol Biochem Zool* **77**, 149–160.

Karten, H. J. (1969). The organization of the avian telencephalon and some speculations on the phylogeny of the amniote telencephalon. In J. Petras, ed. *Comparative and Evolutionary Aspects of the Vertebrate Central Nervous System*, pp.164–179. *Annals of the New York Academy of Sciences*, New York, NY.

Karten, H. J. (1991). Homology and evolutionary origins of the "neocortex". *Brain Behav Evol* **38**, 264–272.

Kaseloo, P. A. and Lovvorn, J. R. (2003). Heat increment of feeding and thermal substitution in mallard ducks feeding voluntarily on grain. *J Comp Physiol B* **173**, 207–213.

Kaseloo, P. A. and Lovvorn, J. R. (2005). Effects of surface activity patterns and dive depth on thermal substitution in fasted and fed lesser scaup (*Aythya affinis*). ducks. *Can J Zool* **83**, 301–311.

Kaseloo, P. A. and Lovvorn, J. R. (2006). Substitution of heat from exercise and digestion by ducks diving for mussels at varying depths and temperatures. *J Comp Physiol B* **176**, 265–275.

Kast, T. L., Ketterson, E. D. and Nolan, V. (1998). Variation in ejaculate quality in dark-eyed juncos according to season, stage of reproduction, and testosterone treatment. *Auk* **115**, 684–693.

Keast, A. and Morton, E. S. (1980). *Migrant Birds in the Neotropics*. Smithsonian Institution Press, Washington, D.C.

Kehoe, F. P. and Ankney, C. D. (1985). Variation in digestive organ size among five species of diving ducks (*Aythya* spp.). *Can J Zool* **63**, 2339–2342.

Kendall, M. D., Ward, P. and Bacchus, S. (1973). A protein reserve in the pectoralis major flight muscle of *Quelea quelea*. *Ibis* **115**, 600–601.

Kenward, R. E. (2001). *A Manual for Wildlife Tagging*. Academic Press, London, UK.

Kerlinger, P. and Moore, F. R. (1989). Atmospheric structure and avian migration. *Curr Ornithol* **6**, 109–142.

Kersten, M. and Visser, W. (1996). The rate of food processing in the oystercatcher: food intake and energy expenditure constrained by digestive bottleneck. *Funct Ecol* **10**, 440–448.

Ketterson, E. D. and King, J. R. (1977). Metabolic and behavioral responses to fasting in the White-crowned Sparrow (*Zonotrichia leucophrys gambelii*). *Physiol Zool* **50**, 115–129.

Ketterson, E. D. and Nolan, V. (1992). Hormones and life histories – an integrative approach. *Am Nat* **140**, S33–S62.

Ketterson, E. D., Nolan, V., Wolf, L., Ziegenfus, C., Dufty, A. M., Ball, G. F., *et al.* (1991). Testosterone and avian life histories – the effect of experimentally elevated testosterone on corticosterone and body-mass in dark-eyed juncos. *Hormones and Behavior* **25**, 489–503.

Ketterson, E. D., Nolan, V., Wolf, L. and Ziegenfus, C. (1992). Testosterone and avian life histories – effects of experimentally elevated testosterone on behaviour and correlates of fitness in the dark-eyed junco (*Junco hyemalis*). *Am Nat* **140**, 980–999.

Kilner, R. M. (2002). The evolution of complex begging displays. In J. Wright and M. L. Leonard, eds. *The Evolution of Begging: Competition, Cooperation and Commnication.* Kulwer Academic Publishers, Dordrecht, Netherlands.

Kimball, R. T. and Ligon, J. D. (1999). Evolution of avian plumage dichromatism from a proximate perspective. *Am Nat* **154**, 182–193.

King, J. R. (1961). The bioenergetics of fattening and starvation in the long-distance migratory garden warbler, *Sylvia borin*, during the migratory stage. *J Comp Physiol B* **164**, 362–371.

King, J. R. (1974). Seasonal allocation of time and energy resources in birds. In R. A. Paynter Jr., ed. *Avian Energetics.* Publ. No **15**, 4–70. Nuttall Ornithological Club, Cambridge, MA.

King, J. R. and Farner, D. S. (1964). Terrestrial animal in humid heat: Birds. In D. B. Dill, ed. *Handbook of Physiology, section 4*: *Adaptation to the Environment*, pp. 603–624. American Physiological Society, Washington, D.C.

King, J. R. and Murphy, M. E. (1985). Periods of nutritional stress in the annual cycles of endotherms: fact or fiction? *Amer Zool* **25**, 955–964.

Kinsky, F. C. (1960). The yearly cycle of the Northern Blue Penguin (*Eudyptula minor novaehollandiae*). in the Wellington Harbour area. *Rec Dominion Mus NZ* **3**, 145–218.

Kirkpatrick, M. (1996). Genes and adaptation: a pocket guide to the theory. In M. R. Rose and G. V. Lauder, eds. *Adaptation*, pp.55–91. Academic Press, San Diego, CA.

Kirkwood, T. B. L. & Holliday, R. (1979). Evolution of aging and longevity. *Proc Roy Soc B* **205**, 531–546.

Klaassen, M. (2004). May dehydration risk govern long-distance migratory behaviour? *J Avian Biol* **35**, 4–6.

Klaassen, M. and Biebach, H. (1994). Energetics of fattening and starvation in the long-distance migratory garden warbler, *Sylvia borin*, during the migratory phase. *J Comp Physiol B* **164**, 362–371.

Klaassen, M. and Biebach, H. (2000). Flight altitude of trans-Sahara migrants in autumn: a comparison of radar observations with predictions from meteorological conditions and water and energy balance models. *J Avian Biol* **31**, 47–55.

Klaassen, M. and Drent, R. H. (1991). An analysis of hatchling resting metabolism: in search of artic tern nestlings and their relations to growth rate. *Condor* **93**, 612–629.

Klaassen, M., Lindstrom, A. And Zijlstra, R. (1997). Composition of fuel stores and digestive limitations to fuel deposition rate in the long-distance migratory Thrush Nightingale, *Luscinia luscinia*. *Physiol Zool* **70**, 125–133.

Klaassen, M., Kvist, A. and Lindström, A. (2000). Flight costs and fuel composition of a bird migrating in a wind tunnel. *Condor* **102(2)**, 444–451.

Klaassen, M., Beekman, J. H. Kontiokorpi, J., Mulder, R. J. W. and Nolet, B. A. (2004). Migrating swans profit from favourable changes in wind conditions at low altitude. *J Ornithol* **145**, 142–151.

Klasing, K. C. (1998). *Comparative Avian Nutrition.* CAB International, New York, NY.

Klasing, K. C. (2007). Nutrition and the immune system. *Brit Poultr Sci* **48**, 525–537.

Kleiber, M. (1932). Body size and metabolism. *Hilgardia* **6**, 315–353.

Klukowski, L. A., Cawthorn, J. M., Ketterson, E. D. and Nolan, V. (1997). Effects of experimentally elevated testosterone on plasma corticosterone and corticosteroid-binding globulin in dark-eyed juncos (*Junco hyemalis*). *Gen Comp Endocrinol* **108**, 141–151.

Koenig, W. D. (1991). The effects of tannins and lipids on digestion of acorns by Acorn Woodpeckers. *Auk* **108**, 79–88.

Komdeur, J., Daan, S., Tinbergen, J and Mateman, C. (1997). Extreme adaptive modification in sex ratio of the Seychelles warbler's eggs. *Nature* **385**, 522–525.

Kontogiannis, J. E. (1968). Effect of temperature and exercise on energy intake and body weight of the white-thorated sparrow, *Zonotrichia albicollis*. *Physiol Zool* **41**, 54–64.

Kooyman, G. L. and Kooyman, T. G. (1995). Diving behavior of emperor penguins nurturing chicks at Coulman Island, Antarctica. *Condor* **97**, 536–549.

Kooyman, G. L. and Ponganis, P. J. (1998). The physiological basis of diving to depth: birds and mammals. *Ann Rev Physiol* **60**, 19–32.

Kooyman, G. L., Cherel, Y., Le Maho, Y., Croxall, J. P., Thorson, P. H., Ridoux, V. and Kooyman, C. A. (1992a). Diving behavior and energetics during foraging cycles in King Penguins. *Ecol Monogr* **62(1)**, 143–163.

Kooyman, G. L., Ponganis, P. J., Castellini, M. A., Ponganis, E. P., Ponganis, K. V., Thorson, P. H., Eckert, S. A. and LeMaho, Y. (1992b). Heart rates and swim speeds of emperor penguins diving under sea ice. *J Exp Biol* **165**, 161–180.

Kowalewsky, S., Dambach, M., Mauck, B. and Dehnhardt, G. (2006). High olfactory sensitivity for dimethyl sulphide in harbour seals. *Biol Lett* **2**, 106–109.

Krag, B. and Skadhauge, E. (1972). Renal salt and water excretion in the budgerygah (*Melopsittacus undulatus*). *Comp Biochem Physiol A* **41**, 667–683.

Krijgsveld K. L., Olson J. M. and Ricklefs R. E. (2001). Catabolic capacity of the muscles of shorebird chicks: Maturation of function in relation to body size. *Physiol and Biochem Zool* **74**, 250–260.

Kroodsma, D. E. and Konishi, M. (1991). A suboscine bird (*eastern phoebe, Sayornis phoebe*). develops normal song without auditory feedback. *Anim Behav* **42**, 477–487.

Kullberg, C., Houston, D. C. and Metcalfe, N. B. (2002). Impaired flight ability – a cost of reproduction in female blue tits. *Behav Ecol* **13**, 575–579.

Kvist, A., Lindström, A., Green, M. and Piersma, T. and Visser, G. H. (2001). Carrying large fuel loads during sustained birdflight is cheaper than expected. *Nature* **413**, 730–732.

Lack, D. (1947). The significance of clutch size. *Ibis* **89**, 302–335.

Lack, D. (1968). *Ecological Adaptation for Breeding in Birds*. Methuen, London, UK.

Laiolo, P. and Tella, J. L. (2005). Habitat fragmentation affects culture transmission: patterns of song matching in Dupont's lark. *J Appl Ecol* **42**, 1183–1193.

Laiolo, P. and Tella, J. L. (2006). Landscape bioacoustics allows detection of the effects of habitat patchiness on population structure. *Ecol* **87**, 1203–1214.

Laiolo, P. and Tella, J. L. (2007). Erosion of animal cultures in fragmented landscapes. *Front Ecol Environ* **5**, 68–72.

Laiolo, P., Vögeli, M., Serrano, D. and Tella, J. L. (2008). Song diversity predicts the viability of fragmented bird populations. *PloS One* **3(3)**, e1822.

Lambert, A. J. and Brand, M. D. (2004). Superoxide production by NADH: Ubiquinone oxidoreductase (complex I). depends on the pH gradient across the mitochondrial inner membrane. *Biochem J* **382**, 511–517.

Land, M. F. and Nilsson, D.-E. (2002). *Animal Eyes*. Oxford University Press, Oxford, UK.

Landys, M. M., Piersma, T. Visser, G. H. Jukema, J. and Wijker, A. (2000). Water balance during real and simulated long-distance migratory flight in the bar-tailed godwit. *Condor* **102**, 645–652.

Landys-Ciannelli, M. M., Piersma, T. and Jukema, J. (2003). Strategic size changes of internal organs and muscle tissue in the bar-tailed godwit during fat storage on a spring stopover site. *Funct Ecol* **17**, 151–159.

Lapillone, A., Clarke, S. and Heird, W. C. (2004). Polyunsaturated fatty acids and gene expression. *Curr Opin Clin Nutr Metab Care* **7**, 151–156.

Larison, J. R., Crock, J. G., Snow, C. M. and Blem, C. (2001). Timing of mineral sequestration in leg bones of white-tailed ptarmigan. *Auk* **118**, 1057–1062.

Larson, A. and Losos, J. B. (1996). Phylogenetic systematics of adaptation. In M. R. Rose and G. V. Lauder, eds. *Adaptation*, pp. 187–220. Academic Press, San Diego, CA.

Lasiewski, R. C. and Dawson, W. R. (1967). A re-examination of the relation between standard metabolic rate and body weight in birds. *Condor* **69**, 13–23.

Lauder, G. (1996). The argument from design. In M. R. Rose and G.V. Lauder, eds. *Adaptation*, pp. 55–91. Academic Press, San Diego, CA.

Laughlin, S. B. (1995). Towards the cost of seeing. In M. Burrows, T. Matthews, P. L. Newland and H. J. Schuppe, eds. *Nervous Systems and Behaviour*, pp. 290. Thieme, Stuttgart, Germany.

Laughlin, S. B., van Steveninck, R. R. D. and Anderson, J. C. (1998). The metabolic cost of neural information. *Nature Neurosci* **1**, 36–41.

Lavin, S. R., McWhorter, T. J. and Karasov, W. H. (2007). Mechanistic bases for differences in passive absorption. *J Exp Biol* **210**, 2754–2764.

Lawless, R. M., Buttemer, W. A., Astheimer, L. B. and Kerry, K. R. (2001). The influence of thermoregulatory demand on contact crèching behaviour in Adélie Penguin chicks. *J Therm Biol* **26**, 555–562.

Laybourne, R. C. (1974). Collision between a vulture and an aircraft at an altitude of 37,000 feet. *Wilson Bull* **86**, 461–462.

Lee, K. A. (2006). Linking immune defenses and life history at the levels of the individual and the species. *Integr Comp Biol* **46**, 1000–1015.

Lee, K. A., Karasov, W. H. and Caviedes-Vidal, E. (2002). Digestive response to restricted feeding in migratory yellow-rumped warblers. *Physiol Biochem Zool* **73(3)**, 314–323.

Lee, K.A., Martin, L.B. and Wikelski, M.C. (2005). Responding to inflammatory challenges is less costly for a successful avian invader, the house sparrow (Passer domesticus) than its less-invasive congener. *Oecologia* **145**, 244–251.

Lee, P. C. (1991). Adaptations to environmental change: An evolutionary perspective. In H. O. Box, ed. *Primate Responses to Environmental Change*, pp. 39–56. Chapman and Hall, London.

Lee, P. C. (2003). Innovation as a behavioural response to environmental challenge. In S. M. Reader and K. N. Laland, eds. *Animal Innovation*, pp. 261–277. Oxford University Press, Oxford, UK.

Lefebvre, L., Whittle, P., Lascaris, E. and Finkelstein, A. (1997). Feeding innovations and forebrain size in birds. *Anim Behav* **53**, 549–560.

Lefebvre, L., Gaxiola, A., Dawson, S., Rosza, L. and Kabai, P. (1998). Feeding innovations and forebrain size in Australasian birds. *Behaviour* **135**, 1077–1097.

Lefebvre, L., Reader, S. M. and Sol, D. (2004). Brains, innovations and evolution in birds and primates. *Brain Behav Evol* **63**, 233–246.

Léger, J. and Larochelle, J. (2006). On the importance of radiative heat exchange during nocturnal flight in birds. *J Exp Biol* **209**, 103–114.

LeMaho, Y., Ka, J. V. V., Koubi, H. (1981). Body composition, energy expenditure, and plasma metabolites in long-term fasting geese. *Am J Physiol* **241**, E342–E354.

Levey, D. J. (1988). Spatial and temporal variation in Costa Rican fruit and fruit-eating bird abundance. *Ecol Monogr* **58**, 251–269.

Levey, D. J. and Cipollini, M. L. (1996). Is most glucose absorbed passively in northern bobwhite? *Comp Biochem Physiol* **113A**, 225–231.

Levey, D. J. and Karasov, W. H. (1989). Digestive responses of temperate birds switched to fruit or insect diets. *The Auk* **106**, 675–686.

Levey, D. J. and Karasov, W. H. (1992). Digestive modulation in a seasonal frugivore, the American robin (*Turdus migratorius*). *Am J Physiol, Gastrointest Liver Physiol* **262(25)**, G711–G718.

Levey, D. J. and F. G. Stiles. (1992). Evolutionary precursors of long-distance migration: resource availability and movement patterns in Neotropical landbirds. *Am Nat* **140**, 447–476.

Levey, D. J., Place, A. R., Rey, P. J. and Martínez del Rio, C. (1999). An experimental test of dietary enzyme modulation in pine warblers *Dendroica pinus*. *Physiol Biochem Zool* **72**, 576–587.

Levine, M. and Tijan, R. (2003). Transcription regulation and animal diversity. *Nature* **424**, 147–151.

Lexer, C., Welch, M. E., Durphy, J. L. and Rieseberg, L. H. (2003). Natural selection for salt tolerance quantitative trait loci (QTLs). in wild sunflower hybrids: implications for the origin of *Helianthus paradoxus*, a diploid hybrid species. *Mol Ecol* **12**, 1225–1235.

Li, C. and Hung Wong, W. (2001). Model-based analysis of oligonucleotide arrays: model validation, design issues and standard error application. *Genome Biol* **2**, research 0032.1–0032.11.

Liechti, F. and Bruderer, B. (1998). The relevance of wind for optimal migration theory. *J Avian Biol* **29**, 561–568.

Liechti, F. and Bruderer, L. (2002). Wingbeat frequency of barn swallows and house martins: a comparison between free flight and wind tunnel experiments. *J Exp Biol* **205**, 2461–2467.

Liechti, F. and Schaller, E. (1999). The use of low-level jets by migrating birds. *Naturwissenschaften* **86**, 549–551.

Liechti, F., Klaassen, M. and Bruderer, B. (2000). Predicting migratory flight altitudes by physiological migration models. *Auk* **117**, 205–214.

Lifjeld, J. T., Dunn, P. O. and Whittingham, L. A. (2002). Short-term fluctuations in cellular immunity of tree swallows feeding nestlings. *Oecologia* **130**, 185–190.

Lifson, N. and McClintock, R. (1966). Theory of use of the turnover rates of body water for measuring energy and material balance. *J Theor Biol* **12**, 46–74.

Lifson, N., Gordon, G. B. and McClintock, R. (1955). Measurement of total carbon dioxide production by means of $D_2{}^{18}O$. *J Appl Physiol* **7**, 704–710.

Lin, H., Deculypere, E. and Buyse, J. (2004a). Oxidative stress induced by corticosterone administration in broiler chickens (*Gallus gallus domesticus*). – 1. Chronic exposure. *Comp Biochem Physiol B, Biochem Mol Biol* **139**, 737–744.

Lin, H., Deculypere, E. and Buyse, J. (2004b). Oxidative stress induced by corticosterone administration in broiler chickens (*Gallus gallus domesticus*). – 2. Short-term effect. *Comp Biochem Physiol B, Biochem Mol Biol* **139**, 745–751.

Linden, R. J. and Mary, D. A. S. G. (1983). The measurement of blood volume. *Cardiovascular Physiol* **P305**, 1–25.

Lindström, A. (1991). Maximum fat deposition rates in migrating birds. *Ornis Scand* **22**, 12–19.

Lingham-Soliar, T., Feduccia, A. and Wang, X. (2007). A new Chinese specimen indicates that "protofeathers" in the Early Cretaceous theropod dinosaur *Sinosauropteryx* are degraded collagen fibres. *Proc R Soc B* **274**, 1823–1829 doi:10.1098/rspb.2007.0352.

Lipar, J. L. and Ketterson, E. D. (2000). Maternally derived yolk testosterone enhances the development of the hatching muscle in the red-winged blackbird *Agelaius phoenicus*. *Proc Roy Soc B* **267**, 2005–2010.

Livezey, B. C. (1993). An ecomorphological review of the dodo (*Raphus cucullatus*) and solitaire (*Pezophaps solitaria*), Flightless Columbiformes of the Mascerene Islands. *J Zool Lond* **230**, 247–292.

Lloyd, P. (1991). Feeding responses of captive Double-collared sunbirds (*Nectarinia afra*). to changes in sucrose food concentrations, and their relation to optimal foraging models. *Suid-Afrikaanse Tydskrif vir Wetenskap* **87**, 67–68.

Lochmiller, R. L., Vestey, M. R. and Boren, J. C. (1993). Relationship between protein nutritional-status and immunocompetence in northern bobwhite chicks. *Auk* **110**, 503–510.

Lockhart, D. J. et al. (1996). Expression monitoring by hybridization to high-density oligonucleotide arrays. *Nature Biotechnol* **14**, 1675–1680.

Lomholt, J. P. (1976). The development of the oxygen permeability of the avian egg shell and its membranes during incubation. *J Exp Zool* **198**, 177–184.

Long, C. A., Zhang, G. P., George, T. F. and Long, C. F. (2003). Physical theory, origin of flight, and a synthesis proposed for birds. *J Theor Biol* **224**, 9–26.

Lotem, A. (1998). Manipulative begging calls by parasitic cuckoo chicks: Why should true offspring not do the same? *TREE* **13**, 342–343.

Lotz, C. N. and Nicolson, S. W. (1996). Sugar preferences of a nectarivorous passerine bird, the Lesser Double-Collared Sunbird (*Nectarinia chalybea*). *Funct Ecol* **10**, 360–365.

Lotz, C. N. and Schondube, J. E. (2006). Sugar preferences in nectar-and fruit-eating birds: Behavioral patterns and physiological causes. *Biotropica* **38(1)**, 3–15.

Love, O. P., Chin, E. H., Wynne-Edwards, K. E. and Williams, T. D. (2005). Stress hormones: a link between maternal condition and sex-biased reproductive investment. *Am Nat* **166**, 751–766.

Love, O. P., Wynne-Edwards, K. E., Bond, L. and Williams, T. D. (2008). Determinants of within- and among-clutch variation in yolk corticosterone in the European starling. *Hormones and Behavior* **53**, 104–111.

Lovegrove, B. G. (2000). The zoogeography of mammalian basal metabolic rate. *Am Nat* **156**, 201–219.

Lovegrove, B. G. (2003). The influence of climate on the basal metabolic rate of small mammals: a slow-fast metabolic continuum. *J Comp Physiol B* **173**, 87–112.

Lovvorn, J. R. (2001). Upstroke thrust, drag effects, and stroke-glide cycles in wing-propelled swimming by birds. *Amer Zool* **41**, 154–165.

Lovvorn, J. R. (2007). Thermal substitution and aerobic efficiency: measuring and predicting effects of heat balance on endotherm diving energetics. *Phil Trans R Soc B* **362**, 2079–2093.

Lundy, H. (1969). A review of the effects of temperature, humidity, turning and gaseous environment in the incubator on hatchability of hen's eggs. In T. C. Carter and B. M. Freeman, eds. *The Fertility and Hatchability of the Hen's Egg*. Oliver and Boyd, Edinburgh, UK.

Lynn, S. E. and Wingfield, J. C. (2003). Male chestnut-collared longspurs are essential for nestling survival: a removal study. *Condor* **105**, 154–158.

Lynn, S. E., Walker, B. G. and Wingfield, J. C. (2005). A phylogenetically controlled test of hypotheses for behavioral insensitivity to testosterone in birds. *Hormones and Behavior* **47**, 170–177.

Machin, M., Simoyi, M. F., Blemings, K. P. and Klandorf, H. (2004). Increased dietary protein elevates plasma uric acid and is associated with decreased oxidative stress in rapidly-growing broilers. *Comp Biochem Physio B, Biochem Mol Biol* **137**, 383–390.

Maclean, G. L. (1996). *Ecophysiology of Desert Birds*. Springer-Verlag, Berlin, Germany.

Madden, J. (2001). Sex, bowers and brains. *Proc R Soc B* **268**, 833–838.

Maeda, K., Henbest, K. B., Cintolesi, F., Kuprov, I., Rodgers, C. T., Liddell, P. A., Gust, D., Timmel, C. R. and Hore, P. J. (2008). Chemical compass model of avian magnetoreception. *Nature* **453**, 387–391.

Maillet, D. and Weber, J.-M. (2007). Relationship between n-3 PUFA content and energy metabolism in the flight muscles of a migrating shorebird: evidence for natural doping. *J Exp Biol* **210**, 413–420.

Maina, J. N. 2000. What it takes to fly: the structural and functional respiratory refinements in birds and bats. *J Exp Biol* **203**, 3045–3064.

Maloney, S. K. and Dawson, T. J. (1998). Changes in pattern of heat loss at high ambient temperature caused by water deprivation in a large flightless bird, the emu. *Physiol Zool* **71(6)**, 712–719.

Manville, R. H. (1963). Altitude record for a mallard. *Wilson Bull* **75**, 92.

Marder, J. and Arieli, Y. (1988). Heat balance of acclimated pigeons exposed to temperatures up to 60°C T$_a$. *Comp Biochem Physiol* **91A**, 165–170.

Marder, J. and Ben-Asher, J. (1983). Cutaneous water evaporation. I. Its significance in heat-stressed birds. *Comp Biochem Physiol* **75A**, 425–431.

Marino, L. (1996). What can dolphins tell us about primate evolution? *Evol Anthropol* **5**, 81–88.

Marsh, R. L. (1983). Adaptations of the gray catbird *Dumetella carolinensis* to long-distance migration: energy stores and blood substrates. *Auk* **100**, 170–179.

Marsh, R. L. (1984). Adaptations of the gray catbird *Dumetella carolinensis* to long-distance migration: fight muscle hypertrophy associated with elevated body mass. *Physiol Zool* **57**, 105–117.

Marshall, D. J. and Uller, T. (2007). When is maternal effect adaptive? *Oikos* **116**, 1957–1963.

Martin, A. C., Zim, J. S. and Nelson, A. L. (1951). *American Wildlife and Plants – A Guide to Wildlife Food Habits*. Dover, New York, NY.

Martin, G. R. (1985). Eye. In A. S. King and J. McClelland, eds. *Form and Function in Birds*, *Vol. 3*, pp. 311–373. Academic Press, London, UK.

Martin, G. R. (1990). *Birds by Night*. Poyser, London, UK.

Martin, G. R. (1993). Producing the image. In H.P. Zeigler and H.-J. Bischoff, eds. *Vision, Brain and Behaviour in Birds*, pp. 5–24. MIT Press, Cambridge, MA.

Martin, L. B., Han, P., Lewittes, J., Kuhlman, J. R., Klasing, K. C. and Wikelski, M. (2006). Phytohemaglutinin-induced skin swelling in birds: Histological support for a classic immunoecological technique. *Funct Ecol* **20**, 290–299.

Martin, R. A. (2007). A review of behavioural ecology of whale sharks (Rhincodontypus). *Fish Res* **84**, 10–16.

Martin, T. E. (2008). Egg size variation among tropical and temperate songbirds: an embryonic temperature hypothesis. *PNAS* **105**, 9268–9271.

Martin, T. E., Auer, S. K., Bassar, R. D., Niklison, A. M. and Lloyd, P. (2007). Geographic variation in avian incubation periods and parental influences on embryonic temperature. *Evolution* **61**, 2558–2569.

Martínez del Rio, C. (1990). Dietary, phylogenetic, and ecological correlates of intestinal sucrase and maltase activity in birds. *Physiol Zool* **63**, 987–1011.

Martínez del Rio, C. and Karasov, W. H. (1990). Digestion strategies in nectar- and fruit-eating birds and the sugar composition of plant rewards. *Am Nat* **136**, 618–637.

Martínez del Rio, C., Karasov, W. H. and Levey, D. J. (1989). Physiological basis and ecological consequences of sugar preferences in cedar waxwings. *Auk* **106**, 67–71.

Martínez del Rio, C., Baker, H. G. and Baker, I. (1992). Ecological and evolutionary implications of digestive processes: bird preferences and the sugar constituents of floral nectar and fruit pulp. *Experientia* **48**, 544–551.

Martínez del Rio, C., Schondube, J. E., McWhorter, T. J. and Herrera, L. G. (2001). Intake responses in nectar feeding birds: digestive and metabolic causes, osmoregulatory consequences, and coevolutionary effects. *Amer Zool* **41**, 902–915.

Masman, D., Daan, S. and Beldhuis, H. J. A. (1988). Ecological energetics of the Kestrel: daily energy expenditure throughout the year based on time-energy budget, food intake and doubly labeled water methods. *Ardea* **76**, 64–81.

Mason, J. R. and Clark, L. (2000). The chemical senses in birds. In G. C. Whittow, ed. *Sturkie's Avian Physiology*, pp. 39–56. Academic Press, New York, NY.

Mathieu-Costello, O., Agey, P. J., Wu, L. Szewczak, J. M. and MacMillen, R. E. (1998). Increased fiber capillarization in flight muscle of finch at altitude. *Respir Physiol* **111**, 189–199.

Maxson, S. J. and Oring, L. W. (1980). Breeding season time and energy budgets of the polyandrous spotted sandpiper. *Behaviour* **74**, 200–263.

McClatchie, S., Greene, C. H., Macaulay, M. C. and Sturley, D. R. M. (1994). Spatial and temporal variability of Antarctic krill: implications for stock assessment. *ICES J Mar Sci* **51**, 11–18.

McDonald, P. G., Buttemer, W. A. & Astheimer, L. B. (2001). The influence of testosterone on territorial defence and parental behavior in male free-living rufous whistlers, *Pachycephala rufiventris*. *Hormones and Behavior* **39**, 185–194.

McKechnie, A. E. (2008). Phenotypic flexibility in basal metabolic rate and changing view of avian physiological diversity: a review. *J Comp Physiol B* **178**, 235–247.

McKechnie, A. E. and Lovegrove, B. G. (2002). Avian facultative hypothermic responses: a review. *Condor* **104**, 705–724.

McKechnie, A. E. and Wolf, B. O. (2004a). The allometry of avian basal metabolic rate: good predictions need good data. *Physiol Biochem Zool* **77**(3), 502–521.

McKechnie, A. E. and Wolf, B. O. (2004b). Partitioning of evaporative water loss in white-winged doves: plasticity in response to short-term thermal acclimation. *J Exp Biol* **207**, 203–210.

McKechnie, A. E., Wolf, B. O. and Martínez del Rio, C. (2004). Deuterium stable isotope ratios as tracers of water resource use: an experimental test with rock doves. *Oecologia* **140**, 191–200.

McKenna, M. C. (1975). Fossil mammals and early Eocene North Atlantic land continuity. *Ann Mo Bot Gard* **62**, 335–353.

McLandress, M. R. and Raveling, D. G. (1981). Changes in diet and body composition of Canada geese before spring migration. *Auk* **98**, 65–79.

McLelland, J. (1979). Digestive system. In A. S. King and J. McLelland, eds. *Form and Function in Birds*, vol. 1, pp. 69–181. Academic Press, New York.

McLelland, J. (1989). Anatomy of the avian cecum. *J Exp Zool* **3**, 2–9.

McMahon T. A. and Bonner, J. T. (1983). *On Size and Life*. Scientific American Library, New York, NY.

McNab, B. K. (2002). *The Physiological Ecology of Vertebrates*. Cornell University Press, Ithaca, NY.

McNab, B. K. (2009). Ecological factors affect the level and scaling of avian BMR. *Comp Biochem Physiol A* **152**, 22–45.

McNeil, R. (1991). Nocturnality in shorebirds. In B. D. Bell, ed. *Acta XX Congressua Internationalia Ornithologici*, Christchurch, New Zealand, 1990, pp. 1098–1104. New Zealand Ornithological Congress Trust Board, Wellington.

McPhail, L. T. and Jones, D. R. (1998). The relationship between power output and heart rate in ducks diving voluntarily. *Comp Biochem Physiol A* **120**, 219–225.

McTaggart, A. R. and Burton, H. (1992). Dimethyl sulfide concentrations in the surface waters of the Australasian Antarctic and Sub-Antarctic oceans during an austral summer. *J Geophys Res Oceans* **97**, 14407–14412.

McWhorter, T. J. (2005). Paracellular intestinal absorption of carbohydrates in mammals and birds. In Starck J. M., T. Wang, eds. *Physiological and Ecological Adaptations to Feeding in Vertebrates*, pp113–140. Science Publishers, Enfield, NH.

McWhorter, T. J. and Martínez del Rio, C. (1999). Food ingestion and water turnover in hummingbirds: how much dietary water is absorbed? *J Exp Biol* **202**, 2851–2858.

McWhorter, T. J. and Martínez del Rio, C. (2000). Does gut function limit hummingbird food intake? *Physiol Biochem Zool* **73**, 313–324.

McWhorter, T. J., Bakken, B. H., Karasov, W. H. and Martínez del Rio, C. (2006). Hummingbirds rely on both paracellular and carrier-mediated intestinal glucose absorption to fuel high metabolism. *Biol Lett* **2**, 131–134.

McWilliams, S. R. and Karasov, W. H. (1998a). Test of a digestion optimization model: effect of a variable-reward feeding schedules on digestive performance of a migratory bird. *Oecologia* **114**, 160–169.

McWilliams, S. R. and Karasov, W. H. (1998b). Test of a digestion optimization model: effects of costs of feeding on digestive parameters. *Physiol Zool* **71(2)**, 168–178.

McWilliams, S. R. and Karasov, W. H. (2001). Phenotypic flexibility in digestive system structure and function in migratory birds and its ecological significance. *Comp Biochem Physiol A* **128**, 579–593.

McWilliams, S. R., Afik, D. and Secor, S. (1997). Patterns and processes in the vertebrate digestive system: implications for the study of ecology and evolution. *TREE* **12**, 420–422.

McWilliams, S. R., Caviedes-Vidal, E. and Karasov, W. H. (1999). Digestive adjustments in cedar waxwings to high feeding rate. *J Exp Zool* **283**, 394–407.

Meier, A. H. and Fivizzani, A. J. (1980). Physiology of migration. In S. A. Gauthreaux, ed. *Animal Migration, Orientation, and Navigation*, pp. 225–282. Academic Press, New York, NY.

Meijer, T. and Drent, R. (1999). Re-examination of the capital and income dichotomy in breeding birds. *Ibis* **141**, 399–414.

Mello, C. V. and Clayton, D. F. (1995). Differential induction of the ZENK gene in the avian forebrain and song control circuit after metrazole-induced depolarization. *J Neurobiol* **26**, 145–161.

Merino, S., Moreno, J., Tomas, G., Martinez, J., Morales, J., Martinez-De La Puente, J. and Osorno, J. L. (2006). Effects of parental effort on blood stress protein HSP60 and immunoglobulins in female blue tits: a brood size manipulation experiment. *J Anim Ecol* **75**, 1147–1153.

Mezentseva, N. V., Kumaratilake, J. S. and Newman, S. A. (2008). The brown adipocyte differentiation pathway in birds: an evolutionary road not taken. *BMC Evol Biol* **6**, 17.

Midtgård, U. (1981). The rete tibiotarsale and arterio-venous association in the hind limb of birds: a comparative morphological study on counter-current heat exchange systems. *Acta Zool* **62**, 67–87.

Midtgård, U., Sejrsen, P. and Johansen, K. (1985). Blood flow in the brood patch of Bantam hens: evidence of cold vasodilatation. *J Comp Physiol B, Biochem Syst Environ Physiol* **155**, 703–709.

Mills, S. C., Hazard, L., Lancaster, L., Mappes, T., Miles, D., Oksanen, T. A.,and Sinervo, B. (2008). Gonadotropin hormone modulation of testosterone, immune function, performance, and behavioral trade-offs among male morphs of the lizard, *Uta stansburiana*. *Am Nat* **171**, 339–357.

Miwa, S., St-Pierre, J., Partridge, L. and Brand, M. D. (2003). Superoxide and hydrogen peroxide production by drosophila mitochondria. *Free Rad Biol Med* **35**, 938–948.

Mock, D. W. and Ploger, B. J. (1987). Parental manipulation of optimal hatch asynchrony in cattle egrets: an experimental study. *Anim Behav* **35**, 150–160.

Monaghan, P., Nager, R. G. and Houston, D. C. (1998). The price of eggs: increased investment in egg production reduces the offspring rearing capacity of parents. *Proc Roy Soc B* **265**, 1731–1735.

Monge, C. C., Léon-Velarde, F. and Gómez de la Torre, G. (1988). Laying eggs at high altitude. *News Physiol Sci* **3**, 69–71.

Montgomery, J. C., Diebel, C., Halstead, M. B. D. and Downer, J. (1999). Olfactory search tracks in Antarctic fish *Trematomus bernacchii*. *Polar Biol* **21**, 151–154.

Moore, M. J. (2005). From birth to death: the complex lives of eukaryotic mRNAs. *Science* **309**, 1514–1518.

Moore, P. and Crimaldi, J. (2004). Odor landscapes and animal behavior: tracking odor plumes in different physical worlds. *J Mar Syst* **49**, 55–64.

Moreno, J., Sanz, J. J. and Arriero, E. (1999a). Reproductive effort and T-lymphocyte cell-mediated immunocompetence in female pied flycatchers, *Ficedula hypoleuca*. *Proc Roy Soc B* **266**, 1105–1109.

Moreno, J., Veiga, J. P., Cordero, P. J. and Minguez, E. (1999b). Effects of paternal care on reproductive success in the polygynous spotless starling, *Sturnus unicolor*. *Behav Ecol Sociobiol* **47**, 47–53.

Moreno, J., Sanz, J. J., Merino, S. and Arriero, E. (2001). Daily energy expenditure and cell-mediated immunity in pied flycatchers while feeding nestlings: interaction with moult. *Oecologia* **129**, 492–497.

Morrison, R. I. G. and Hobson, K. A. (2004). Use of body stores in shorebirds after arrival on high-arctic breeding grounds. *Auk* **121**, 333–344.

Moss, R. (1974). Winter diet, gut length, and interspecific competition in Alaskan ptarmigan. *Auk* **91**, 737–746.

Motani, R., Rothschild, B. M. and Wahl, W. (1999). Large eyeballs in diving ichthyosaurs. *Nature* **402**, 747.

Mourer-Chauviré, C., Hugueney, M. and Jonet, P. (1989). Découverte de Passeriformes dans l'Oligocene supérieur de France. *C R Acad Sci Sér II* **309**, 843–849.

Mouritsen, H. (1998). Redstarts, *Phoenicurus phoenicurus*, can orient in a true zero magnetic field. *Anim Behav* **55**, 1311–1324.

Mouritsen, H. and Larsen, O. N. (2001). Migrating songbirds tested in computer-controlled Emlen funnels use stellar cues for a time-independent compass. *J Exp Biol* **204**, 3855–3865.

Mouritsen, H. And Ritz, T. (2005). Magnetoreception and its use in bird navigation. *Curr Opinion in Neurobiol* **15**, 406–415.

Mouritsen, H., Janssen-Bienhold, U., Liedvogel, M., Feenders, G., Stalleicken, J., Dirks, P. and Weiler, R. (2004). Night-vision brain area in migratory songbirds. *PNAS* **101**, 14294–14299.

Mouritsen, H., Feenders, G., Liedvogel, M., Wada, K. and Jarvis, E. D. (2005). Night-vision brain area in migratory songbirds. *PNAS* **102(23)**, 8339–8344.

Mugaas, J. N. and King, J. R. (1981). Annual variation of daily energy expenditure by the Black-billed Magpie. A study of thermal and behavioral energetics. *Stud Avian Biol* **5**, 1–78.

Muheim, R., Bäckman, J. and Akesson, S. (2002). Magnetic compass orientation in European robins is dependent on both wavelength and intensity of light. *J. Exp. Biol.* **205**, 3845–3856.

Mujahid, A., Sato, K., Akiba, Y. and Toyomizu, M. (2006). Acute heat stress stimulates mitochondrial superoxide production in broiler skeletal muscle, possibly via downregulation of uncoupling protein content. *Poultr Sci* **85**, 1259–1265.

Müller, C., Sendler, M. and Hildebrandt, J.-P. (2006). Downregulation of aquaporins 1 and 5 in nasal gland by osmotic stress in ducklings, *Anas platyrhynchos*: implications for the production of hypertonic fluid. *J Exp Biol* **209**, 4067–4076.

Muñoz-Garcia, A. and Williams, J. B. (2008). Developmental plasticity of cutaneous water loss and lipid composition in stratum corneum of desert and mesic nestling house sparrows. *PNAS* **105**, 15611–15616.

Murphy, M. E. (1994). Amino-acid compositions of avian eggs and tissues – nutritional implications. *J Avian Biol* **25**, 27–38.

Murray, K. G., Russel, S., Picone, C. M., Winnett-Murray, K., Sherwood, W. and Kuhlmann, M. L. (1994). Fruit laxatives and seed passage rates in frugivores: consequences for plant reproductive success. *Ecology* **75**, 989–994.

Nager, R. G., Ruegger, C. and Vannoordwijk, A. J. (1997). Nutrient or energy limitation on egg formation: a feeding experiment in great tits. *J Anim Ecol* **66**, 495–507.

Nager, R. G., Monaghan, P. and Houston, D. C. (2001). The cost of egg production: Increased egg production reduces future fitness in gulls. *J Avian Biol* **32**, 159–166.

Nagy, K. A. (1980). CO_2 production in animals: analysis of potential errors in the doubly labeled water method. *Am J Physiol.* **238**, R466–R473.

Nagy, K. A., Siegfried, W. R. and Wilson, R. P. (1984). Energy utilisation by free-ranging Jackass Penguins, *Spheniscus demersus*. *Ecology* **65**, 1648–1655.

Nagy, K. A., Girard, I. A. and Brown, T. K. (1999). Energetics of freeranging mammals, reptiles and birds. *Ann Rev Nutr* **19**, 247–277.

Nevitt, G. A. (1999a). Foraging by seabirds on an olfactory landscape. *Am Sci* **87**, 46–53.

Nevitt, G. A. (1999b). Olfactory foraging in Antarctic seabirds: a species-specific attraction to krill odors. *Mar Ecol Prog Ser* **177**, 235–241.

Nevitt, G. A. (2000). Olfactory foraging by Antarctic procellariiform seabirds: life at high Reynolds numbers. *Biol Bull* **198**, 245–253.

Nevitt, G. A. (2008). Sensory ecology on the high seas: the odor world of the procellariiform seabirds. *J Exp Biol* **211**, 1706–1713.

Nevitt, G. A. and Bonadonna, F. (2005). Sensitivity to dimethyl sulphide suggests a mechanism for olfactory navigation by seabirds. *Biol Lett* **1**, 303–305.

Nevitt, G. A. and Haberman, K. (2003). Behavioral attraction of Leach's storm-petrels (*Oceanodroma leucorhoa*). to dimethyl sulfide. *J Exp Biol* **206**, 1497–1501.

Nevitt, G. A., Veit, R. R. and Kareiva, P. (1995). Dimethyl sulphide as a foraging cue for Antarctic Procellariiform seabirds. *Nature* **376**, 681–682.

Nevitt, G. A., Reid, K. and Trathan, P. (2004). Testing olfactory foraging strategies in an Antarctic seabird assemblage. *J Exp Biol* **207**, 3537–3544.

Nevitt, G. A., Bergstrom, D. M. and Bonadonna, F. (2006). The potential role of ammonia as a signal molecule for procellariiform seabirds. *Mar Ecol Progr Ser* **315**, 271–277.

Nevitt, G. A., Losekoot, M. and Weimerskirch, H. (2008). Evidence for olfactory search in wandering albatross (*Diomedea exulans*). *PNAS* **105**, 4576–4581.

Nicolakakis, N. and Lefebvre, L. (2000). Forebrain size and innovation rate in European birds: feeding, nesting and confounding variables. *Behaviour* **137**, 1415–1429.

Nicolson, S. W. (2002). Pollination by passerine birds: why are the nectars so dilute. *Comp Biochem Physiol B* **131**, 645–652.

Nicolson, S. W. and Van Wick, B. E. (1998). Nectar Sugar in Proteaceae: patterns and processes. *Austr J Bot* **46**, 489–504.

Nicolson, S. W., Hoffmann, D. and Fleming, P. A. (2005). Short-term energy regulation in nectar-feeding birds: the response of Whitebellied Sunbirds (*Nectarinia talatala*) to a midday fast. *Funct Ecol* **19**, 988–994.

Nilsson J.-Å., Åkesson, M. and Nilsson, J. F. (2009). Heritability of resting metabolic rate in a wild population of blue tits. *J Evol Biol* **22**, 1867–1874.

Nisbet, I. C. T. (1963). Measurements with radar of the height of nocturnal migration over Cape Cod, Massachussetts. *Bird-Banding* **34(2)**, 57–67.

Norberg, U. M. (1990). *Vertebrate flight: Mechanics, physiology, morphology, ecology and evolution.* Springer-Verlag, Berlin, Germany.

Norberg, U. M. L. (1995). How a long tail and changes in mass and wing shape affect the cost for flight in animals. *Funct Ecol* **9**, 48–54.

Norberg, U. M. L. (2002). Structure, form, and function of flight in engineering and the living world. *J Morphol* **252**, 52–81.

Nottebohm, F. (1972). The origins of vocal learning. *Amer Nat* **106**, 116–140.

Nottebohm, F. (1980). Brain pathways for vocal learning in birds: a review of the first 10 years. *Prog Psychobio Physio Psych* **9**, 85–124.

Nottebohm, F., Stokes, T. M. and Leonard, C. M. (1976). Central control of song in the canary, *Serinus canarius*. *J Comp Neurol* **165**, 457–486.

Nur, N. (1984). The consequences of brood size for breeding blue tits: 1. Adult survival, weight change and the cost of reproduction. *J Anim Ecol* **53**, 479–496.

Obst, B. and Nagy, K. A. (1993). Stomach oil and the energy budget of Wilson's Storm-petrel nestlings. *Condor* **95**, 792–805.

O'Connor, R. J. (1984). *The Growth and Development of Birds.* John Wiley & Sons, London, UK.

Odum, E. P., Rogers, D. T. and Hicks, D. L. (1964). Homeostasis of nonfat components of migrating birds. *Science* **143**, 1037–1039.

O'Dwyer, T. W., Buttemer, W. A. and Priddel, D. M. (2007). Differential rates of offspring provisioning in Gould's petrels: are better feeders better breeders? *Austr J Zool* **55**, 155–160.

O'Dwyer, T. W., Buttemer, W. A., Priddel, D. M. and Downing, J. A. (2006). Prolactin, body condition and the cost of good parenting: An interyear study in a long-lived seabird, Gould's petrel (*Pterodroma leucoptera*). *Funct Ecol* **20**, 806–811.

Ohmart, R. D. (1972). Physiological and ecological observations concerning salt-secreting glands of the roadrunner. *Comp Biochem Physiol A* **43**, 311–316.

Ohmart, R. D. and Lasiewski, R. C. (1971). Roadrunners: energy conservation by hypothermia and absorption of sunlight. *Science* **172**, 67–69.

Olsen, N. J. and Kovacs, W. J. (1996). Gonadal steroids and immunity. *Endocr Rev* **17**, 369–384.

Olson, S. L. (1973a). Evolution of the rails of the South Atlantic Islands (Aves: Rallidae) *Smithsonian Contribs Zool* **152**, 1–53.

Olson, S. L. (1973b). A classification of the Rallidae. *Wilson Bull* **85**, 381–416.

Olson, S. L. and James, H. 1991 Descriptions of thirty-two new species of birds from the Hawaiian Islands: part 1. Non-passeriformes. *Ornithol Monogr* **45**, 1–88.

Olson, C. R., Vleck, C. M. and Vleck, D. (2006). Periodic cooling of bird eggs reduces embryonic growth efficiency. *Physiol Biochem Zool* **79**, 927–936.

Olson, V. A., Davies, R. G., Orme, C. D. L., Thomas, G. H., Gaston, K. J., Owens, I. P.F. and Bennett, P. M. (2009). Global biogeography and ecology of body size in birds. *Ecol Letters* **12**, 249–259.

Ophir, E., Arieli, Y., Marder, J. and Horowitz, M. (2002). Cutaneous blood flow in the pigeon *Columba livia*: its possible relevance to cutaneous water evaporation. *J Exp Biol* **205**, 2627–2636.

Ornelas, J. F., González, C. and Uribe, J. (2002). Complex vocalizations and aerial displays of the Amethyst-throated Hummingbird (*Lampornis amethystinus*). *Auk* **119**, 1141–1149.

Ott, A., Bouchaud, J. P., Langevin, D. and Urbach, W. (1990). Anomalous diffusion in "living polymers": a genuine Lévy flight? *Phys Rev Lett* **65**, 2201–2204.

Ottinger, M. A., Kubakawa, K., Kikuchi, M., Thompson, N. and Ishii, S. (2002). Effects of exogenous testosterone on testicular luteinizing hormone and follicle-stimulating hormone receptors during aging. *Exp Biol Med* **227**, 830–836.

Owen-Ashley, N. T., Hasselquist, D. and Wingfield, J. C. (2004). Androgens and the immunocompetence handicap hypothesis: Unraveling direct and indirect pathways of immunosuppression in song sparrows. *Am Nat* **164**, 490–505.

Owens, I. P. F. and Short, R. V. (1995). Hormonal basis of sexual dimorphism in birds – implications for new theories of sexual selection. *TREE* **10**, 44–47.

Padian, K. (1986). *Origin of Birds and the Evolution of Flight*. California Academy of Sciences, San Francisco, CA.

Padian, K. and Dial, K. P. (2005). Origin of flight: could "four-winged" dinosaurs fly? *Nature* **438**, 3–4.

Paladino, F. and King, J. R. (1984). Thermoregulation and oxygen consumption during terrestrial locomotion by white-crowned sparrows *Zonotricia leucophrys gambelii*. *Physiol Zool* **57**, 226–236.

Pamplona, R., Portero-Otin, M., Requena, J. R., Thorpe, S. R., Herrero, A. and Barja, G. (1999a). A low degree of fatty acid unsaturation leads to lower lipid peroxidation and lipoxidation-derived protein modification in heart mitochondria of the longevous pigeon than in the short-lived rat. *Mech Ageing Develop* **106**, 283–296.

Pamplona, R., Portero-Otin, M., Riba, D., Ledo, F., Gredilla, R., Herrero, A., *et al.* (1999b). Heart fatty acid unsaturation and lipid peroxidation, and aging rate, are lower in the canary and the parakeet than in the mouse. *Aging-Clin Exp Res* **11**, 44–49.

Pamplona, R., Barja, G. and Portero-Otin, M. (2002). Membrane fatty acid unsaturation, protection against oxidative stress, and maximum life span – a homeoviscous-longevity adaptation? *Increasing healthy life span: Conventional measures and slowing the innate aging process* **959**, 475–490.

Pap, P. L. and Markus, R. (2003). Cost of reproduction, t-lymphocyte mediated immunocompetence and health status in female and nestling barn swallows, *Hirundo rustica. J Avian Biol* **34**, 428–434.

Pappenheimer, J. R. (1993). On the coupling of membrane digestion with intestinal absorption of sugars and amino acids. *Am J Physiol, Gastrointest Liver Physiol* **265(28)**, G409–G417.

Parejo, D. and Danchin, E. (2006). Brood size manipulation affects frequency of second clutches in the blue tit. *Behav Ecol Sociobiol* **60**, 184–194.

Parmesan, C. and Yohe, G. (2003). A globally coherent fingerprint of climate changes impacts across natural systems. *Nature* **421**, 37–42.

Paton, T., Haddrath, O. and Baker, A. J. (2002). Complete mitochondrial DNA genome sequences show that modern birds are not descended from transitional shorebirds. *Proc. R. Soc. Ser. B* **269**, 839.

Pearson, J. T. (1994). Oxygen consumption rates of adults and chicks during brooding in king quail (*Coturnix chinensis*). *J Comp Physiol B, Biochem Syst Environ Physiol* **164**, 415–424.

Pearson, J. T. (1998). Development of thermoregulation and posthatching growth in the altricial cockatiel *Nymphicus hollandicus*. *Physiol Zool* **71**, 237–244.

Pearson, J. T. (1999). Energetics of embryonic development in the cockatiel (*Nymphicus hollandicus*). and the king quail (*Coturnix chinensis*). *Austr J Zool* **47**, 565–577.

Pennycuick, C. J. (1975). Mechanics of flight. In D. S. Farner and J. R. King, eds. *Avian Biology*, Vol. 5, pp.1–75. Academic Press, New York, NY.

Pennycuick, C. J. (2002). Gust soaring as a basis for the flight of petrels and albatrosses (Procellariiformes). *Avian Sci* **2**, 1–12.

Pennycuick, C. J. (2008). *Modelling the Flying Bird*. Academic Press, New York, NY.

Pennycuick, C. J., Einarsson, O. Bradbury, T. A. M. and Owen, M. (1996). Migrating whooper swans *Cygnus cygnus*: Satellite tracks and flight performance calculations. *J Avian Biol* **27**, 118–134.

Penry, D. L. and Jumars, P. A. (1987). Modeling animal guts as chemical reactors. *Am Nat* **129**, 69–96.

Phalan, B., Phillips, R. A., Silk, J. R. D., Afanasyev, V., Fukuda, A., Fox, J., Catry, P., Higuchi, H. and Croxall, J. P. (2007). Foraging behaviour of four albatross species by night and day. *Mar Ecol Progr Ser* **340**, 271–286.

Piersma, T. (1998). Phenotypic flexibility during migration: optimization of organ size contingent on the risks and rewards of fueling and flight. *J Avian Biol* **29**, 551–520.

Piersma, T. and Drent, J. (2003). Phenotypic flexibility and the evolution of organismal design. *TREE* **18**, 228–233.

Piersma, T. and Gill, R. E. J. (1998). Guts don't fly: small digestive organs in obese bar-tailed godwits. *Auk* **115**, 196–203.

Piersma, T. and Lindström, A. (1997). Rapid reversible changes in organ size as a component of adaptive behavior. *TREE* **12**, 134–138.

Piersma, T. Koolhaas, A. And Dekinga, A. (1993). Interactions between stomach structure and diet choice in shorebirds. *Auk* **110**, 552–564.

Piersma, T., van Gils, J., De Goeij, P. and Van Der Meer, J. (1995). Holling's functional response model as a tool to link the food-finding mechanism of a probing shorebird with its spatial distribution. *J Animal Ecol* **64**, 493–504.

Piersma, T., Dietz, M. W., Dekinga, A., Nebel, S., van Gils, J., Battley, P. F. and Spaans, B. (1999a). Reversible size-changes in stomachs of shorebirds: when, to what extent, and why? *Acta Ornithol* **34**, 175–181.

Piersma, T., Gudmundsson, G. A. and Lilliendahl, K. (1999b). Rapid changes in the size of different functional organ and muscle groups during refueling in a long-distance migrating shorebird. *Physiol Biochem Zool* **72**, 405–415.

Piersma, T., Perez-Tris, J., Mouritsen, H., Bauchinger, U. and Bairlein, F. (2005). Is there a "migratory syndrome" common to all migrant birds? *Ann N Y Acad Sci* **1046**, 282–293.

Pigliucci, M. (1996). How organisms respond to environmental changes: from phenotypes to molecules (and vice versa). *TREE* **11**, 168–173.

Pigliucci, M. (2001). *Phenotypic Plasticity: Beyond Nature and Nurture*. Johns Hopkins University Press, Baltimore.

Pihlaja, M., Siitari, H. and Alatalo, R. V. (2006). Maternal antibodies in a wild altricial bird: effects on offspring immunity, growth and survival. *J Anim Ecol* **75**, 1154–1164.

Pike, T. W. and Petrie, M. (2005). Maternal body condition and plasma hormones affect offspring sex ratio in peafowl. *Anim Behav* **70**, 745–751.

Pimm, S. L., Moulton, M. P. and Justice, L. J. (1994). Bird extinctions in the central Pacific. *Phil Trans R Soc Lond B* **344**, 27–33.

Pitelka, F. A. (1942). Territoriality and related problems in North American hummingbirds. *Condor* **44**, 189–204.

Place, A. R., Stoyan, N. C., Ricklefs, R. E. and Butler, R. G. (1989). Physiological basis of stomach oil formation in Leach's Storm-petrel (*Oceanodroma leucorhoa*). *Auk* **106**, 687–699.

Pohnert, G., Steinke, M. and Tollrian, R. (2007). Chemical cues, defence metabolites and the shaping of pelagic interspecific interactions. *TREE* **22**, 198–204.

Ponganis, P. J., Kooyman, G. L., Starke, L. N., Kooyman, C. A. and Kooyman, T. G. (1997). Post-dive blood lactate concentrations in emperor penguins, *Aptenodytes forsteri*. *J Exp Biol* **200**, 1623–1626.

Ponganis, P. J., Van Dam, R. P., Marshall, G., Knower, T. and Levenson, D. H. (2000). Sub-ice foraging behavior of emperor penguins. *J Exp Biol* **203**, 3275–3278.

Ponganis, P. J., Van Dam, R. P., Knower, T. and Levenson, D. H. (2001). Temperature regulation in emperor penguins foraging under sea ice. *Comp Biochem Physiol* **129A**, 811–820.

Ponganis, P. J., Van Dam, R. P., Levenson, D. H., Knower, T., Ponganis, K. V. and Marshall, G. (2003). Regional heterothermy and conservation of core temperature in emperor penguins diving under sea ice. *Comp Biochem Physiol* **135A**, 477–487.

Ponganis, P. J., van Dam, R. P., Knower, T., Levenson, D. H. and Ponganis, K. V. (2004). Deep dives and aortic temperatures of emperor penguins: new directions for bio-logging at the isolated dive hole. *Mem Natl Inst Polar Res Special Issue* **58**, 155–161.

Ponganis, P. J., Stockard, T. K., Meir, J. U., Williams, C. L., Ponganis, K. V., van Dam R. P. and Howard, R. (2007). Returning on empty: extreme blood O_2 depletion underlies dive capacity of emperor penguins. *J Exp Biol* **210**, 4279–4285.

Poole, J. H., Tyack, P. L., Stoeger-Horwath, A. S. and Watwood, S. (2005). Animal behaviour: elephants are capable of vocal learning. *Nature* **434**, 455–456.

Potts, R. (1998). Variability selection in hominid evolution. *Evol Anthropol* **7**(3), 81–96.

Powell, F. L. (2000). Respiration. In G. C. Whittow, ed. *Sturkie's Avian Physiology*, pp. 233–264. Academic Press, New York, NY.

Powers, D. R. and Nagy, K. A. (1988). Field metabolic rate and food consumption by free-living Anna's hummingbirds *Calypte anna*. *Physiol Zool* **61**, 500–506.

Preest, M. R. and Beuchat, C. A. (1997). Ammonia excretion by hummingbirds. *Nature* **386**, 561–562.

Prinzinger, R. and Hänssler, I. (1980). Metabolism-weight relationship in some small nonpasserine birds. *Experientia* **36**, 1299–1300.

Prinzinger, R., Preßmar, A. and Schleucher, E. (1991). Body temperature in birds. *Comp Biochem Physiol* **99A**, 499–506.

Prum, P. O. (1999). Development and evolutionary origin of feathers. *J Exp Zool* **285**(4), 291–306.

Prum, R. O. and Brush, A. H. (2002). The evolutionary origin and diversification of feathers. *Q Rev Biol* **77**(3), 261–295.

Pryke, S. R. and Griffith, S. C. (2006). Red dominates black: agonistic signalling among head morphs in the colour polymorphic Gouldian finch. *Proc Roy Soc B* **273**, 949–957.

Pryke, S. R., Astheimer, L. B., Buttemer, W. A. and Griffith, S. C. (2007). Frequency-dependent physiological trade-offs between competing colour morphs. *Biol Lett* **3**, 494–497.

Pulido, F. and Berthold, P. (2004). Microevolutionary response to climate change. In A. P. Møller, W. Fiedler and P. Berthold, eds. *Birds and Climate Change*, pp. 151–183. Academic Press, London, UK.

Pyke, G. H. and Waser, N. M. (1981). The production of dilute nectars by hummingbirds and honeyeater flowers. *Biotropica* **13**, 260–270.

Quintana, F., Wilson, R. P. and Yorio, P. (2007). Dive depth and plumage air in wettable birds: the extraordinary case of the imperial cormorant. *Mar Ecol Progr Ser* **334**, 299–310.

Rahn, H. (1984). Factors controlling the rate of incubation water loss in bird eggs. In R. S. Seymour, ed. *Respiration and Metabolism of Embryonic Vertebrates*, pp. 271–288. Dr W. Junk Publishers, Dordrecht, Netherlands.

Rahn, H., Paganelli, C. V. and Ar, A. (1974). The avian egg: air-cell gas tension, metabolism and incubation time. *Respir Physiol* **22**, 297–309.

Rahn, H., Ackermann, R. A. and Paganelli, C. V. (1977a). Humidity in the avian nest and egg water loss during incubation. *Physiol Zool* **50**, 269–283.

Rahn, H., Carey, C., Balmas, K., Bahtia, B. and Paganelli, C. V. (1977b). Reduction of pore area of the avian eggshell as an adaptation to altitude. *PNAS* **74**, 3095–3098.

Rahn, H., Krog, J. and Mehlum, F. (1983). Microclimate of the nest and egg water loss of the eider *Somateria mollissima* and other waterfowl in Spitsbergen. *Polar Res* **1**, 171–183.

Raimbault, S., Dridi, S., Denjean, F., Lachuer, J., Couplan, E., Bouillaud, F., *et al.* (2001). An uncoupling protein homologue putatively involved in facultative muscle thermogenesis in birds. *Biochem J* **353**, 441–444.

Ramdas, L., Coombes, K. R., Baggerly, K., Abruzzo, L., Highsmith, W. E., Krogmann, T., Hamilton, S. R. and Zhang, W. (2001). Sources of nonlinearity in cDNA microarray expression measurements. *Genome Biol* **2**, research0047.1–0047.7.

Ramsay, S. L. and Houston, D. C. (1998). The effect of dietary amino acid composition on egg production in blue tits. *Proc Roy Soc B* **265**, 1401–1405.

Rand, R. W. (1960). The distribution, abundance and feeding habits of the Cormorants *Phalacrocoracidae* off the south-western coast of the Cape province. *The Biology of Guano-producing Seabirds*, chapter 3, *Division of Fisheries, Investigational Report* **42**, 32pp.

Raouf, S. A., Parker, P. G., Ketterson, E. D., Nolan, V. and Ziegenfus, C. (1997). Testosterone affects reproductive success by influencing extra-pair fertilizations in male dark-eyed juncos (aves: *Junco hyemalis*). *Proc Roy Soc B* **264**, 1599–1603.

Rappole, J. H. (1995). *The Ecology of Migrant Birds: A Neotropical Perspective*. Smithsonian Institution Press, Washington, D.C.

Rappole, J. H. and Jones, P. (2002). Evolution of Old and New World migration systems. *Ardea* **90**, 525–537.

Rashotte, M. E., Saarela, S., Henderson, R. P. and Hohtola, E. (1999). Shivering and digestion-related thermogenesis in pigeons during dark phase. *Am J Physiol* **277**, R1579–R1587.

Raubenheimer, D. and Simpson, S. J. (1998). Nutrient transfer functions: the site of integration between feeding behaviour and nutritional physiology. *Chemoecology* **8**, 61–68.

Rautenberg, W. (1952–53). Korpergewicht und grundumsatz beim kastrierten mannlichen vogel. *Wissenschaftliche Zeitschrift der Universitat Greifswald Jahrgang* **2**, 230–236.

Rayner, J. M. V. (1990). The mechanics of bird migration performance. In E. Gwinner, ed. *Bird Migration: Physiology and Ecophysiology*, pp.283–327. Springer-Verlag, Berlin, Germany.

Rayner, J. M. V. (1999). Estimating power curves of flying vertebrates. *J Exp Biol* **202**, 3449–3461.

Reader, S. M. (2004). Don't call me clever. *New Sci* **183**, 34–37.

Reader, S. M. and MacDonald, K. N. (2003). Environmental variability and primate behavioural flexibility. In S. M. Reader and K. N. Laland, eds. *Animal Innovation*, pp. 83–116. Oxford University Press, Oxford, UK.

Reader, S. M. and Laland, K. N. (2002). Social intelligence, innovation and enhanced brain size in primates. *PNAS* **99**, 4436–4441.

Reid, K., Croxall, J. P. and Prince, P. A. (1996). The fish diet of black-browed albatross *Diomedea melanophris* and grey-headed albatross *D. chrysostoma* at South Georgia. *Polar Biol* **16**, 469–477.

Reid, K., Croxall, J. P., Edwards, T. M., Hill, H. J. and Prince, P. A. (1997). Diet and feeding ecology of the diving petrels *Pelecanoides georgicus* and *P. urinatrix* at South Georgia. *Polar Biol* **17**, 17–24.

Reid, R. C., Prausnitz, J. M. and Poling B. E. (1987). *The Properties of Gases and Liquids*. McGraw-Hill, New York, NY.

Reid, W. V. (1987). The cost of reproduction in the glaucous winged gull. *Oecologia* **74**, 458–467.

Reinertsen, R. E. (1996). Physiological and ecological aspects of hypothermia. In C. Carey, ed. *Avian Energetics and Nutritional Ecology*, pp. 125–157. Chapmann and Hall, New York.

Reisz, R. R. and Müller, J. (2004). Molecular timescales and the fossil record: a paleontological perspective. *Trends in Genetics* **20(5)**, 237–241.

Reyer, H. U. (1984). Investment and relatedness: a cost benefit analysis of breeding and helping in the pied kingfisher (*Ceryle rudis*). *Anim Behav* **32**, 1163–1178.

Reynolds, P. S. and Lee, R. M. (1996). Phylogenetic analysis of avian energetics: passerines and non-passerines do not differ. *Am Nat* **147**, 735–759.

Rezende, E. L., Swanson, D. L. Novoa, F. F. and Bozinovic, F. (2002). Passerines versus non-passerines: so far, no statistical differences in the scaling of avian energetics. *J Exp Biol* **205**, 101–107.

Ribak, G., Weihs, D. and Arad, Z. (2005a). Submerged swimming of the great cormorant *Phalacrocorax carbo sinensis* is a variant of the burst-and-glide gait. *J Exp Biol* **208**, 3835–3849.

Ribak, G., Weihs, D. and Arad, Z. (2005b). Water retention in the plumage of diving great cormorants *Phalacrocorax carbo sinensis*. *J Avian Biol* **36**, 89–95.

Rice, G. E. and Skadhauge, E. (1982). Caecal water and electrolyte absorption and the effects of acetate and glucose, in dehydrated, low-NaCl diet hens. *J Comp Physiol B* **147**, 61–64.

Ricklefs, R. E. (1974). Energetics of reproduction in birds. In R. A. Paynter, Jr., ed. *Avian Energetics*, Publ. no. 15, pp. 152–292. Nuttal Ornithological Club, Cambridge, MA.

Ricklefs, R. E. (1976). Growth rate of birds in the humid New World tropics. *Ibis* **118**, 176–207.

Ricklefs, R. E. (1979). Adaptation, constraint, and compromise in avian postnatal development. *Biol Rev* **54**, 269–290.

Ricklefs, R. E. (1983). Some considerations on the reproductive energetics of pelagic seabirds. *Stud Avian Biol* **8**, 84–94.

Ricklefs, R. E. (1984). Prolonged incubation in pelagic seabirds: a comment on Boersma's paper. *Amer Nat* **123**, 710–720.

Ricklefs, R. E. (1996). Avian energetics, ecology, and evolution. In C. Carey, ed. *Avian Energetics and Nutritional Ecology*, pp. 1–30. Chapman and Hall, New York, NY.

Ricklefs, R. E. (2000). Density dependence, evolutionary optimization, and the diversification of avian life histories. *Condor* **102**, 9–22.

Ricklefs, R. E. and Hainsworth, F. R. (1968). Temperature dependent behavior of the Cactus Wren. *Ecology* **49**, 227–233.

Ricklefs, R. E., Roby, D. D. and Williams, J. B. (1986). Daily energy expenditure of adult Leach's Storm-petrels during the nesting cycle. *Physiol Zool* **59**, 649–660.

Ricklefs, R. E., Konarzewski, M. and Daan, S. (1996). The relationship between basal metabolic rate and daily energy expenditure in birds and mammals. *Am Nat* **147**, 1047–1071.

Rijke, A. M. (1968). The water repellency and feather structure of cormorants, Phalacrocoracidae. *J Exp Biol* **48**, 185–189.

Ripley, S. D. 1977. *Rails of the World*. Godine, Boston.

Ritz, T., Adem, S. and Schulten, K. (2000). A model for photoreceptor-based magnetoreception in birds. *Biophys J* **78**, 707–718.

Robbins, C. T. (1993). *Wildlife Feeding and Nutrition*. Academic Press, New York, NY.

Robbins, C. T., Hagerman, A. E., Austin, P. J., McArthur, C. and Hanley, T. A. (1991). Variation in mammalian physiological responses to a condensed tannin and its ecological implications. *J Mamm* **72**, 480–486.

Roberts, M. L., Buchanan, K. L. and Evans, M. R. (2004). Testing the immunocompetence handicap hypothesis: a review of the evidence. *Anim Behav* **68**, 227–239.

Robertson, B. C., Elliott, G. P., Eason, D. K., Clout, M. N. and Gemmell, N. J. (2006). Sex allocation theory aids species conservation. *Biology Letters* **2**, 229–231.

Robertson, G. R. (1995). *The Foraging Ecology of Emperor Penguins (Aptenodytes forsteri) at two Mawson Coast Colonies, Antarctica*. Australia: Antarctic Division, Department of Environment, Sport, and Territories.

Rodhouse, P. G. and Prince, P. A. (1993). Cephalopod prey of the black-browed albatross *Diomedea melanophrys* at South Georgia. *Polar Biol* **13**, 373–376.

Rønning, B., Jensen, H., Moe, B. and Bech, C. (2007). Basal metabolic rate: heritability and genetic correlations with morphological traits in the zebra finch. *J Evol Biol* **20**, 1815–1822.

Rooke, I. J., Bradshaw, S. D. and Langworthy, R. A. (1983). Aspects of water, electrolyte, and carbohydrate physiology of the silvereye, *Zosterops lateralis* (Aves). *Aust J Zool* **31**, 695–704.

Rottenberg, H. (2007). Exceptional longevity in songbirds is associated with high rates of evolution of cytochrome b, suggesting selection for reduced generation of free radicals. *J Exp Biol* **210**, 2170–2180.

Rubner, M. (1902). *Die Gesetze des Energieverbrauchs bei der Ernährung*. Franz Deuticke, Lepizig, Germany.

Rusch, K. M., Pytte, C. L. and Ficken, M. S. (1996). Organization of agonistic vocalizations in Black-chinned Hummingbirds. *Condor* **98**, 557–566.

Rusch, K. M., Thusius, K. and Ficken, M. S. (2001). The organization of agonistic vocalizations in Ruby-throated Hummingbirds with a comparison to Black-chinned Hummingbirds. *Wilson Bulletin* **113**, 425–430.

Russell, R. W., Hunt, G. L., Coyle, K. O. and Cooney, R. T. (1992). Foraging in a fractal environment: spatial patterns in a marine predator-prey system. *Landscape Ecol* **7**, 195–209.

Sabat, P., Cavieres, G., Veloso, C. and Canals, M. (2006a). Water and energy economy of an omnivorous bird: population differences in the Rufous-collared sparrow (*Zonotrichia capensis*). *Comp Biochem Physiol* **144A**, 485–490.

Sabat, P., Maldonado, K., Canals, M. and Martínez del Rio, C. (2006b). Osmoregulation and adaptive radiation in the ovenbird genus *Cinclodes* (Passeriformes: Furnariidae). *Funct Ecol* **20**, 799–805.

Sabat, P., Novoa, F. Bozinovic, F. and Martínez del Rio, C. (1998). Dietary flexibility and intestinal plasticity in birds: a field and laboratory study. *Physiol Zool* **71**, 226–236.

Saino, N. and Moller, A. P. (1995). Testosterone correlates of mate guarding, singing and aggressive-behaviour in male barn swallows, *Hirundo rustica*. *Anim Behav* **49**, 465–472.

Saino, N., Calza, S., Ninni, P. and Moller, A. P. (1999). Barn swallows trade survival against offspring condition and immunocompetence. *J Anim Ecol* **68**, 999–1009.

Saino, N., Moller, A. P. and Bolzern, A. M. (1995). Testosterone effects on the immune system and parasite infestations in the barn swallow (*Hirundo rustica*): an experimental test of the immunocompetence hypothesis. *Behav Ecol* **6**, 397–404.

Salt, G. W. (1964). Respiratory evaporation in birds. *Biol Rev* **39**, 113–136.

Sancar, A. (2000). Cryptochrome: the second photoactive pigment in the eye and its role in circadian photoreception. *Annu Rev Biochem* **69**, 31–67.

Sancar, A. (2003). Structure and function of DNA photolyase and cryptochrome blue-light photoreceptors. *Chem Rev* **103**, 2203–2237.

Sanmartín, I., Enghoff, H. and Ronquist, F. (2001). Pattern of animal dispersal, vicariance and diversification in the Holarctic. *Biol J Linn Soc* **73**, 345–390.

Sancar, A. (2004). Regulation of the mammalian circadian clock by cryptochrome. *J Biol Chem* **279(33)**, 34079–34082.

Sanvito, S., Galimberti, F. and Miller, E. H. (2007). Observational evidences of vocal learning in Southern elephant seals: a longitudinal study. *Ethology* **113**, 137–146.

Sanz, A., Pamplona, R. and Barja, G. (2006). Is the mitochondrial free radical theory of aging intact? *Antioxidants & Redox Signaling* **8**, 582–599.

Sapolsky, R. M., Romero, L. M. and Munck, A. U. (2000). How do glucocorticoids influence stress responses? Integrating permissive, suppressive, stimulatory, and preparative actions. *Endocr Rev* **21**, 55–89.

Sato, K., Mitani, Y., Cameron, M. F., Siniff, D. B. and Naito, Y. (2003). Factors affecting stroking patterns and body angle in diving Weddell seals under natural conditions. *J Exp Biol* **206**, 1461–1470.

Sato, K., Ponganis, P. J., Habara, Y. and Naito, Y. (2005). Emperor penguins adjust swim speed according to the above-water height of ice holes through which they exit. *J Exp Biol* **208**, 2549–2554.

Saunders, D. K. and Fedde, M. R. (1991). Physical conditioning: effect on the myoglobin concentration in skeletal and cardiac muscle of bar-headed geese. *Comp Biochem Physiol A* **100**, 349–352.

Scharff, C. and Nottebohm, F. (1991). A comparative study of the behavioral deficits following lesions of various parts of the zebra finch song system: implications for vocal learning. *J Neurosci* **11**, 2896–2913.

Schaub, M., Liechti, F. and Jenni, L. (2004). Departure of migrating European robins, *Erithacus rubecula*, from a stopover site in relation to wind and rain. *Anim Behav* **67**, 229–237.

Scheid, P. and Piiper, J. (1971). Direct measurement of the pathway of respired gas in duck lungs. *Respir Physiol* **11**, 308–314.

Scheid, P. and Piiper, J. (1972). Cross-current gas exchange in avian lungs: effects of reversed parabronchial air flow in ducks. *Respir Physiol* **16**, 304–312.

Schena, M., Shalon, D., Davis, R. W. and Brown, P. O. (1995). Quantitative monitoring of gene expression patterns with a complementary DNA microarray. *Science* **270**, 467–470.

Schlichting, C. D. and Pigliucci, M. (1998). *Phenotypic Evolution: A Reaction Norm Perspective*. Sinauer Associates, Inc., Sunderland.

Schmid, D., Grémillet, D. J. H. and Culik, B. M. (1995). Energetics of underwater swimming in the great cormorant (*Phalacrocorax carbo sinensis*). *Mar Biol* **123**, 875–881.

Schmidt-Nielsen, B. and Schmidt-Nielsen, K. (1950). Evaporative water loss in desert rodents in their natural habitat. *Ecology* **31**, 75–85.

Schmidt-Nielsen, K. (1964). *Desert Animals: Physiological Problems of Heat and Water*. Clarendon Press, Oxford, UK. Reprinted (1979) Dover Publications Inc, New York, NY.

Schmidt-Nielsen, K. (1984). *Scaling. Why is Animal Size so Important!* Cambridge University Press, Cambridge, UK.

Schmidt-Nielsen, K. (1997). *Animal Physiology: Adaptation and Environment*. Cambridge University Press, Cambridge.

Schmidt-Nielsen, K., Barker-Jørgensen, C. and Osaki, H. (1958). Extrarenal salt excretion in birds. *Am J Physiol* **193**, 101–107.

Schmidt-Nielsen, K., Hainsworth, F. R. and Murrish, D. E. (1970). Counter-current heat exchange in the respiratory passages: effect on water and heat balance. *Respir Physiol* **9**, 263–276.

Schneider, R. A. & Helms, J. A. (2003). The cellular and molecular origins of beak morphology. *Science* **299**, 565–568.

Schoech, S. J., Ketterson, E. D. and Nolan, V. (1999). Exogenous testosterone and the adrenocortical response in dark-eyed juncos. *Auk* **116**, 64–72.

Scholander, P. F. (1940). Experimental investigations on the respiratory function in diving mammals and birds. *Hvalradets Skrifter* **22**, 1–131.

Scholander, P. F., Hock, R., Walters, V., Johnson, F. and Irving, L. (1950). Heat regulation in some arctic and tropical mammals and birds. *Biol Bull* **99**, 237–258.

Schondube, J. E. and Martínez del Rio, C. (2003). Concentration-dependent sugar prefrences in nectar-feeding birds: Mechanisms and consequences. *Funct Ecol* **17**, 445–453.

Schreiber, E. A. and Burger, J. (2002). Seabirds in the marine environment. In E. A. Schreiber and J. Burger, eds. *Biology of Marine Birds*, pp. 1–15. CRC Press, Boca Raton, FL.

Schulten, K., Swenberg, C. and Weller, A. (1978). A biomagnetic sensory mechanism based on magnetic field modulated coherent electron spin motion. *Z Phys Chem* **111**, 1–5.

Schwabl, H. (1996). Maternal testosterone in the avian egg enhances postnatal growth. *Comp Biochem Physiol A* **114**, 271–276.

Scott, G. R. and Milsom, W. K. (2007). Control of breathing and adaptation to high altitude in the bar-headed goose. *Am J Physiol, Regul Integr Comp Physiol* **293**, R379–R391.

Secor, S. M. (2009). Specific dynamic action: a review of the postprandial metabolic response. *J Comp Physiol B* **179**, 1–56.

Sell, J. L., Kolodovsky, O. and Reid, B. L. (1989). Intestinal disaccharidases of young turkeys: temporal development and influence of diet composition. *Poult Sci* **68**, 265–277.

Selman, R. G. and Houston, D. C. (1996a). A technique for measuring lean pectoral muscle mass in live small birds. *Ibis* **138**, 348–350.

Selman, R. G. and Houston, D. C. (1996b). The effect of prebreeding diet on reproductive output in zebra finches. *Proc Roy Soc B* **263**, 1585–1588.

Semm, P. and Demaine, C. (1986). Neurophysiological properties of magnetic cells in the pigeon's visual system. *J Comp Physiol A* **159**, 619–625.

Seyfarth, R. M. and Cheney, D. L. (2002). What are big brains for? *PNAS* **99**, 4141–4142.

Seymour, R. S. and Rahn, H. (1978). Gas conductance in the eggshell of the mound-building brush turkey. In J. Piiper ed. *Respiratory Function in Birds, Adult and Embryonic*, pp. 243–246. Springer-Verlag, New York, NY.

Shaffer, S. A., Costa, D. P., and Weimerskirch. (2003). Foraging effort in relation to the constraints of reproduction in free-ranging albatrosses. *Funct Ecol* **17**, 66–74.

Shaffer, S. A., Tremblay, Y., Awkerman, J. A., Henry, R. W., Teo, S. L. H., Anderson, D. J., Croll, D. A., Block, B. A. and Costa, D. P. (2005). Comparison of light- and SST-based geolocation with satellite telemetry in free-ranging albatrosses. *Mar Biol* **147**, 833–843.

Shaffer, S. A., Tremblay, Y., Weimerskirch, H., Scott, D., Thompson, D. R., Sagar, P. M., Moller, H., Taylor, G. A., Foley, D. G., Block, B. A. and Costa, D. P. (2006). Migratory shearwaters integrate oceanic resources across the Pacific Ocean in an endless summer. *PNAS* **103**, 12799–12802.

Shams, H. and Scheid, P. (1993). Effects of hypobaria on parabronchial gas exchange in normoxic and hypoxic ducks. *Respir Physiol* **91**, 155–163.

Sheldon, B. C. and Verhulst, S. (1996). Ecological immunology: costly parasite defences and trade-offs in evolutionary ecology. *TREE* **11**, 317–321.

Shepherd, G. M., Chen, W. R., Willhite, D., Migliore, M. and Greer, C. A. (2007). The olfactory granule cell: from classical enigma to central role in olfactory processing. *Brain Res Rev* **55**, 373–382.

Shiu, S.-H. and Borevitz, J. O. (2008). The next generation of microarray research: applications in evolutionary and ecological genomics. *Heredity* **100**, 141–149.

Shlesinger, M. F. and Klafter, J. (1986). Lévy walks versus Lévy flights. In H. E. Stanley and N. Ostrowsky, eds. *On Growth and Form: Fractal and Non-Fractal Patterns in Physics*, pp. 279–283. Martinus Nijhoff Publishers, Dordrecht, Netherlands.

Shlesinger, M. F., Zaslavsky, G. and Klafter, J. (1993). Strange kinetics. *Nature* **363**, 31–37.

Shlesinger, M. F., Zaslavsky, G. and Frisch, U. (1995). *Lévy Flights and Related Topics in Physics*. Springer, Berlin, Germany.

Sibley, C. G. and Ahlquist, J. E. (1990). *Phylogeny and Classification of Birds: a Study in Molecular Evolution*. Yale University Press, New Haven, CT.

Silby, R. M. (1981). Strategies of digestion and defecation. In C. R. Townsend and P. Calow, eds. *Physiological Ecology*, pp. 109–139. Sinauer, Sunderland, MA.

Silverin, B. (1980). Effects of long-acting testosterone treatment on free-living pied flycatchers, *Ficedula hypoleuca*, during the breeding period. *Anim Behav* **28**, 906–912.

Silverin, B. (1982). Endocrine correlates of brood size in adult pied flycatchers, *Ficedula hypoleuca*. *Gen Comp Endocrinol* **47**, 18–23.

Silverin, B. (1993). Territorial aggressiveness and its relation to the endocrine system in the pied flycatcher. *Gen Comp Endocrinol* **89**, 206–213.

Silverin, B., Baillien, M. and Balthazart, J. (2004). Territorial aggression, circulating levels of testosterone, and brain aromatase activity in free-living pied flycatchers. *Hormones and Behavior* **45**, 225–234.

Silverman, E., Veit, R. R. and Nevitt, G. A. (2004). Nearest neighbors as foraging cues: information transfer in a patchy environment. *Mar Ecol Progr Ser* **277**, 25–36.

Simoyi, M. F., Van Dyke, K. and Klandorf, H. (2002). Manipulation of plasma uric acid in broiler chicks and its effect on leukocyte oxidative activity. *Am J Physiol, Reg Integr Comp Physiol* **282**, R791–R796.

Simpson, H. B. and Vicario, D. S. (1990). Brain pathways for learned and unlearned vocalizations differ in *zebra finches*. *J Neurosci* **10**, 1541–1556.

Simpson, S. J. and Raubenheimer, D. (1995). The geometric analysis of feeding and nutrition: a user's guide. *J Insect Physiol* **41**, 545–553.

Skadhauge, E. (1981). *Osmoregulaton in Birds*. Springer-Verlag, New York, NY.

Skadhauge, E. (1983). Ionic and osmotic regulation in birds. *Verh Dtsch Zool Ges*, 69–81.

Skadhauge, E. (1993). Basic characteristics and hormonal regulation of ion transport in avian hindguts. *Adv Comp Environ Physiol* **16**, 67–93.

Skadhauge, E. and Bradshaw, S. D. (1974). Saline drinking and cloacal excretion of salt and water in the zebra finch. *Am J Physiol* **227**, 1263–1267.

Skadhauge, E., Thomas, D. H., Chadwick, A. and Jallageas, M. (1983). Time course of adaptation to low and high NaCl diets in the domestic fowl: effects on electrolyte excretion and on the plasma hormone levels (aldosterone, corticosterone and prolactin). *Pflügers Arch Eur J Physiol* **396**, 301–307.

Skadhauge, E., Warüi, C. N., Kamau, J. M. Z. and Maloiy, G. M. O. (1984). Function of the lower intestine and osmoregulation in the ostrich: preliminary anatomical and physiological observations. *Quart J Exp Physiol* **69**, 809–818.

Skadhauge, E., Maloney, S. K. and Dawson, T. J. (1991). Osmotic adaptation of the emu (*Dromaius novaehollandiae*). *J Comp Physiol B* **161**, 173–178.

Skadhauge, E., Dawson, T. J., Prys-Jones, R. and Warüi, C. N. (1996). The role of the kidney and gut in osmoregulation. In D. C. Deeming, ed. *Improving our Understanding of Ratites in a Farming Environment*, pp. 115–122. Ratite Conference, Oxfordshire, UK.

Skopec, M. M., Hagerman, A. E. and Karasov, W. H. (2004). Do salivary proline-rich proteins counteract dietary hyrolyzable tannin in laboratory rats? *J Chem Ecol* **30(9)**, 1679–1692.

Smith, A. D. and Dufty, A. M., Jr. (2005). Variation in the stable-hydrogen isotope composition of northern goshawk feathers: relevance to the study of migratory origins. *Condor* **107**, 547–558.

Snow, B. K. (1973). Social organization of the Hairy Hermit *Glaucis hirsuta*. *Ardea* **61**, 94–105.

Snow, B. K. and Snow, D. W. (1972). Feeding niches of hummingbirds in a Trinidad valley. *J Anim Ecol* **41**, 471–485.

Snow, D. W. (1962). The natural history of the Oilbird, *Steatornis caripensis*, in Trinidad, W. I. II. Population, breeding ecology and food. *Zoologica* **47**, 199–221.

Snow, D. W. (1968). The singing assemblies of Little Hermits. *Living Bird* **7**, 47–55.

Sockman, K. W., Sharp, P. J. and Schwabl, H. (2006). Orchestration of avian reproductive effort: an integration of the ultimate and proximate bases for flexibility in clutch size, incubation behaviour, and yolk androgen deposition. *Biol Rev* **81**, 629–666.

Sohal, R. S. and Weindruch, R. (1996). Oxidative stress, caloric restriction, and aging. *Science* **273**, 59–63.

Sol, D., Duncan, R. P., Blackburn, T. M., Cassey, P. and Lefebvre, L. (2005). Big brains, enhanced cognition, and response of birds to novel environments. *PNAS* **12(15)**, 5460–5465.

Solomon, M. E. (1949). The natural control of animal populations. *J Animal Ecol* **18**, 1–35.

Sorenson, J. G., Kristensen, T. N. and Loeschke, V. (2003). The evolutionary and ecological role of heat shock proteins. *Ecol Lett* **6**, 1025–1037.

Spanier, E., Weihs, D. and Almog-Shtayer, G. (1991). Swimming of the Mediterranean slipper lobster. *J Exp Mar Biol Ecol* **145**, 15–31.

Speakman, J. R. (1997). *Doubly Labelled Water: Theory and Practice*. Chapman & Hall, London, UK.

Staaland, H. (1967). Anatomical and physiological adaptations of the nasal glands in charadriiformes birds. *Comp Biochem Physiol* **23**, 933–944.

Stark, J. M. (1996). Intestinal growth in the altricial European starling (*Sturnus vulgaris*) and the precocial Japanese quail (*Coturnix coturnix japonica*). A morphometric and cytokinetic study. *Acta Anatomica* **156**, 289–306.

Steadman, D. W. 1995 Prehistoric extinctions of Pacific island birds: biodiversity meets zooarcheology. *Science* **267**, 1123–1131.

Steadman, D. W. and Olson, S. L. (1985). Bird remains from an archaeological site on Henderson Island, South Pacific: Man-caused extinctions on an 'uninhabited' island. *PNAS* **82**, 6191–6195.

Stearns, S. C. (1976). Life history tactics: a review of the ideas. *Quart Rev Biol* **51**, 3–47.

Stearns, S. C. (1989). The evolutionary significance of reaction norms. *Bioscience* **39**, 436–446.

Steiger, S. S., Fidler, A. E., Valcu, M. and Kempenaers, B. (2008). Avian olfactory receptor gene repertoires: evidence for a well-developed sense of smell in birds? *Proc R Soc B* **275**, 2309–2317.

Stein, E. D. and Diamond, J. M. (1989). Do dietary levels of pantothenic acid regulate its intestinal uptake in mice? *J Nutr* **119**, 1973–1983.

Stephens, D. W. and Krebs, J. R. (1986). *Monographs in Behavior and Ecology: Foraging Theory*. Princeton University Press, Princeton, NJ.

Stephenson, R., Lovvorn, J. R., Heieis, M. R. A., Jones, D. R. and Blake, R. W. (1989). A hydromechanical estimate of the power requirements of diving and surface swimming in lesser scaup (*Aythya affinis*). *J Exp Biol* **147**, 507–519.

Stewart, A. G. (1978). Swans flying at 8000 meters. *Brit Birds* **71**, 459–460.

Stiles, F. G. (1982). Aggressive and courtship displays of the male Anna's Hummingbird. *Condor* **84**, 208–225.

Stiles, F. G. (1995). Behavioral, ecological and morphological correlates of foraging for arthropods by the hummingbirds of a tropical wet forest. *Condor* **97**, 853–878.

Stiles, F. G. and Wolf, L. L. (1979). Ecology and evolution of lek mating behavior in the Long-Tailed Hermit Hummingbird. *Ornithol Monogr* **27**.

Stockard, T. K., Heil, J., Meir, J. U., Sato, K., Ponganis, K. V. and Ponganis, P. J. (2005). Air sac PO_2 and oxygen depletion during dives of emperor penguins. *J Exp Biol* **208**, 2973–2980.

Stoehr, A. M. and Hill, G. E. (2000). Testosterone and the allocation of reproductive effort in male house finches (*Carpodacus mexicanus*). *Behav Ecol Sociobiol* **48**, 407–411.

Storer, R. W. (1960). Adaptive radiation in birds. In A. J. Marshall, ed. *Biology and Comparative Physiology of Birds*, vol. 1, pp. 15–55, Academic Press, New York, NY.

Stull, R. B. (2000). *Meteorology for Scientists and Engineers*. Brooks/Cole, Pacific Grove, California, CA.

Sumida, S.S. and Brochu, C. A. (2000). Phylogenetic context for the origin of feathers. *Amer Zool* **40**, 486–303.

Surai, P. F. (2002). *Natural Antioxidants in Avian Nutrition and Reproduction*. Nottingham University Press, Nottingham, UK.

Swan, L. W. (1961). The ecology of the high Himalayas. *Sci Am* **205**, 68–78.

Swan, L. W. (1970). Goose of the Himalayas. *Nat Hist* **70**, 68–75.

Szabo, A., Febel, H., Mezes, M., Balogh, K., Horn, P. and Romvari, R. (2006). Body size related adaptations of the avian myocardial phospholipid fatty acyl chain composition. *Comp Biochem Physiol B, Biochem Mol Biol* **144**, 496–502.

Talbot, D. A., Duchamp, C., Rey, B., Hanuise, N., Rouanet, J. L., Sibille, B. and Brand, M. D. (2004). Uncoupling protein and ATP/ADP carrier increase mitochondrial proton conductance after cold adaptation of king penguins. *J Physiol* **558**, 123–135.

Tattersal, G. J., Andrade, D. V. and Abe, A. S. (2009). Heat exchange from the toucan bill reveals a controllable vascular thermal radiator. *Science* **325**, 468–470.

Thomas, D. H. and Robin, A. P. (1977). Comparative studies of thermoregulatory and osmoregulatory behaviour and physiology of five species of sandgrouse (*Aves: Pterocliidae*). in Morocco. *J Zool* **183**, 229–249.

Thomas, D. W., Bosque, C. and Arends, A. (1993). Development of thermoregulation and the energetics of nestling Oilbirds (*Steatornis caripensis*). *Physiol Zool* **66**, 422–448.

Thomas, R. J., Székely, T., Powell, R. F. and Cuthill, I. C. (2006). Eye size, foraging methods and the timing of foraging in shorebirds. *Funct Ecol* **20**, 157–165.

Thompson, D. (1961). *On Growth and Form.* Cambridge University Press, Cambridge, UK.

Thomson, D. L., Monaghan, P. and Furness, R. W. (1998). The demands of incubation and avian clutch size. *Biol Rev* **73**, 293–304.

Thorup, K, Bisson, I.-A., Bowlin, M. S., Holland, R. A., Wingfield, J. C., Ramenofsky, M. and Wikelski, M. (2007). Evidence for a navigational map stretching across the continental U.S. in a migratory songbird. *PNAS* **104(46)**, 18115–18119.

Tieleman, B. I. and Williams, J. B. (1999). The role of hyperthermia in the water economy of desert birds. *Physiol Biochem Zool* **72(1)**, 87–100.

Tieleman, B. I. and Williams J. B. (2000). The adjustment of avian metabolic rates and water fluxes to desert environments. *Physiol Biochem Zool* **73**, 461–479.

Tieleman, B. I., Williams, J. B. (2002). Cutaneous and respiratory water loss in larks from arid and mesic environments. *Physiol Biochem Zool* **75(6)**, 590–599.

Tieleman, B. I., Williams, J. B. and Buschur, M. E. (2002). Physiological adjustments to arid and mesic environments in larks (*Alaudidae*). *Physiol Biochem Zool* **75(3)**, 305–313.

Tieleman, B. I., Williams, J. B. and Bloomer, P. (2003). Adaptation of metabolism and evaporative water loss along an aridity gradient. *Proc R Soc B* **270**, 207–214.

Timmel, C. R. and Henbest, K. B. (2004). A study of spin chemistry in weak magnetic fields. *Phil Trans R Soc A* **362**, 2573–2589.

Timmermans, S., Lefebvre, L., Boire, D. and Basu, P. (2001). Relative size of the hyperstriatum ventrale is the best predictor of feeding innovation rate in birds. *Brain Behav Evol* **56**, 196–203.

Tinbergen, J. M. (1987). Costs of reproduction in the great tit: intraseasonal costs associated with brood size. *Ardea* **75**, 111–122.

Tinbergen, J. M. and Williams, J. B. (2002). Energetics of incubation. In D. C. Deeming, ed. *Avian Incubation: Behaviour, Environment and Evolution.* Oxford University Press, Oxford, UK.

Tobalske, B. W., Hedrick, T. L., Dial, K. P. and Biewener, A. A. (2003). Comparative power curves in bird flight. *Nature* **421**, 363–366.

Tøien, Ø. (1993). Control of shivering and heart rate in incubating bantam hens upon sudden exposure to cold eggs. *Acta Physiol Scand* **149**, 205–214.

Tøien, Ø., Aulie, A. and Steen, J. B. (1986). Thermoregulatory responses to egg cooling in incubating bantam hens. *J Comp Physiol B, Biochem Syst Environ Physiol* **156**, 303–307.

Torre-Bueno, J. R. (1978). Evaporative cooling and water-balance during flight in birds. *J Exp Biol* **75**, 231–236.

Toyomizu, M., Ueda, M., Sato, S., Seki, Y., Sato, K. and Akiba, Y. (2002). Cold-induced mitochondrial uncoupling and expression of chicken UCP and ant mRNA in chicken skeletal muscle. *FEBS Lett* **529**, 313–318.

Tracy, C. R. (1972). Newton's law: its application for expressing heat losses from homeotherms. *Bioscience* **22**, 656–659.

Travis, J. (1994). Evaluating the adaptive role of morphological plasticity. In P. C. Wainwright and S. M. Reilly, eds. *Ecological Morphology: Integrative Organismal Biology*, pp. 99–122. University of Chicago Press, Chicago, IL.

Trewick, S. A. (1996). Morphology and evolution of two takahe: Flightless rails of New Zealand. *J Zool Lond* **238**, 221–237.

Trewick, S. A. (1997). Flightlessness and phylogeny amongst endemic rails (Aves: Rallidae). of the New Zealand region. *Phil Trans R Soc B* **352**, 429–446.

Tseng, G. C., Oh, M. K., Rohlin, L., Liao, J. C. and Wong, W. H. (2001). Issues in cDNA microarray analysis: quality filtering, channel normalization, models of variations and assessment of gene effects. *Nucleic Acids Res* **29**, 2549–2557.

Tucker, V. A. (1968a). Respiratory physiology of house sparrows in relation to high-altitude flight. *J Exp Biol* **48**, 55–66.

Tucker, V. A. (1968b). Respiratory exchange and evaporative water loss in the flying Budgerigar. *J Exp Biol* **75**, 223–229.

Tucker, V. A. (1972). Metabolism during flight in the laughing gull, *Larus atricilla*. *Am J Physiol* **222(2)**, 237–245.

Tucker, V. A. (1973). Bird metabolism during flight: evaluation of a theory. *J Exp Biol* **58**, 689–709.

Turner J. S. (1987). Blood circulation and the flows of heat in an incubated egg. *J Exp Zool Supple* **1**, 99–104.

van Buskirk, R. W. and Nevitt, G. A. (2008). The influence of developmental environment on the evolution of olfactory foraging behaviour in procellariiform seabirds. *J Evol Biol* **21**, 67–76.

van Dam, R. P., Ponganis, P. J., Ponganis, K. V., Levenson, D. H. and Marshall, G. (2002). Stroke frequencies of emperor penguins diving under sea ice. *J Exp Biol* **205**, 3769–3774.

van der Meer, J. and Piersma, T. (1994). Physiologically inspired regression models for estimating and predicting nutrient stores and their composition in birds. *Physiol Zool* **67**, 305–329.

van Duyse, E., Pinxten, R. and Eens, M. (2002). Effects of testosterone on song, aggression, and nestling feeding behavior in male great tits, *Parus major*. *Hormones and Behavior* **41**, 178–186.

van Gils, J. A., Battley, P. F., Piersma, T. and Drent, R. (2005). Reinterpretation of gizzard sizes of red knots world-wide emphasises overriding importance of prey quality at migratory stopover sites. *Proc R Soc B* **272**, 2609–2618.

van Gils, J. A., Piersma, T., Dekinga, A. and Dietz, M. W. (2003). Cost-benefit analysis of mollusc-eating in a shorebird II. Optimizing gizzard size in the face of seasonal demands. *J Exp Biol* **206**, 369–380.

Vanderwerf, E. (1992). Lack clutch size hypothesis – an examination of the evidence using meta-analysis. *Ecology* **73**, 1699–1705.

Vates, G. E. and Nottebohm, F. (1995). Feedback circuitry within a song learning pathway. *PNAS* **92**, 5139–5143.

Vates, G. E., Broome, B. M., Mello, C. V. and Nottebohm, F. (1996). Auditory pathways of caudal telencephalon and their relation to the song system of adult male zebra finches. *J Comp Neurol* **366**, 613–642.

Veasey, J. S., Houston, D. C. and Metcalfe, N. B. (2000). Flight muscle atrophy and predation risk in breeding birds. *Funct Ecol* **14**, 115–121.

Velculescu, V. E., Zhang, L., Vogelstein, B. and Kinzler, K. W. (1995). Serial analysis of gene expression. *Science* **270**, 484–487.

Vezina, F. and Williams, T. D. (2002). Metabolic costs of egg production in the European starling (*Sturnus vulgaris*). *Physiol Biochem Zool* **75**, 377–385.

Vianna, C. R., Hagen, T., Zhang, C. Y., Bachman, E., Boss, O., Gereben, B., Moriscot, A. S., Lowell, B. B., Bicudo, J. E. P. W. and Bianco, A. C. (2001). Cloning and functional characterization of an uncoupling protein homologue in hummingbirds. *Physiol Genomics* **5**, 137–145.

Videler, J. J. (2006). *Avian Flight*. Oxford Ornithological Series. Oxford University Press, Oxford, UK.

Videler, J. J. and Weihs, D. (1982). Energetic advantages of burst-and-coast swimming of fish at high speeds. *J Exp Biol* **97**, 169–178.

Viney, M. E., Riley, E. M. and Buchanan, K. L. (2005). Optimal immune responses: Immunocompetence revisited. *TREE* **20**, 665–669.

Visser, G. H. and Ricklefs, R. E. (1993). Temperature regulation in neonates of shorebirds. *Auk* **110**, 445–457.

Visser, M. E. and Lessells, C. M. (2001). The costs of egg production and incubation in great tits (*Parus major*). *Proc Roy Soc B* **268**, 1271–1277.

Viswanathan, G. M., Afanasyev, V., Buldyrev, S. V., Murphy, E. J., Prince, P. A. and Stanley, H. E. (1996). Lévy flight search patterns of wandering albatrosses. *Nature* **381**, 413–415.

Vleck, C. M. and Hoyt, D. F. (1991). Metabolism and energetics of reptilian and avian embryos. In D. C. Deeming and M. W. J. Ferguson, eds. *Egg Incubation: its Effects on Embryonic Development in Birds and Reptiles*. Cambridge University Press, Cambridge, UK.

Vleck, C. M. and Vleck, D. (1987). Metabolism and energetics of avian embryos. *J Exp Zool Suppl* **1**, 111–25.

Vogel, S. (1994). *Life in Moving Fluids*. Princeton University Press, Princeton, NJ.

Vogel, S. (2003). *Comparative Biomechanics: Life's Physical World*. Princeton University Press, Princeton, NJ.

Vrba, E. S. (1985). Ecological and adaptive changes associated with early hominid evolution. In E. Delson, ed. *Ancestors: the Hard Evidence*, pp.63–71. Alan R. Liss, Inc., New York, NY.

Wada, K., Sakaguchi, H., Jarvis, E. D. and Hagiwara, M. (2004). Differential expression of glutamate receptors in avian neural pathways for learned vocalization. *J Comp Neurol* **476**, 44–64.

Wainwright, P. C. and Reilly, S. M. (1994). *Ecological Morphology: Integrative Organismal Biology*. University of Chicago Press, Chicago, IL.

Waite, T. A. and Strickland, D. (2006). Climate change and the demise of a hoarding bird living on the edge. *Proc Roy Soc B* **273**, 2809–2813.

Walls, G. L. (1942). *The Vertebrate Eye and its Adaptive Radiation*. Hafner, New York, NY.

Walsberg, G. E. (1986). Thermal consequences of roost-site selection: the relative importance of three modes of heat conservation. *Auk* **103**, 1–7.

Walsberg, G. E. (2000). Small mammals in hot deserts: some generalizations revisited. *Bioscience* **50**, 109–120.

Walsberg, G. E. and Weathers, W. W. (1986). A simple technique for estimating operative environmental temperature. *J Therm Biol* **11**, 67–72.

Wang, N., Banzett, R. B., Butler, J. P. and Fredberg, J. J. (1988). Bird lung models show that convective inertia effects inspiratory aerodynamic valving. *Respir Physiol* **73**, 111–124.

Wang, N., Banzett, R. B., Nations, C. S. and Jenkins, F. A. (1992). An aerodynamic valve in the avian primary bronchus. *J Exp Biol* **262**, 441–445.

Wangensteen, O. D., Wilson, D. and Rahn, H. (1971). Diffusion of gases across the shell of the hen's egg. *Respir Physiol* **11**, 16–30.

Wangensteen, O. D., Rahn, H., Burton, R. R. and Smith, A. H. (1974). Respiratory gas exchange of high altitude adapted chick embryos. *Respir Physiol* **21**, 61–70.

Ward, P. and Zahavi, A. (1973). The importance of certain assemblages of birds as "information centres" for food finding. *Ibis* **115**, 517–534.

Ward, S., Rayner, J. M. V., Möller, U., Jackson, D. M., Nachtigall, W. and Speakman, J. R. (1999). Heat transfer from starlings *Sturnus vulgaris* during flight. *J Exp Biol* **202**, 1589–1602.

Ward, S., Bishop, C. M., Woakes, A. J. and Butler, P. J. (2002). Heart rate and the rate of oxygen consumption of flying and walking barnacle geese (*Branta leucopsis*). and bar-headed geese (*Anser indicus*). *J Exp Biol* **205**, 3347–3356.

Warham, J. (1990). *The Petrels: Their Ecology and Breeding Systems*. Academic Press, London, UK.

Warham, J. (1996). *The Behaviour, Population Biology and Physiology of the Petrels*. Academic Press, London, UK.

Warren, W. C., Clayton, D. F., Ellegren, H., Arnold, A. P., Hillier, L. W., Künstner, A., *et al* (2010). The genome of a songbird. *Nature* **464**, 757–762.

Warrick, D. R., Tobalske, B. W., Powers, D. R. (2005). Aerodynamics of the hovering hummingbird. *Nature* **435**, 1094–1097.

Warüi, C. N. (1989). Light microscopic morphometry of the kidneys of fourteen avian species. *J Anat* **162**, 19–31.

Warren, W. C., Clayton, D. F., Ellegren, H., Arnold, A. P., Hillier, L. W., Künstner, A., Searle, S., et al (2010). The genome of a songbird. *Nature* **464**, 757–762.

Wassenaar, L. I. and Hobson, K. A. (2000). Stable-carbon and hydrogen isotope rations reveal breeding origins of red-winged blackbirds. *Ecol Appl* **10**, 911–916.

Watts, L. T., Rathinam, M. L., Schenker, S. and Henderson, G. I. (2005). Astrocytes protect neurons from ethanol-induced oxidative stress and apoptotic death. *J Neurosci Res* **80**, 655–666.

Weathers, W. W. (1981). Physiological thermoregulation in heat stressed birds: consequences of body size. *Physiol Zool* **54**, 345–361.

Weathers, W. W. and Schoenbaechler, D. C. (1976). Regulation of body temperature in the budgerygah, *Melopsittacus undulates*. *Aust J Zool* **24**, 39–47.

Weathers, W. W. and Sullivan, K. A. (1989). Juvenile foraging proficiency, parental effort, and avian reproductive success. *Ecol Monogr* **59**, 223–246.

Weathers, W. W. and Sullivan, K. A. (1993). Seasonal patterns of time and energy allocation by birds. *Physiol Zool* **66**, 511–536.

Weathers, W. W., Davidson, C. L., Olson, C. R., Morton, M. L., Nur, N. and Famula, T. R. (2002). Altitudinal variation in parental energy expenditure by white-crowned sparrows. *J Exp Biol* **205**, 2915–2924.

Webb, D. R. (1987). Thermal tolerance of avian embryos: a review. *Condor* **89**, 874–898.

Webb, D. R. (1993). Maternal-nestling contact geometry and heat transfer in an altricial bird. *J Therm Biol* **18**, 117–124.

Webb, P. W. and Fairchild, A. G. (2001). Performance and maneuverability of three species of teleostean fishes. *Can J Zool* **79**, 1866–1877.

Webb, S. D. and Opdyke, N. D. (1995). Global climatic influence of Cenozoic land mammal faunas. In *Effects of Past Global Change on Life*, pp. 184–208, National Academy Press, Washington, DC.

Webster, M. D. and Weathers, W. W. (1989). Validation of single-sample doubly labeled water method. *Am J Physiol* **256**, R572–R576.

Webster, M. D. and Weathers, W. W. (1990). Heat produced as a by-product of foraging activity contributes to thermoregulation by verdins, *Auriparus flaviceps*. *Physiol Zool* **63**, 777–794.

Weihs, D. (1974). Energetic advantages of burst swimming of fish. *J Theor Biol* **48**, 215–229.

Weihs, D. (2002). Dynamics of dolphin porpoising revisited. *Integr Comp Biol* **42**, 1071–1078.

Weimerskirch, H. (1998). Foraging strategies of southern albatrosses and their relationship with fisheries. In G. Robertson and R. Gales, ed. *Albatross Biology and Conservation*, pp. 168–179. Surrey Beatty, Sydney Australia.

Weimerskirch, H., Ancel, A., Caloin, M., Zahariev, A., Spagiari, J., Kersten, M. and Chastel, O. (2003). Foraging efficiency and adjustment of energy expenditure in a pelagic seabird provisioning its chick. *J Animal Ecol* **72**, 500–508.

Weimerskirch, H., Chastel, O., Barbraud, C. and Tostain, O. (2003). Frigatebirds ride high on thermals. *Nature* **421**, 333–334.

Weinig, C. et al. (2002). Novel loci control variation in reproductive timing in *Arabidopsis thaliana* in natural environments. *Genetics* **162**, 1875–1884.

Weinstein, Y., Bernstein, M. H., Bickler, P. E., Gonzalez, D. V., Samaniego, F. C. and Escobedo, M. A. (1985). Blood respiratory properties in pigeons at high altitudes: effects of acclimation. *Am J Physiol* **249**, R765–R775.

Weis-Fogh, T. (1972). Energetics of hovering flight in hummingbirds and in *Drosophila*. *J Exp Biol* **56**, 79–104.

Weiss, S. L., Lee, E. A. and Diamond, J. M. (1998). Evolutionary matches of enzymes and transporter capacities to dietary substrate loads in the intestinal brush border. *PNAS* **95**, 2117–2121.

Wenzel, B. M. and Meisami, E. (1987). Number, size, and density of mitral cells in the olfactory bulbs of the northern fulmar and rock dove. *Ann N Y Acad Sci* **510**, 700–702.

White, C.R. and Seymour, R.S. (2003). Mammalian basal metabolic rate is proportional to body mass$^{2/3}$. *PNAS* **100**, 4046–4049.

White, C. R., Martin, G. R. and Butler, P. J. (2008). Wing-spreading, wing-drying and food-warming in great cormorants *Phalacrocorax carbo*. *J Avian Biol* **39**, 576–578.

White, F. N., Bartholomew, G. A. and Kinney, J. L. (1978). Physiological and ecological correlates of tunnel nesting in the European bee-eater, *Merops apiaster*. *Physiol Zool* **51**, 140–154.

White, S. C. (1974). Ecological aspects of growth and nutrition in tropical fruit-eating birds. Ph.D. diss., University of Pennsylvania, Philadelphia.

Whitfield C. W., Cziko, A. M. and Robinson, G. E. (2003). Gene expression profiles in the brain predict behavior in individual honey bees. *Science* **302**, 296–299.

Whittow, G. C. and Tazawa, H. (1991). The early development of temperature regulation in birds. *Physiol Zool* **64**, 1371–1390.

Wideman, R. F., Jr. (1988). Avian kidney anatomy and physiology. In *CRC Critical Reviews in Poultry Biology*. Vol. 1, pp. 133–176. CRC Press, Boca Raton, FL.

Wiersma, P., Muñoz-Garcia, A., Walker, A. and Williams, J. B. (2007). Tropical birds have a slow pace of life. *PNAS* **104**, 9340–9345.

Wiersma, P., Selman, C., Speakman, J. R. and Verhulst, S. (2004). Birds sacrifice oxidative protection for reproduction. *Proc Roy Soc B* **271**, S360–S363.

Wikelski, M. Tarlow, E. M., Raim, A., Diehl, R. H., Larkin, R. P. and Visser, G.H. (2003). Cost of migration in free-flying songbirds. *Nature* **423**, 704.

Wikelski, M., Hau, M. and Wingfield, J. C. (1999). Social instability increases plasma testosterone in a year-round territorial neotropical bird. *Proc Roy Soc B* **266**, 551–556.

Wild, J. M. (1997). Neural pathways for the control of birdsong production. *J Neurobiol* **33**, 653–670.

Wiley, C. J. and Goldizen, A. W. (2003). Testosterone is correlated with courtship but not aggression in the buff-banded rail, *Gallirallus philippensis*. *Hormones and Behavior* **43**, 554–560.

Wiley, R. H. (1971). Song groups in a singing assembly of Little Hermits. *Condor* **73**, 28–35.

Williams, G. C. (1966). Natural selection, the cost of reproduction, and a refinement of lack's principle. *Am Nat* **100**, 687–690.

Williams, J. B. (1996a). A phylogenetic perspective of evaporative water loss in birds. *Auk* **113**, 457–472.

Williams, J. B. (1996b). Energetics of avian incubation. In C. Carey, ed. *Avian Energetics and Nutritional Ecology*. Chapman Hall, New York, NY.

Williams, J. B. (1985). Validation of the doubly labeled water technique for measuring energy metabolism in starlings and sparrows. *Comp Biochem Physiol* **80A**, 349–353.

Williams, J. B. (1993). Energetics of incubation in free-living orange-breasted sunbirds in South Africa. *Condor* **95**, 115–126.

Williams, J. B. and Nagy, K. A. (1984). Daily energy expenditure of Savannah sparrows: comparison of time-energy budget and doubly-labeled water estimates. *Auk* **101**, 221–229.

Williams, J. B. and Tieleman, B. I. (2002). Ecological and evolutionary physiology of desert birds: a progress report. *Integr Comp Biol* **42**, 68–75.

Williams, J. B. and Tieleman, B. I. (2005). Physiological adaptation in desert birds. *Bioscience* **55(5)**, 416–425.

Williams, J. B., Anderson, M. D. and Richardson, P. R. K. (1997). Seasonal differences in field metabolism, water requirements, and foraging behavior of free-living aardwolves in South Africa. *Ecology*, **78**, 2588–2602.

Williams, J. B., Tieleman, B. I., Visser, G. H. and Ricklefs, R. E. (2007). Does growth rate determine the rate of metabolism in shorebird chicks living in the Arctic? *Physiol Biochem Zool* **80**, 500–513.

Williams, T. D. (1994). Intraspecific variation in egg size and egg composition in birds – effects on offspring fitness. *Biological Reviews of the Cambridge Philosophical Society* **69**, 35–59.

Williams, T. D. (2005). Mechanisms underlying the costs of egg production. *Bioscience* **55**, 39–48.

Williams, T. D. and Martyniuk, C. J. (2000). Tissue mass dynamics during egg-production in female zebra finches, *Taeniopygia guttata*: dietary and hormonal manipulations. *J Avian Biol* **31**, 87–95.

Willis, M. A. (2005). Odor-modulated navigation in insects and artificial systems. *Chem Senses* **30**, 1287–1288.

Wilson, J. X. (1997). Antioxidant defense of the brain: a role for astrocytes. *Can J Physiol Pharmacol* **75**, 1149–1163.

Wilson, R. P. and Grémillet, D. (1996). Body temperatures of free-living African penguins (*Spheniscus demersus*). and bank cormorants (*Phalacrocorax neglectus*). *J Exp Biol* **199**, 2215–2223.

Wilson, R. P., Cooper, J. and Plötz, J. (1992a). Can we determine when marine endotherms feed? A case study with seabirds. *J Exp Biol* **167**, 267–275.

Wilson, R. P., Hustler, K., Ryan, P. G., Burger, A. E. and Nöldeke, E. C. (1992b). Diving birds in cold water: do Archimedes and Boyle determine energetic costs? *Am Nat* **140**, 179–200.

Wilson, R. P., White, C. R., Quintana, F., Halsey, L. G., Liebsch, N., Martin, G. R. and Butler, P. J. (2006). Moving towards acceleration for estimate of activity-specific metabolic rate in free-living animals. *J Animal Ecol* **75**, 1081–1090.

Wiltschko, R. and Wiltschko, W. (1995a). *Magnetic Orientation in Animals*. Springer, Berlin, Germany.

Wiltschko, W. and Wiltschko R. (1995b). Migratory orientation of European robins is affected by the wavelength of light as well as by a magnetic pulse. *J Comp Physiol A* **177**, 363–369.

Wiltschko, W. and Wiltschko, R. (1972). Magnetic compass of European robins. *Science* **176**, 62–64.

Wiltschko, W. and Wiltschko, R. (2002). Magnetic compass orientation in birds and its physiological basis. *Naturwissenschaften* **89**, 445–452.

Wiltschko, W., Munro, U., Ford, H. and Wiltschko, R. (1993). Red light disrupts magnetic orientation of migratory birds. *Nature* **364**, 525–527.

Wiltschko, W., Traudt, J., Güntürkün, O., Prior, H. and Wiltschko, R. (2002). Lateralisation of magnetic compass orientation in a migratory bird. *Nature* **419**, 467–470.

Wingfield, J. C. (1984a). Androgens and mating systems: testosterone-induced polygyny in normally monogamous birds. *Auk* **101**, 665–671.

Wingfield, J. C. (1984b). Environmental and endocrine control of reproduction in the song sparrow, *Melospiza melodia*. 1. Temporal organisation of the breeding cycle. *Gen Comp Endocrinol* **56**, 406–416.

Wingfield, J. C. and Farner, D. S. (1978). Annual cycles of luteinizing hormone and sex steroid hormones in the plasma of the white-crowned sparrow, *Zonotrichia leucophyrs gambelii*. *Biol Reprod* **19**, 1046–1056.

Wingfield, J. C. and Marler, P. (1988). Endocrine basis of communication in reproduction and aggression. In E. Knobil and J. D. Neill, eds. *The Physiology of Reproduction*. Raven Press, New York, NY.

Wingfield, J. C., Hegner, R. E., Dufty, A. M. and Ball, G. F. (1990). The challenge hypothesis – theoretical implications for patterns of testosterone secretion, mating systems, and breeding strategies. *Am Nat* **136**, 829–846.

Wingfield, J. C., Lynn, S. E. and Soma, K. K. (2001a). Avoiding the "costs" of testosterone: ecological bases of hormone-behavior interactions. *Brain Behav Evol* **57**, 239–251.

Wingfield, J. C., Soma, K. K., Wikelski, M., Meddle, S. L. and Hau, M. (2001b). Life cycles, behavioural traits and endocrine mechanisms. In A. Dawson and C. M. Chaturvedi, eds. *Avian Endocrinology*. Narosa Publishing House, New Delhi, India.

Withers, P. C. and Williams, J. B. (1990). Metabolic and respiratory physiology of an arid-adapted Australasian bird, the Spinifex pigeon. *Condor* **92**, 961–969.

Woakes, A. J. and Butler, P. J. (1983). Swimming and diving in tufted ducks, *Aythya fuligula*, with particular reference to heart rate and gas exchange. *J Exp Biol* **107**, 311–329.

Wolf, B. O. and Walsberg, G. E. (1996). Respiratory and cutaneous evaporative water loss at high environmental temperatures in a small bird. *J Exp Biol* **199**, 451–457.

Wolf, L. L. and Stiles, F. G. (1970). Evolution of pair cooperation in a tropical hummingbird. *Evolution* **24**, 759–773.

Wu, B. J., Else, P. L., Storlien, L. H. and Hulbert, A. J. (2001). Molecular activity of Na$^+$/K$^+$-ATPase from different sources is related to the packing of membrane lipids. *J Exp Biol* **204**, 4271–4280.

Wu, T. D. (2001). Analysing gene expression data from DNA microarrays to identify candidate genes. *J Pathol* **195**, 53–65.

Wunder, M. B., Kester, C. L, Knopf, F. L and Rye, R. O. (2005). A test of geographical assignment using isotope tracers in feathers of known origin. *Oecologia* **144**, 607–617.

Wylie, D. R. W., Bischof, W. F. and Frost, B. J. (1998). Common reference frame for neural coding of translational and rotational optic flow. *Nature* **392**, 278–282.

Xu, P., Vernooy, S. Y., Guo, M. and Hay, B. A. (2003). The *Drosophila* microRNA mir-14 supresses cell death and is required for normal fat metabolism. *Curr Biol* **13**, 790–795.

Xu, X., Norell, M. A, Kuang, X., Wang, X., Zhao, Q. and Jia, C. (2004). Basal tyrannosauroids from China and evidence for protofeathers in tyrannosauroids. *Nature* **431**, 680–684.

Xu, X., Zhou, Z. and Prum, R. O. (2001). Branched integumental structures in Sinornithosaurus and the origin of feathers. *Nature* **410**, 1036–1037.

Xu, X., Zhou, Z., Wang, X., Kuang, X., Zhang, F. and Du, X. (2003). Four-winged dinosaurs from China. *Nature* **421**, 335–340.

Yang, Y., Cui, Y., Wang, W., Zhang, L., Bufford, L., Sasaki, S., Fan, Z. and Nishimura, H. (2004). Molecular and functional characterization of a vasotocin-sensitive aquaporin water channel in quail kidney. *Am J Physiol, Regul Integr Comp Physiol* **287**, R915–924.

Yarbrough, C. G. and Johnston, D. W. (1965). Lipid deposition in wintering and premigratory myrtle warblers. *Wilson Bull* **77**, 175–191.

Ydenburg, R. C. and Bertram, D. F. (1989). Lack's clutch size hypothesis and brood enlargement studies on colonial seabirds. *Colonial Waterbirds* **12**, 134–137.

Yoda, K. and Ropert-Coudert, Y. (2004). Decision-rules for leaping Adélie penguins (*Pygoscelis adeliae*). *J Zool.* **263**, 1–5.

Yon, H. L., Lamanna, M. C., Harris, H. D., Chiappe, L. M., O'Connor, J., Ji, S. A., Lu, S. C., Yuan, C. X., Li, D. Q., Zhang, X., Lacovara, K. J., Dodson, P. and Ji, Q. (2006). A nearly modern amphibious bird from the Early Cretaceous of northwestern China. *Science* **312**, 1640–1643.

Yu, M., Wu, P., Widelitz, R. B. and Chuong, C. M. (2002). The morphogenesis of feathers. *Nature* **420**, 308–312.

Yue, Z., Jiang, T. X., Widelitz, R. B. and Chuong, C. M. (2006). Wnt3a gradient converts radial to bilateral feather symmetry via topological arrangement of epithelia. *PNAS* **103**, 951–955.

Zar, J. H. (1968). Standard metabolism comparisons between orders of birds. *Condor* **70**, 278.

Zhou, Z. (2004). The origin and early evolution of birds: discoveries, disputes and perspectives from fossil evidence. *Naturwissenchaften* **91**, 455–471.

Zhou, Z. and Zhang, F. (2005). Discovery of an ornithurine bird and its implications for Early Cretaceous avian radiation. *PNAS* **102 (52)**, 18998–19002.

Zhou, Z., Barrett, P. M. and Hilton, J. (2003). An exceptionally preserved Lower Cretaceous ecosystem. *Nature* **421**, 807–814.

Zien, A., Aigner, T., Zimmer, R. and Lengauer, T. (2001). Centralization: a new method for the normalization of gene expression data. *Bioinformatics* **17(Suppl. 1)**, S323–S331.

Zimmer-Faust, R. K., Finelli, C. M., Pentcheff, N. D. and Wethey, D. S. (1995). Odor plumes and animal navigation in turbulent water flow: a field study. *Biol Bull* **188**, 111–116.

Zuk, M. and Stoehr, A. M. (2002). Immune defense and host life history. *Am Nat* **160**, S9–S22.

Zwarts, L. and Blomert, A. M. (1992). Why knot *Calidris canutus* take take medium-sized *Macoma balthica* when six prey species are available. *Mar Ecol Progr Ser* **83**, 113–128.

Zwarts, L., Blomert, A. M. and Wanink, J. H. (1992). Annual and seasonal variation in the food supply harvestable by knot *Calidris canutus* staging in the Wadden Sea in late summer. *Mar Ecol Progr Ser* **83**, 129–139.

Index

Ingram Content Group UK Ltd.
Milton Keynes UK
UKHW022352130323
418540UK00004B/252